CISM Courses and Lectures

The series presents lecture notes, monographs, edited works and proceedings in the field of Mechanics, Engineering, Computer Science and Applied Mathematics.
Purpose of the series is to make known in the international scientific and technical community results obtained in some of the activities organized by CISM, the International Centre for Mechanical Sciences.

International Centre for Mechanical Sciences

Courses and Lectures Vol. 542

For further volumes:
www.springer.com/series/76

Wieslaw Ostachowicz · J. Alfredo Güemes
Editors

New Trends in Structural Health Monitoring

Springer

Editors

Wieslaw Ostachowicz
Polish Academy Of Sciences
Gdansk, Poland

J. Alfredo Güemes
Universidad Politecnica De Madrid
Madrid, Spain

ISSN 0254-1971
ISBN 978-3-7091-1389-9 ISBN 978-3-7091-1390-5 (eBook)
DOI 10.1007/978-3-7091-1390-5
Springer Wien Heidelberg New York Dordrecht London

All contributions have been typeset by the authors
Printed in Italy

Printed on acid-free paper

Springer is part of Springer Science+Business Media (www.springer.com)

PREFACE

This book is the outcome of a European Project sponsored by the European Commission (Specific Support Action ASA6-CT-2006-044636), entitled "Structural Health Monitoring (SHM) in Action". It all began in May 2007 at University Politcnica de Madrid, Spain, during the first meeting where we had the first lectures on SHM in Action. It was clear from the beginning that we wanted to continue the lectures, where the partners had complimentary expertise. Following the first meeting in Madrid we organized eight additional Advanced Lecture Courses at the University of Siegen, Germany (Nov '07) — University of Sheffield, U.K. (Apr '08) — Polish Academy of Sciences, IFFM, Poland (Jun / July '08) — Instituto Superior Tecnico (IST), Portugal (Nov '08) — University of Patras, Greece (Apr '09) — RISOE National Laboratory, Denmark (Jul '09) — University of Catalunya, Spain (Nov. '09) — Deutsche Gesellschaft fur Zerstorungsfreie Prufung E.V., Germany (Nov '10). Structural Health Monitoring (SHM) is a truly multidisciplinary topic. SHM in its own has now been around for more than two decades. Much of the science developed has been generated in either the academic or industrial environment. Numerous institutions and industries dealing with aerospace and civil engineering have been involved. The large number of publications, high attendance in conferences and workshops, research projects and R&D groups dealing with SHM provides significant proof that this technology has gained wide interest, not only in the engineering world but also in physics, electronics or computer science, to just name a few. The objective of the course on "New Trends in SHM" (Udine, Jun '11) was to prepare future experts on SHM and control methods and systems for civil, aerospace and mechanical structures. Its aim was to describe and explain current research, areas of application and tendencies for the future of this technology including various design issues involved. During the course the theory and techniques important for understanding the covered topics were addressed. The Course gathered experts actively working on various SHM and control techniques in order to cover the main disciplines. This book, with the same name, is an additional logical output of this course. In designing this book, we wanted it to be a reference for professionals and students in the area of engineering, applied natural sciences and also engineering management. The course was not only be based on lecturing but also was illustrated through examples using some of the latest hardware and software tools, experimental data from small scale laboratory demonstrators and measurements made on real structures. It is our hope that the reader will find this book interesting and will share the enthusiasm of many scientists who have contributed to the research work presented in it.

Wieslaw Ostachowicz and Alfredo Guemes

CONTENTS

Structural Health Monitoring – Its Association and Use

Christian Boller [*‡]

[*] Chair in NDT and Quality Assurance, Saarland University, Saarbrücken/Germany

[‡] Fraunhofer Institute for Nondestructive Testing, Dresden & Saarbrücken/Germany

Abstract Structural Health Monitoring (SHM) is an expression being around for more than two decades now. Triggered by damage tolerant design where damage has to be inspected at defined intervals based on non-destructive testing (NDT) techniques modern sensing hardware may now be integrated into structural materials and combined with advanced signal processing software making NDT to become an integral part of materials and structures and hence conventional inspection processes to be automated. This chapter provides an insight into the background of and the motivation for SHM by describing needs and assumptions made when designing a damage tolerant structure. It explains the need for loads monitoring and the implications those loads have on damage such as fatigue propagating during a structure's operational life. NDT techniques with a specific potential for SHM are addressed including their impact on monitoring carbon-fibre reinforced composites being one of the different material types where SHM plays a significant role. Finally some outlook is made with regard to SHM implementation and the benefits to be gained where examples have been taken from aviation and some outlook is provided considering SHM applications in wind energy generation.

1 A Motivation for Structural Health Monitoring

Engineering structures today are continuously ageing, would those be in civil engineering, energy generation or aeronautics, to just name a few. Civil engineering buildings such as houses or bridges are not truly designed for a finite life, although one is aware that those will not last forever. In nuclear energy generation there is currently a significant discussion ongoing with regard to life extension of existing power plants which also stems from the fact that many of the power plants have been used less than they were expected to be used. This is characteristic with regard to many of the high asset values a society has, where any extension of an initially assumed operational life is a potential benefit to the operator.

An area where usage is often higher than expected can be seen with road transportation and its related infrastructure. The increased number of vehicles as well as vehicle loads in general has increased deterioration of road tarmacs and bridges, to just name a few. This has again resulted in increased inspection and maintenance effort required to keep those structures operational since cost for replacement of those buildings will rather exceed resources being available, would those be related to material, people or funding in general. The big challenge in managing the criticality of a structure is therefore the determination of a structure's current damage condition even in a quantifiable manner. This quantification requires an in depth understanding of a material's and structure's damage condition, which is currently still quite difficult although a lot of know how being present.

An observation similar to the civil engineering sector can be made with aircraft. Aircraft are engineering structures well designed from an engineering point of view. Due to their complexity in structural design as well as in operation aircraft do contain a variety of components as well as loading conditions which can lead to damage to occur during in service operation. This damage is quantitatively covered during roughly the first half of an aircraft's operational life, specifically in terms of damage tolerant design. However when it comes to the second half of those aircrafts' lives a lot of damages do occur which have not been anticipated in the aircraft's initial design and which have to be covered by a variety of additional inspections and possibly even modifications. One of the most challenging aircraft types in that regard is the Boeing B-52 bomber, which has been designed and built shortly after World War II and is due to stay in operation until the year 2045. Any modification to be done on this type of aircraft today – where there are still sufficient to be expected – does suffer from the lack of knowledge in the structural design of those old structures as well as in the degree an individual structure has truly aged. Means of identifying such a structure's degree of damage are therefore very much in need.

Aircraft are also very much prone to accidental damage, mainly with ground vehicles loading the aircraft. Some of those damages can be clearly recognised such as the one shown in Figure 1. However there are many damages which are hardly to be seen. A critical area among those is frames around the main cargo door of large aircraft such as the Boeing 747 where heavy loading trucks may collide with the lower frame and the locking mechanisms frequently. This has led to accumulated damage and resulted in some

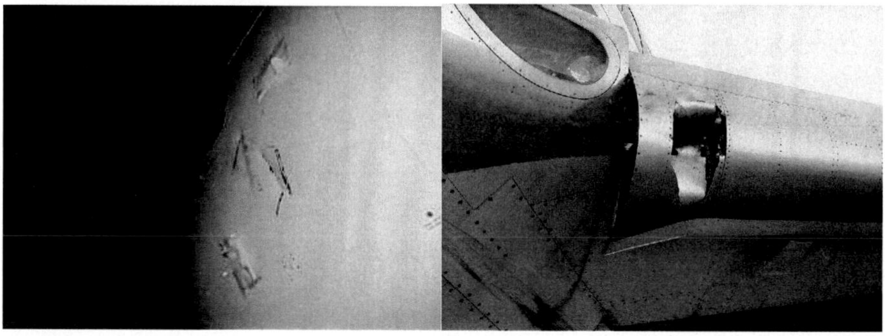

Figure 1. Accidental damage in aircraft structures resulting from ground vehicles

cargo door lock failures in the early days of the Boeing 747 that triggered enhanced inspection of cargo door locks and has now become an issue for continuous monitoring. Aircraft do also have to withstand a variety of other types of repeating operational loads. One of them is hard landings where judgement of the hard landing is mainly referred to a pilot's – fairly subjective – judgement. Although landing gears are structures designed safe life and are hence replaced after defined intervals, it may be difficult to judge what effect the hard landing loads might have on a landing gear's adjacent structural components such as fittings, spars or frames to just name a few. Aircraft may however also be operated to their design limits which is specifically true with military aircraft. Figure 2 shows three examples in that regard a) a Panavia Tornado fighter aircraft, which has been initially designed in the late 1960ies for East-West attack in central Europe and is nowadays used for reconnaissance missions in countries such as Afghanistan and other places in the world, b) load spectra of a F-16 and CF-18 respectively, where the true operational spectrum has exceeded the design spectrum after a portion of its operational life. If those changes in load spectra would not have been monitored significant damage would have occurred to those types of aircraft and the air forces respectively. It has to be considered that a change in the operational conditions such as payloads, manoeuvres or environment will result in a higher amount of damage since the structure having been designed initially for specific operational conditions cannot be changed anymore. Besides metals carbon fibre reinforced polymers (CFRP) play an increasing role in aircraft structures nowadays where barely visible impact damage (BVID) is a source of significant concern and hence an option for monitoring.

Aircraft structures, specifically when it comes to commercial aircraft, are designed damage tolerant which allows structures to be designed lighter weight. Damage tolerance means that the structure considered is able to withstand damages up to a defined size above which a structure will then fail badly. Allowable damages in aircraft structures can achieve a significant amount in length ranging from a few millimetres up to even a metre or more in size, depending on where a crack is due to initiate. Figure 3 shows an example for a hidden corrosion along a front spar web that starts around the rivets at the inner side and gradually grows into two directions before it finally appears on the surface above the fillet seal. Assuming that a crack of 5 mm in length and more might be reliably detectable by an inspector the true crack length would already be in the order of centimetres or more when being finally detected.

Once damaged, a structure has to be repaired. Repairs in aircraft structures are standardised up to a certain degree. However there are many conditions where standardisation in repair does not apply such as with accidental damage. In those cases often tailored repair solutions have to be determined which might be monitored best continuously if technical conditions do allow. So far this has unfortunately not happened and in a few cases some major aircraft failed catastrophically from repairs such as with the Japan Airlines B-747 in 1985 or the China airlines B-747 in 2001.

Damage in aircraft structures can have a variety of reasons which have been summarised in Table 1 below. All of this damage is detected and quantified through non-destructive inspection (NDI) where most of the inspection is done on a visual inspection basis supported by NDI techniques such as ultrasonic and eddy current testing and is further supported by specific damage tolerance design criteria and/or loads monitoring if applicable at all. Inspection however requires a significant amount of effort specifically when the location to be inspected is located at a hidden place where a significant amount of components therefore has to be dismantled and reassembled again afterwards. This requires an aircraft to be out of operation for a substantial amount of time and it would be advantageous if this amount of time could be significantly reduced due to automation of the inspection process. Furthermore many of the subjective judgements done today by pilots such as with regard to hard landings could be made more objective through a loads monitoring system.

If structural health monitoring (SHM) is considered as the integration of sensors and possibly also actuators then the incident which triggered the birth of SHM in aeronautics has been the accidents of the de Havilland Comet aircraft in the early 1950ies (2). This aircraft as shown in Figure 4 below included the three significant engineering features of damage tolerant

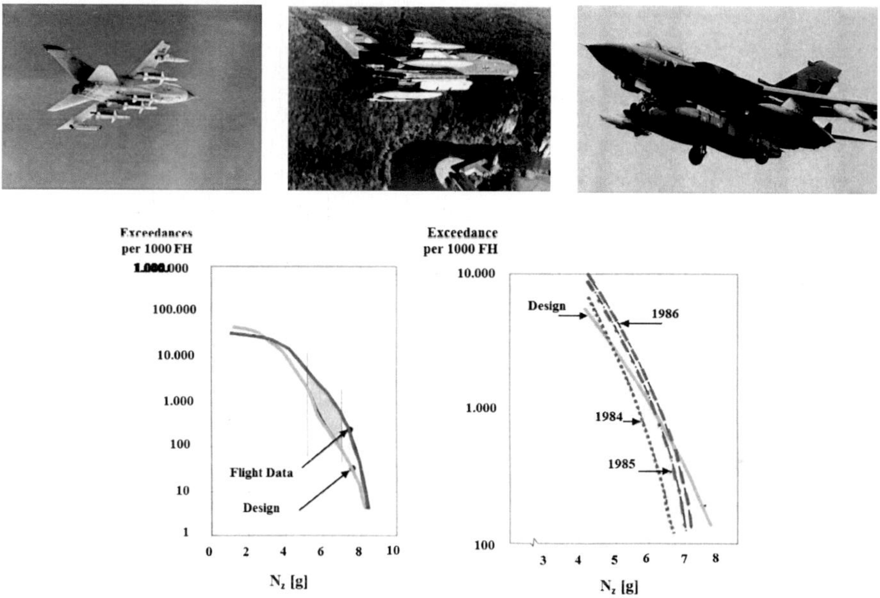

Figure 2. Changes in operational conditions of fighter airplanes of Panavia Tornado (top) and F-16 (bottom left) and CF-18 (bottom right) (1)

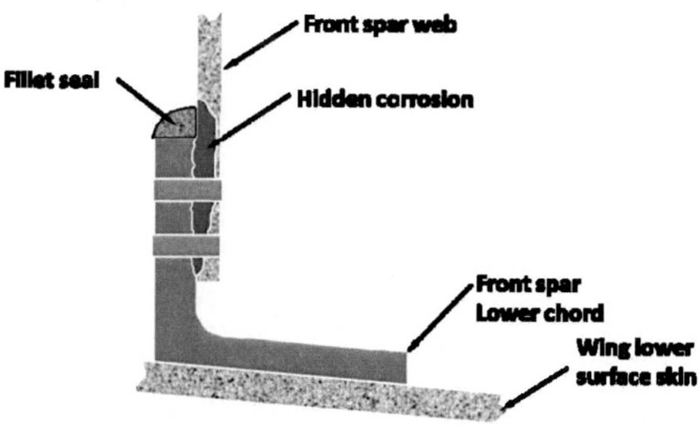

Figure 3. Wing front spar web with hidden corrosion of significant tolerable size

<table>
<tr><th>Issue</th><th>Solution</th><th>Comment</th></tr>
<tr><td>Accident</td><td>Visual and/or Non-Destructive Inspection (NDI); Repair</td><td>Barely visible damage can be overseen</td></tr>
<tr><td>Loads</td><td>Design spectrum; pilot judgement followed by possible NDI</td><td>Very subjective judgement</td></tr>
<tr><td>Fatigue & Fracture</td><td>Damage Tolerance Design; Major Structural Testing; NDI; Inspection Interval</td><td rowspan="4">Can be time consuming and labour intensive</td></tr>
<tr><td>Corrosion</td><td>Design; Corrosion Protection Plan; NDI; Inspection Interval</td></tr>
<tr><td>Multi-Site Damage</td><td>NDI; Inspection Interval</td></tr>
<tr><td>Repair</td><td>Quality Assurance; NDI; Inspection Interval</td></tr>
</table>

Table 1. Issues of damage in aircraft structures and current means of solution

design, integrated jet engines and a pressurised fuselage as a novelty and has possibly been the most dramatic technology push which ever happened with an aircraft type in aviation history. However this technology push resulted in different significant aircraft crashes and hence losses within a very short period of time which triggered immediate measures to be taken on how to enhance aircraft safety in general. One of those measures taken has been the major airframe fatigue test (MAFT) which has nowadays become mandatory to be performed with regard to any new aircraft type going into operation. Figure 5 schematically shows such a type of test for an Airbus A300 and the Eurofighter Typhoon respectively. Along this test all operational loads are applied to the structure on ground with structural design to be validated with regard to its fatigue performance.

Another measure introduced also as a consequence of the Comet accident has been loads monitoring systems of the type shown in Figure 6. This system consists of eight accelerometers set at different trigger levels where each accelerometer measures the number the respective trigger level has been exceeded. Acceleration can be correlated with an aircraft's mass and is therefore a measure to characterise an aircraft's individual load and hence the load spectrum. The spectrum measured has then been a basis to determine a fatigue index FI which has been used to characterise an individual aircraft's usage and hence damage.

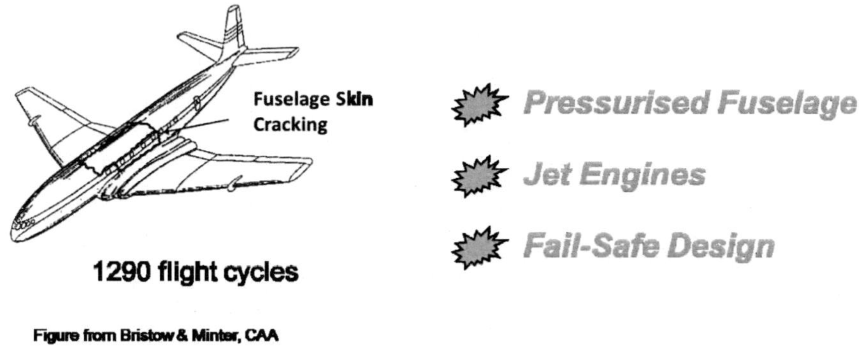

Figure 4. De Havilland Comet aircraft indicating the form of structural damage as well as the innovations made

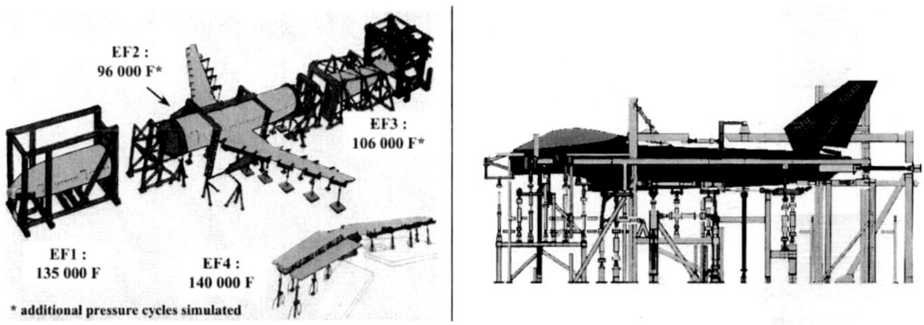

Figure 5. Major airframe fatigue tests performed on Airbus A300 (left) and Eurofighter Typhoon (right) (figures from H.-J. Schmidt/Airbus (left) and DASA (now Cassidian) (right))

Unluckily a trigger of many of the structural monitoring technologies introduced has been accidents. One of the milestones in that regard has been the Aloha Airlines Flight 243 accident in 1988 (Figure 7), where fracture of fuselage panels resulted from multi-site damage (MSD) along rivet lines due to ageing. This accident led to divide the life of aircraft structures into two phases. The first phase is conventionally around the first 15 years of an aircraft's life, where no specific measures have to be taken with regard to MSD and any measures with regard to ageing. It is only beyond this age where an aircraft is considered to age and an increasing in-depth analysis has to be done with regard to MSD and other issues of ageing resulting

$$FatigueIndex(FI) = K_2 \times S_1 \times (2.31A + 0.03B + 0.001C + 0.001D + 0.28E + 3.43F + 10.36G + 18.63H + 1.16WL)$$

K_2 = Mission Coefficient
S_1 = Stores Configuration Coefficient
WL = Landing Coefficient
A bis H = Fatigue Meter Readings

Figure 6. Acceleration exceedance measurement system to be used for loads and fatigue monitoring in aircraft of the Royal Air Force since 1954 (3)

in an increased amount for inspection. It is also during this phase where damage monitoring and hence SHM becomes relevant.

Due to the fact that aircraft are designed damage tolerant and are therefore able to tolerate structural damage of significant size the amount of hull losses due to structural failure in aircraft has significantly decreased over the past decades, as shown in Figure 8. Today no more than 5% of the hull losses can be related to structural failure which is a consequence of enhanced structural understanding in terms of damage accumulation and a resulting effort in structural inspection. This inspection effort increases with an aircraft's increasing age and is hence a source for automation of the inspection process in case SHM can provide that option accordingly. Maintenance errors are a much higher source of error where SHM might be able to have a positive impact of alleviation too.

Implementing SHM into engineering structures in general requires the engineering structural design process to be understood. This has to include the operational loads to be applied on the respective structure. Those loads have to be assumed either in general when nothing else is known or have to be derived from past experience where they might have been measured on equivalent typed structures. Based on the structures' geometric shape

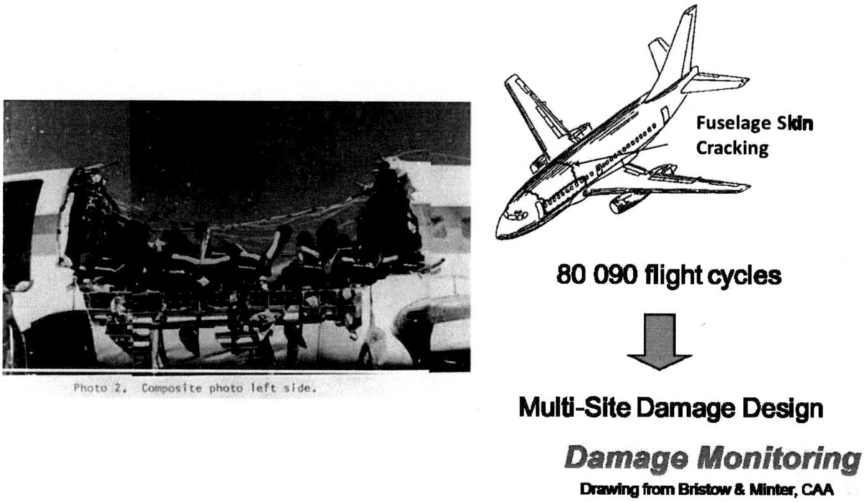

Figure 7. The Aloha Airlines Flight 243 accident as an impact to the ageing aircraft discussion

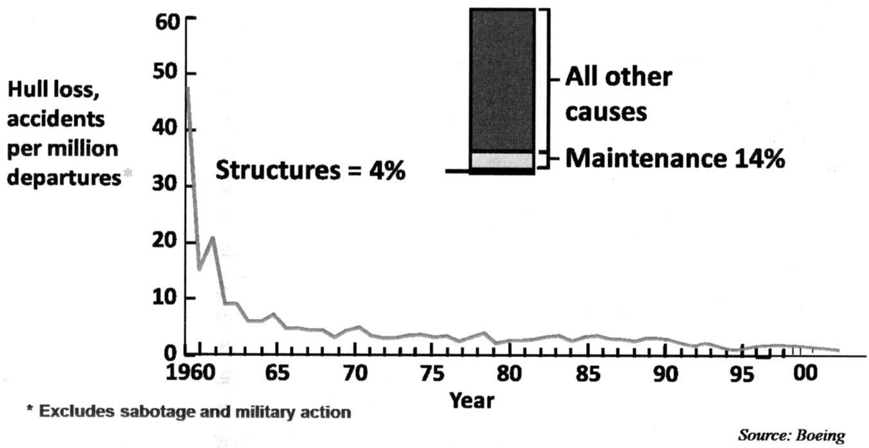

Figure 8. Jet fleet safety records and causes over the past 50 years

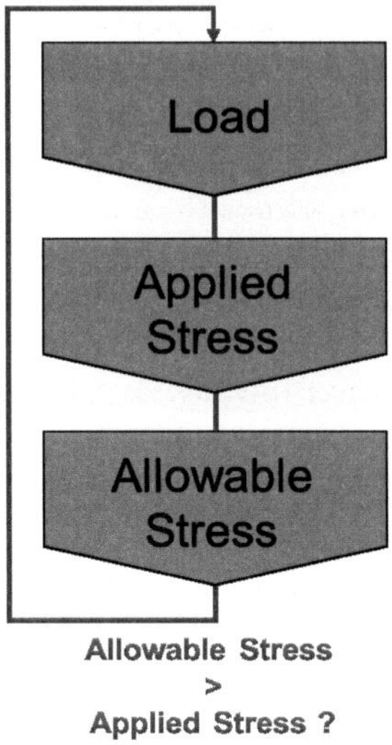

Figure 9. The structural design process

resulting in notches and the material chosen to build the structure itself, stresses and strains within the structure are calculated which results in the determination of local stresses and strains to be compared with the allowable stresses and strains resulting from the properties of the materials chosen. In case applied stresses are larger than allowable stresses a change in the geometric design might be required to reduce any stress concentration factors or a different material may have to be considered alternatively that would have to provide higher allowable stresses respectively.

A first parameter to be monitored in terms of SHM in that regard is therefore loads. The stress information obtained can then be used to evaluate fatigue damage in terms of a fatigue life evaluation and does allow locations on a structure to be determined where damage is most likely to occur. Those locations would then be the locations where damage monitoring may be performed best in terms of SHM. This approach in looking at loads monitoring first, combining this with structural assessment and finally performing damage monitoring at damage critical locations only is the most efficient way to minimise the number of sensors required in terms of SHM and which is therefore recommended to be pursued here. This approach is even applicable to multiaxial loading conditions.

Along this structural design process a central question arises with regard to making a structure better available, lighter weight, more cost efficient and more reliable by making sensors and possibly actuators an integral part of the structure considered. What about looking at advanced sensors which are continuously becoming smaller, lighter and lower cost to become an integral part of a material and structure and further combining the sensors through advanced microelectronics and possibly wireless technology including advanced microprocessors and advanced signal processing? An answer to this question is SHM which is considered to be the integration of sensing and possibly also actuation devices to allow the loading and damaging conditions of a structure to be recorded, analysed, localised and predicted in a way that non-destructive testing becomes an integral part of the structure. SHM encompasses therefore of multitude of disciplines would those be structural strength including fatigue and fracture, structural dynamics, non-destructive testing, signal processing, and possibly much more.

What we therefore need to know to handle a structure adequately in terms of SHM is a structure's behaviour and performance which is described by the loads as well as the design and maintenance principles being applied. Depending on whether only loads or additionally also damage play a role, adequate sensors will have to be integrated which will have to work with appropriate assessment procedures would those be sensor signal processing as well as structural simulation tools. In that regard it is worth looking at a variety of emerging technologies where sensor and signal processing hardware as well as software and also manufacturing technologies that do allow for realisation of the SHM system have to be considered.

2 Loads and Overloads

Loads are the source of any material deterioration would this deterioration happen at the micro or at the macro scale. Loads are usually associated with mechanical loads where those are due to generate fatigue and fracture as damage. Loads may however also be environmental loads such as generated by temperature, humidity, irradiation or any other type of parameter that might lead to deterioration. Fatigue and resulting fracture is generated in a more complex way when loads are repetitive. Hence the time domain signal of a load needs to be known and therefore may have to be monitored. Since mechanical loads are usually the major point of concern in fatigue analysis considerations will be limited to mechanical loads here only. However ideas being expressed below may be synonymously transferred to other types of loads such as those being generated from environment. Loads are a fingerprint of a structure's operational conditions. Many of the different structural components therefore do not follow the same pattern of a load sequence, specifically when they are exposed to a stochastic process and even combined with additional environmental loads. Only a very limited number of structures operate under a sequence of loads where the height of the load is permanently constant and where the spectrum will be called 'constant amplitude'. Most of the loads sequences have their specific spectra. Just imagine an aircraft and the load at the wing attachment box as shown schematically in Figure 10. While the aircraft is on the ground the load might be slightly compressive or close to zero while once the aircraft has taken off the load increases significantly to a high tensile load being directly related to the aircraft's flight load. This load cycle is also called the ground-to-air cycle. While in the air, the aircraft will manoeuvre and face different gusts which will result in dynamic loads cycling around the mean load carrying the aircraft. Once the aircraft lands again the tensile load will vanish or even turn into compressive. Different other applications will show different time domain loading patterns where a few examples are provided in Figure 11.

Reasons for a load-time function may be various and the result of a superposition of different effects as shown schematically in Figure 12. Besides a main load being composed of the structure's weight and a potential payload which can be both mainly considered to be static there are additional loads being generally of a fatigue load nature with manoeuvre loads usually occurring at lower frequencies when compared to higher frequency loads being generated from vibrations.

Load sequences themselves are difficult to be characterised from a time domain point of view such as shown in Figure 11. Different load cycle counting procedures have been developed during the 1930ies to 1970ies where the rainflow cycle counting method (5) has turned out to be the recently most widely accepted. Figure 13 shows a time-domain sequence with its various peaks and troughs where the shape of the time-domain signal can be imagined as the bottom of a water basin reservoir. Now imagine this reservoir to be filled to its top with water (Figure 13 left) and we are gradually shaving the bottom of the water basin from its lowest trough upwards. The first trough which will be hit will be load level -2. Once cut water will flow out from an upper level of +4.5 down to a lower level of -2. This is considered as the cycle of the largest loading span and has been registered as cycle 3 in the table shown in the upper right of Figure 13. The next trough to be hit is at load level -1.5 where water will flow out from load level 3.5 down to -1.5 and is considered as cycle 1 in the table in Figure 13. This is followed then by the trough at load level -1 where water now only flows out from load level 2.5 down to -1 because water level was already reduced from 4.5 to 2.5 due to load cycle 3. Finally a trough at load level +1 is hit where

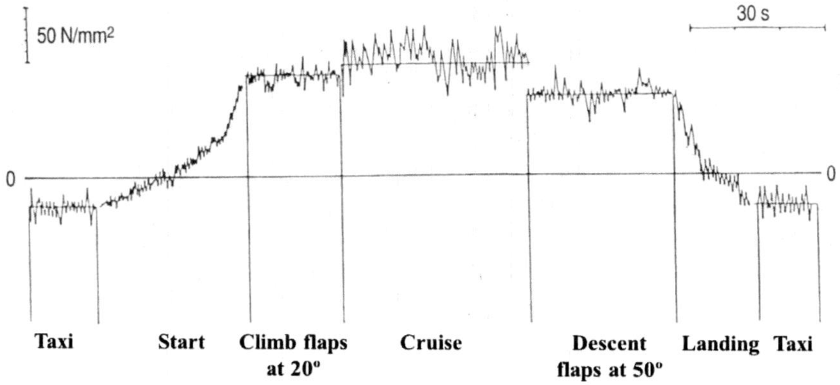

Figure 10. Stress Sequence in Tension Girder of the Wing Root of a Transport Aircraft (4)

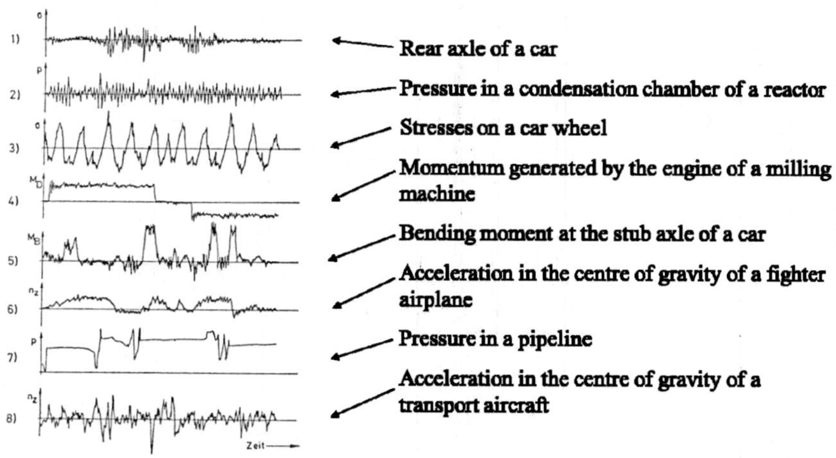

Figure 11. Examples of load sequences for different structures and systems (Figure from (4))

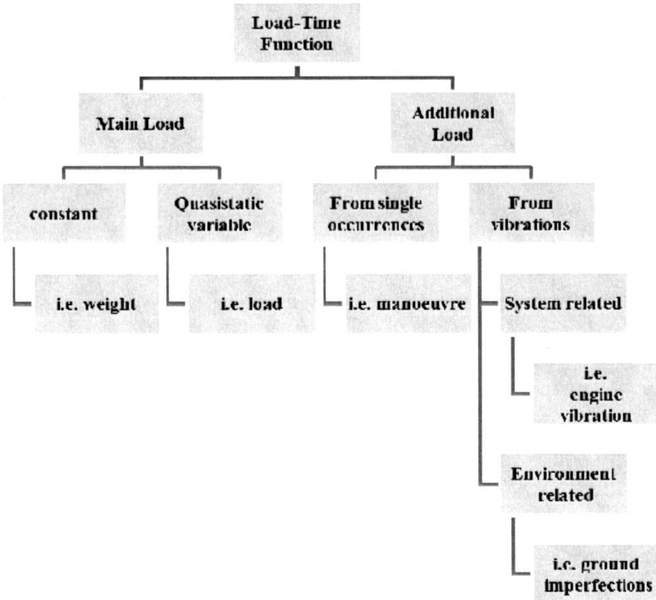

Figure 12. Reasons for load-time functions

water drops from load level 2.5 down to 1 due to the water level reduction already resulting from cycle 1. There is a residual remaining with a trough a load level +2 which cannot be considered since it is unknown where the load sequence is going to continue. With conventional load sequences and their generally large number of cycles this residual can be considered to be negligible. Each cycle can now be represented either in terms of the maximum and minimum load/stress or the amplitude and the mean as shown in the table in Figure 13. The cycles can be further plotted in terms of their loading range as exceedances as shown on the lower left of Figure 13. The shape of this plot can be also considered as a characteristic of a spectrum. Just imagine the spectrum would be constant amplitude. In that case the spectrum would turn out to be of squared shape as shown schematically in Figure 14. Every spectrum being of a random service load nature would turn out to be of a non-squared shape when all spectra are normalised on the same maximum load. In that case the smaller area the spectrum shape covers the weaker the spectrum is considered to be in terms of damaging.

Load spectra may be a combination of different discrete events and this at different levels. Assume an aircraft performing a series of different flights.

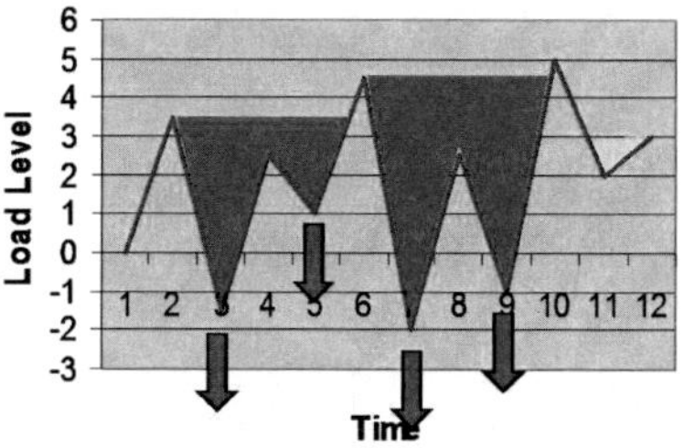

Cycle	σ max	σ min	σ_a	σ_m
1	3.5	-1.5	2.5	1.0
2	2.5	1	0.75	1.75
3	4.5	-2	3.25	1.25
4	2.5	-1	1.75	0.75

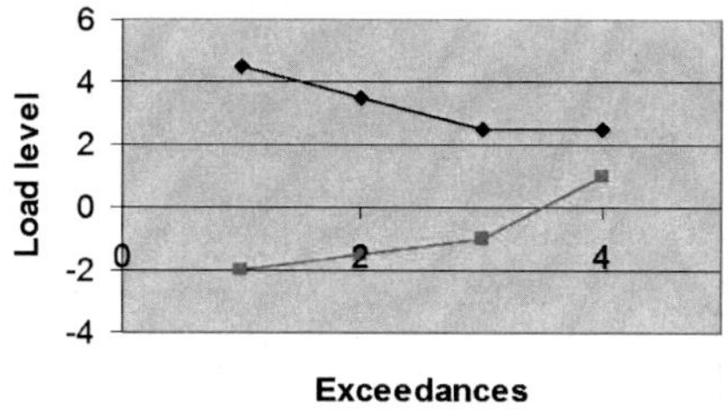

Figure 13. Rainflow cycle counting method (5)

Each flight consists of a variety of different load cycles where the ground to air cycle might be the dominant one added by a variety of smaller load cycles being characterised by flight manoeuvres as well as gusts. Load spectra for civil aircraft may be of a mainly linear shape nature in accordance to the examples provided in Figure 14. Each flight again may be of a different characteristic. Most of the flights are of a smooth character with a lower maximum load. However there is also a limited number of flights under turbulent conditions where loads may be more severe and occasionally an aircraft might even have to fly through a thunderstorm where the maximum load might even be higher. Flight spectra have been characterised and as a consequence standardised for a variety of applications where some examples of a mainly aeronautical nature can be found in (6). One of the most popular aeronautical load spectra is the Transport WIng Standard (TWIST) (7) shown in Figure 15. The spectrum consists of principally two sub spectra being a flight spectrum and a ground taxiing spectrum respectively. The ground taxiing spectrum is again divided into two steps (load levels) while the flight spectrum is divided into ten steps, with both resulting in a mainly linear shaped spectrum.

Detailed numbers for the TWIST spectrum are provided in Table 2. Besides being divided into the load steps I to X there is also a division into flight types A to J, where flight type A represents the most severe flight type with the highest loads and the lowest likelihood to occur during an aircraft's operational life while flight type J is the flight type to occur most and consisting of the lowest loads distribution.

Load cycles as described for the TWIST spectrum in Table 2 may also

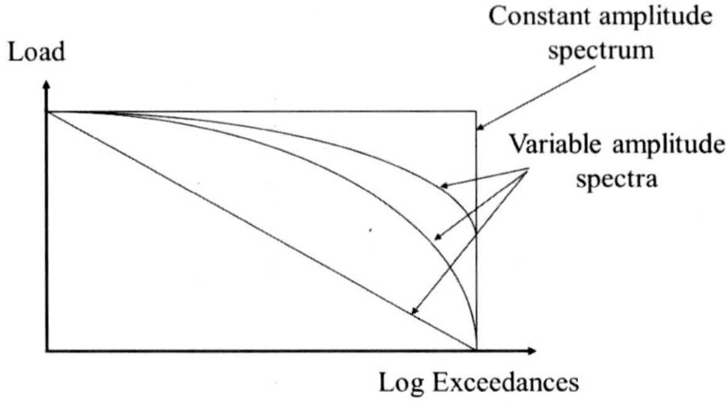

Figure 14. Shape of different loading spectra

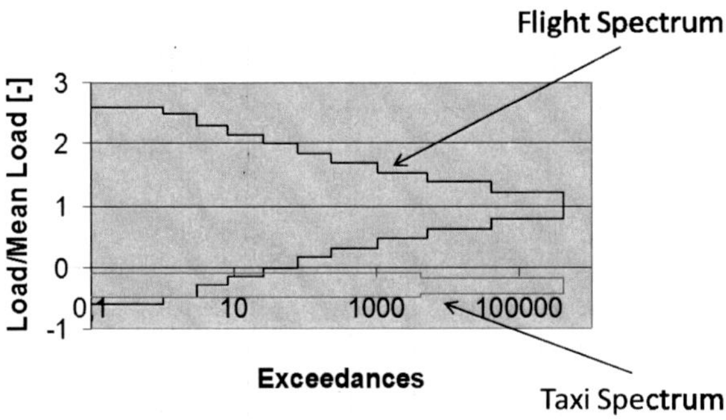

Figure 15. Transport Wing Standard (TWIST) spectrum (7)

Flight type	Number of Flights	Cycles per step										Cycles per flight
		I	II	III	IV	V	VI	VII	VIII	IX	X	
A	1	1	1	1	4	8	18	64	112	391	900	1500
B	1		1	1	2	5	11	39	76	366	899	1400
C	3			1	1	2	7	22	61	277	879	1250
D	9				1	1	2	14	44	208	680	950
E	24					1	1	6	24	165	603	800
F	60						1	3	19	115	512	650
G	181							1	7	70	412	490
H	420								1	16	233	250
I	1090									1	69	70
J	2211										25	25
Total cycles in all		1	2	5	18	52	152	800	4170	34800	358665	
Exceedances		1	3	8	26	78	230	1030	5200	40000	398665	

Table 2. Defined load spectra descriptors of the TWIST spectrum (7)

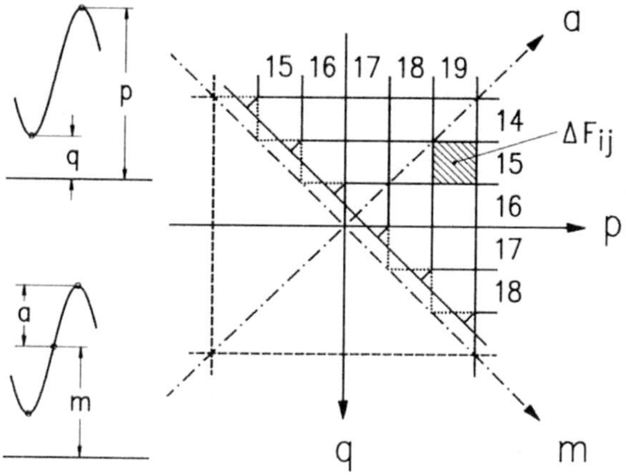

Figure 16. Load matrix descriptions and definitions (figure from (4))

be summarised and described in a matrix as the number of load half cycles going from a minimum load level q (trough) to a maximum load level p (peak) within a complete load spectrum determined in accordance to the rainflow cycle counting method mentioned before. Numbers 14 to 19 indicated as an example in Figure 16 represent arbitrary load levels which have to be decided upon in accordance to the specific needs. In the case of the TWIST load sequence the total number of load levels is 10. Figure 16 hence represents a matrix where each of the elements ΔF_{ij} represents the number of cycles going from load class i to load class j. Going along the horizontal axis and looking at the upper right triangle provides all the peaks p with increasing number while looking at the vertical axis downwards and the matrix's lower left triangle provides all the troughs q respectively. Similar considerations can be made with amplitudes a and means m when converting those as follows:

$$\begin{aligned} q &= m - a & p &= m + a \\ a &= (p - q)/2 & m &= (p + q)/2 \end{aligned}$$

A rainflow matrix as shown in Figure 16 and having been once filled can be pseudo-randomised in accordance to a procedure described in Figure 17. The 32x32 rainflow matrix shown in Figure 17 is filled with numbers a_{ij}. Assume the randomised service load sequence to originate from class α. In this case a term T_α is determined in accordance to $T_\alpha = \sum_{j=\alpha+1}^{n} a_{\alpha j}$ with a random number R_n being determined as:

$$R_n = A \cdot R_{n-1} + (B + 1 + W)\,(MOD\,M)$$

With $M = 2^r$ and $r = 1 + INT\,(\ln T_\alpha / \ln 2)$

$$\begin{aligned} D &= M - T_\alpha \\ A &= (5)\,MAX\,(M/2 - 3) \\ B &= (3)\,MAX\,(M/4 - 1) \end{aligned}$$

$W = +1$ when towards a peak and $W = -1$ when towards a trough.

For the very first cycle when loading turns towards a peak R_{0i} turns out to be:

$$R_{01} = D_i + \sum_{j=i+1}^{16} a_{ij} \text{ for } i = 1...15 \text{ and } R_{0i} = D_i \text{ for } R_{0i} = D_i \text{ for } i = 16...32$$

When the first cycle turns towards a trough R_{0i} is determined as:

$$R_{0i} = D_i \text{ for } i = 1 \ldots 17 \text{ and } R_{0i} = D_i + \sum_{j=17}^{i=1} a_{ij} \text{ for } i = 18 \ldots 32.$$

Target load level β is achieved when $R_n - \sum_{j=\alpha+1}^{\beta} a_{\alpha j} \leq 0$ and $a_{\alpha\beta}$ is now reduced by 1.

If load level β is a peak (as shown in Figure 17) the same procedure is determined the opposite way to determine load level γ as shown in Figure 17. This procedure is then continued until all peaks and troughs within the matrix have been absorbed. A resulting load sequence determined might then look like the one shown as an example in Figure 18 where N_0/N_1 is the ratio of the number of cycles of zero crossings versus the complete number of cycles within the load spectrum.

3 Fatigue, Fracture and Damage Tolerance

3.1 Fatigue

Fatigue and fracture do specifically occur in areas where stresses and strains concentrate. Locations of stress and strain concentrations are notches where stresses and strains do increase significantly by even a factor. Figure 19 left shows such an example for an elliptic notch where the following equation

$$(\sigma_\varphi)_{\max} = S\left(1 + 2\frac{1+c}{1-c}\right) = S\left(1 + 2\frac{a}{b}\right) = S\left(1 + 2\sqrt{\frac{a}{\rho}}\right) \qquad (1)$$

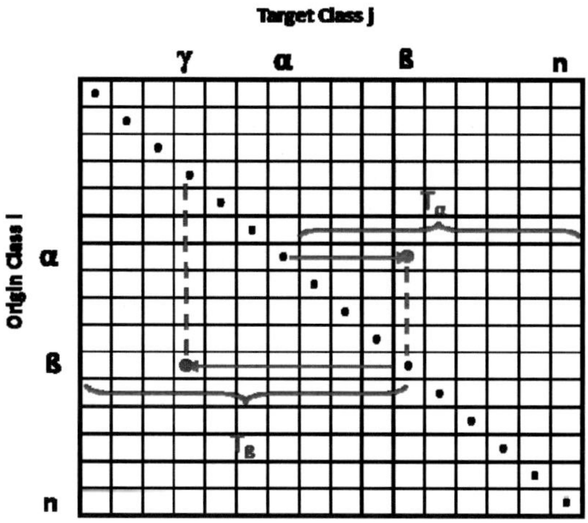

Figure 17. Pseudo-random load sequence generation scheme based on a rainflow matrix (8)

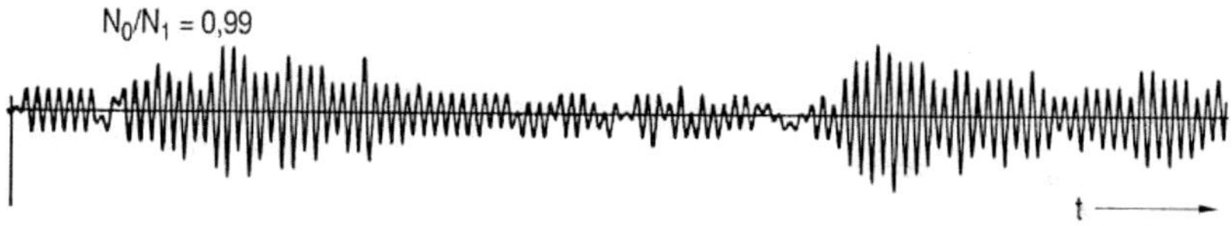

Figure 18. Pseudo-random load sequence generated from a rainflow matrix

does allow the maximum stress in the notch root $(\sigma_\varphi)_{\max}$ to be determined as a function of the nominal stress S, the notch depth a and the notch root radius ρ as shown in Figure 19.

Nominal stress S resulting from an axial load F is usually defined as $S = F/A$ where A is the cross net-section. Similar definitions can be made for bending and torque. It is important that one continuously sticks to the same nominal stress definition. Stress increase due to a notch is defined as

$$K_t = \frac{(\sigma_\varphi)_{\max}}{S} = \frac{\text{maximum stress}}{\text{nominal stress}}$$

where K_t is named the stress concentration factor. K_t is derived from a component's pure elastic behaviour and its definition is therefore only

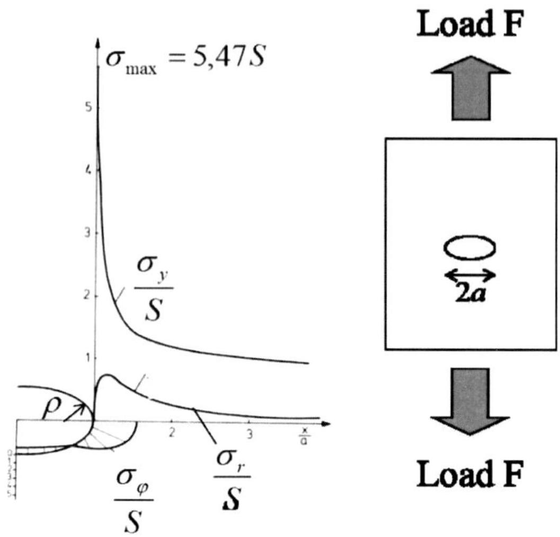

Figure 19. Stress concentration in a notch root

valid when loading is macroscopically elastic with small scale yielding in the notch root being allowed only. The definition is also just valid for one type of loading would this either be tension, bending or torsion. A different type of loading will also result in a different K_t value. K_t values can be either determined from handbooks (i.e. (9)) or might have to be determined through a FE analysis.

To characterise the fatigue behaviour of an unnotched (material) or a notched component (structure) constant amplitude fatigue tests are performed at different load levels as shown in Figure 20. The result obtained is a fatigue life curve (S-N curve) for either crack initiation or complete fracture. Those S-N curves can be linearized when being plotted in a log-log scale, where an endurance limit can be seen at fatigue lives around 10^6 cycles, resulting in a bi-linear relationship in a log-log scale in the end. An S-N curve does depend upon the material, component (geometry/shape), loading and nominal stress definition considered. S-N curves may be determined experimentally or may be taken from handbooks (i.e. (10)).

Once an S-N curve is made available fatigue life of a structure can be estimated. The way such an estimation is done in principle is shown as an example in Figure 21. Assume a load sequence to consist of the following

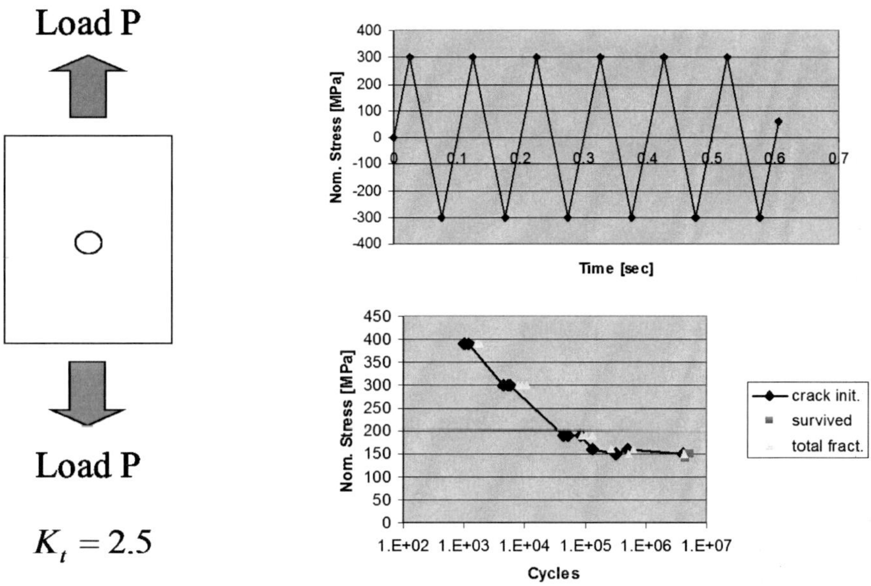

Figure 20. Determination of a constant amplitude S-N curve of a notched specimen

three loading blocks: a) $n_1 = 6$ cycles of a maximum stress of 300 MPa, b) $n_2 = 4$ cycles of a maximum stress of 400 MPa and c) $n_3 = 5$ cycles of a maximum stress of 200 MPa. For the stress range block a) the component would endure around $N_1 = 2 \times 10^5$ cycles, which would result in a linearized damage D_i per load cycle to be $D_i = 1/N_1$ or for the complete block $D_1 = n_1/N_1$. The same is applied for blocks b) and c) where N_3 turns out to be infinite in the case of block c) and hence damage per loading cycle results in being zero. Damage is assumed to accumulate linear in accordance to the Palmgren-Miner [11,12] rule which is defined as shown in Figure 21. Fatigue life is achieved when accumulated damage D turns out to be unity.

With such an approach combined with a FE-analysis accumulated fatigue damage can be virtually calculated for each element allowing areas of high damage accumulation to be determined. These areas are those where damage monitoring then will have to occur first. It therefore turns out that this way of fatigue life evaluation is an essential strategic means for an efficient SHM process. What that means in terms of a SHM strategy can be summarised as follows:

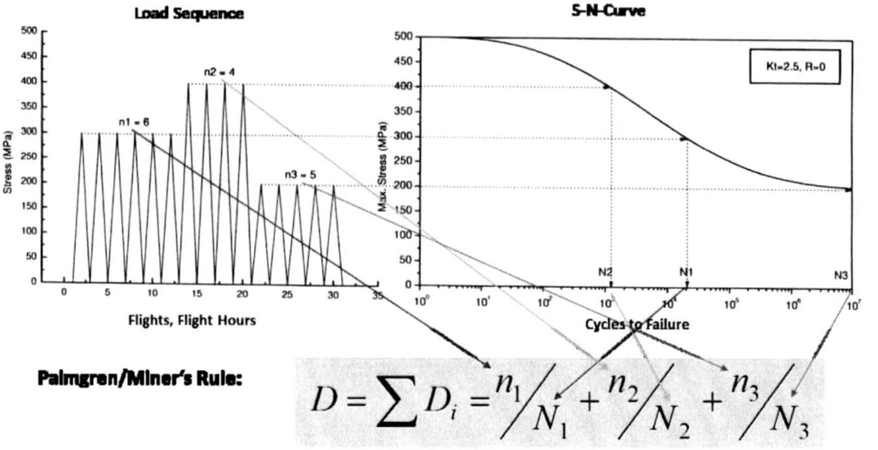

Figure 21. Fatigue life evaluation according to Palmgren Miner's rule

1. Establish a loads monitoring system based on load sequence measurement that allows the operational loads sequence to be monitored, analysed and stored as a rainflow matrix.
2. Use the rainflow spectrum to perform fatigue life evaluation under true service loading conditions and determine the locations on the component considered where damage is most likely to occur (hot spots).
3. Equip hot spots with damage monitoring systems that will possibly allow the initiation of a crack to be detected but will certainly detect an allowable damage such as a defined crack length in accordance to the damage tolerance principle.

3.2 Fracture and Damage Tolerance

Structures when being loaded can withstand a considerable number of load cycles until they fracture. Fracture is considered as a damage condition defined which might be a crack of a defined size. When a component of a material, shape and manufacturing process being well defined is fatigue loaded it will fracture at a specific number of fatigue cycles. If the test is done again on a next component made of the same material, shape and manufacturing process it will most likely fail at a different number of fatigue cycles. Further repetitions of the fatigue test will finally result in a scatter band of fatigue lives as shown as an example in Figure 22. Looking at the resulting S-N curve this does result in a zone where: a) nothing fractures, b) fracture occurs and c) everything has fractured. If a structure is to be

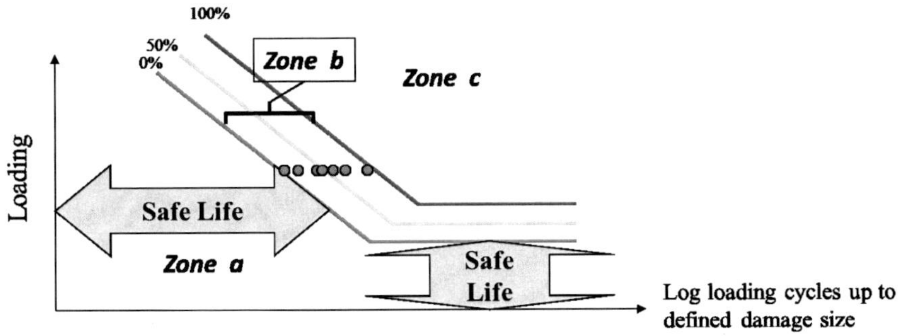

Figure 22. Scatter in S-N curves and the impact on structural design principles

designed such that it is not allowed to fracture it has to be designed such that it falls into zone a). Once the component has achieved the fatigue life defined by the S-N curve of 0% of probability of fracture the component will have to be removed although it might still have a substantial amount of remaining fatigue life. This is what is called *safe life design*. Since the abscissa of an S-N curve is plotted in a logarithmic scale the scatter band of fatigue lives (zone b)) may easily be in the range of a factor of two and more. Hence if the component mentioned before having been removed under safe life design conditions would be run until its true fracture life it might have lasted an additional number of fatigue cycles possibly of the same amount as it would have lasted under safe life design. Allowing a component to be run up to fracture however requires damage mechanisms within the component to be understood and the component to be inspected at well-defined intervals. This is what is called *damage tolerant design* and where SHM comes into play. Damage tolerant design allows a structure to last longer or if this is not desired to apply higher stresses which results in lighter weight design. The latter is what aeronautics has specifically taken advantage of and what now has become standard specifically in the commercial aviation world.

The idea of damage tolerant design can also be extended to engineering structures in general. In that case a component in a structure may fail as long as neighbouring components may be able to take over the loads in the sense of a redundancy and the structure with the resulting fracture may be even controllable under these circumstances. This is what is also generally known as *fail safe design*. Figure 23 shows the range damage tolerance can have within potential fatigue life. If microcracking is not considered

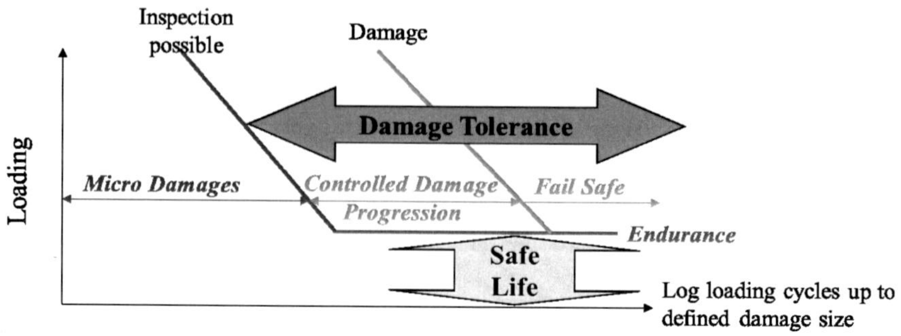

Figure 23. Range of damage tolerance along structural fatigue life

to impair the structure's integrity, damage tolerance starts from where a crack can be reliably observed and damage progression can be controlled until a tolerable damage constellation within the fail safe area with the size of tolerable damage being defined by structural design. A further aspect to be considered is that non-destructive testing concepts have to be further developed along the design phase of a damage tolerant structure (13).

A key issue within damage tolerant design is the knowledge on how cracks propagate. From a very generic principle cracks can even be considered as a notch for which Eq. (1) might be applicable and where the notch depth a may be equivalent to the crack length and hence large when compared to the notch radius ρ being extremely small in a crack leading finally to an incredibly high stress concentration. This would however only be true with a brittle material where material and structural conditions are fully elastic. With ductile materials crack tips however become blunt due to plasticity which then significantly changes the stress intensity and makes Eq. (1) to become invalid in that case.

Cracks can be principally loaded under the three different conditions shown in Figure 24 and which are called the cracking modes or modes simply. Mode 1 represents an axial tensile loading, mode 2 a bending loading and mode 3 a twist loading respectively. Along the following only cases of mode 1 will be considered.

Stresses σ_{kl} around the tip of an arbitrary crack under arbitrary loading can be principally determined on the basis of the following equation:

$$\sigma_{kl} = \frac{1}{\sqrt{2\pi r}}\left[\underbrace{K_I f_{kl}^I(\varphi)}_{\text{Mode I}} + \underbrace{K_{II} f_{kl}^{II}(\varphi)}_{\text{Mode II}} + \underbrace{K_{III} f_{kl}^{III}(\varphi)}_{\text{Mode III}}\right] \quad (2)$$

$$k, l = x, y, z$$

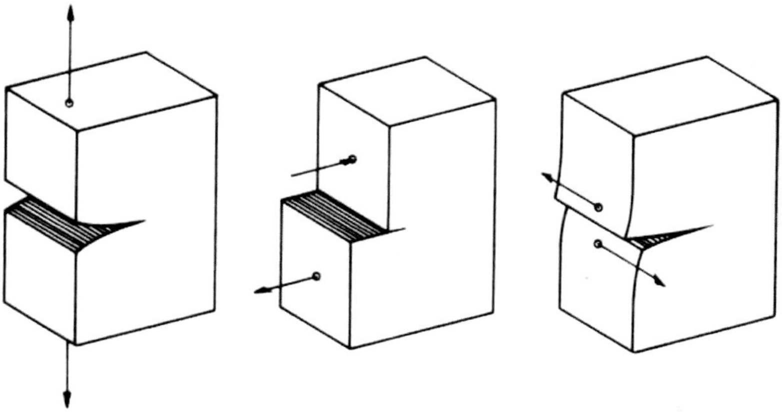

Figure 24. Different modes of fracture

where r is the radius from the centre of the coordinate system as shown in Figure 25 for a crack in an infinite plate under bi-axial loading, K_I , K_{II} and K_{III} are the stress intensities for the different modes and $f(\varphi)$ are dimensionless functions depending on the crack mode. In the case of mode 1 K_I turns out to be:

$$K_I = \sigma\sqrt{\pi a} \quad \left[N\big/\sqrt{mm^3}\right] \tag{3}$$

Since the geometry of the component has an influence on the stress intensity factor Ka geometry factor $Y$is additionally introduced extending Eq. (3) to become

$$K_I = Y\sigma\sqrt{\pi a} \quad \left[N\big/\sqrt{mm^3}\right] \tag{4}$$

Reference for all geometry factors is a crack in an infinite plate where Y is unity while for all others this will differ, for which an example is given in Figure 26 for a finite plate cracked. Further geometry factors can be found in handbooks such as [14-17] and possibly also in other relevant publications or may have to be determined by numerical means.

When a component is fatigue loaded the load alternates between a minimum and a maximum load and hence stress. The stress intensity factor

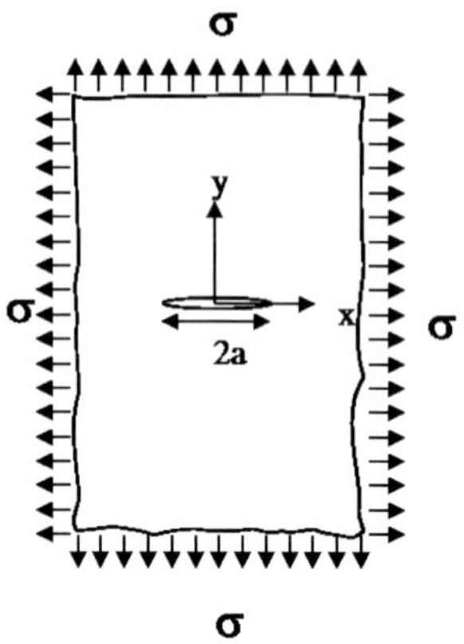

$$\sigma_x = \frac{K_I}{\sqrt{2\pi r}} \cos\frac{\varphi}{2}\left(1 - \sin\frac{\varphi}{2}\sin\frac{3\varphi}{2}\right)$$

$$\sigma_y = \frac{K_I}{\sqrt{2\pi r}} \cos\frac{\varphi}{2}\left(1 + \sin\frac{\varphi}{2}\sin\frac{3\varphi}{2}\right)$$

$$\tau_{xy} = \frac{K_I}{\sqrt{2\pi r}} \sin\frac{\varphi}{2}\cos\frac{\varphi}{2}\cos\frac{3\varphi}{2}$$

Figure 25. Stresses under mode 1 around a crack tip for a crack in an infinite plate under bi-axial loading

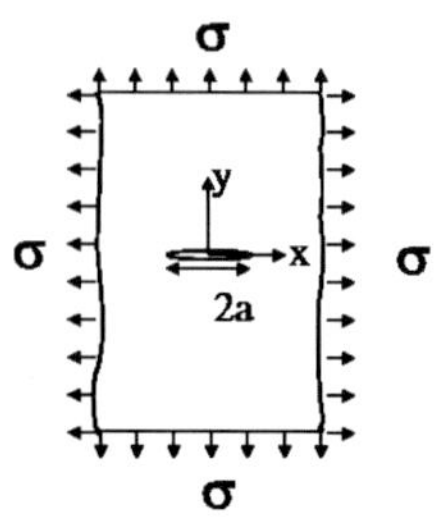

Infinite plate:

$$Y = 1$$

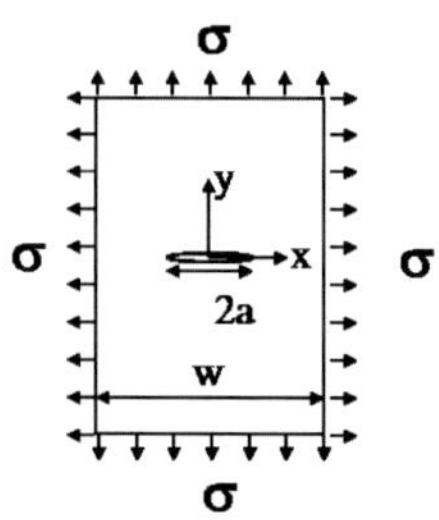

Finite plate:

$$Y = \sqrt{\sec\frac{\pi a}{w}}$$

Figure 26. Geometry factors Y for a cracked infinite and finite plate

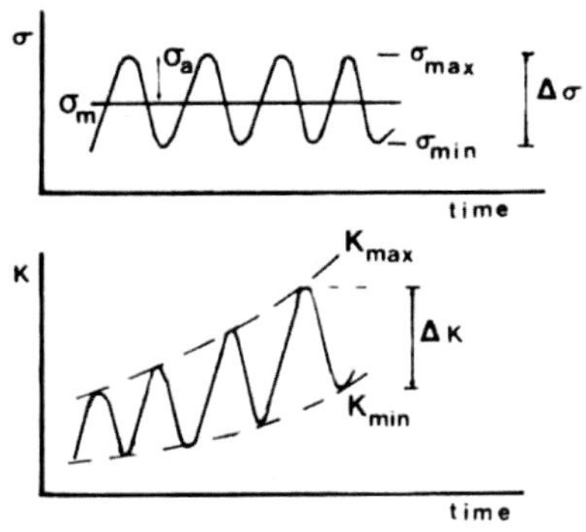

Figure 27. Development of stress intensity for a crack under fatigue loading

K being dependent on stress will therefore alternate as well and can therefore be differentiated between a minimum and a maximum value as to the following:

$$\begin{aligned} K_{\min} &= Y\sigma_{\min}\sqrt{\pi a} \\ K_{\max} &= Y\sigma_{\max}\sqrt{\pi a} \end{aligned} \tag{5}$$

and with a range ΔK being

$$\Delta K = K_{\max} - K_{\min} = Y\left(\sigma_{\max} - \sigma_{\min}\right)\sqrt{\pi a} \tag{6}$$

Since K also depends on the crack length and the influence of the crack length might be less under the minimum load than under the maximum load due to an effect being called crack closure, development of crack intensities along a fatigue test may develop as schematically shown in Figure 27.

What can be indirectly observed from the lower plot in Figure 27 is the nonlinear propagation of a crack length a over fatigue life. When a fatigue test is performed on a cracked plate under constant amplitude loading crack propagation can be observed as to the right hand curve of the left diagram in Figure 28. A similar test done at a higher load level will allow the crack to propagate at higher speed and hence result in the left hand curve of the left diagram in Figure 28. When plotting the change in crack propagation over fatigue life da/dN over the range in stress intensity ΔK in a logarithmic scale, a fairly linear relationship is obtained as can be seen from the right

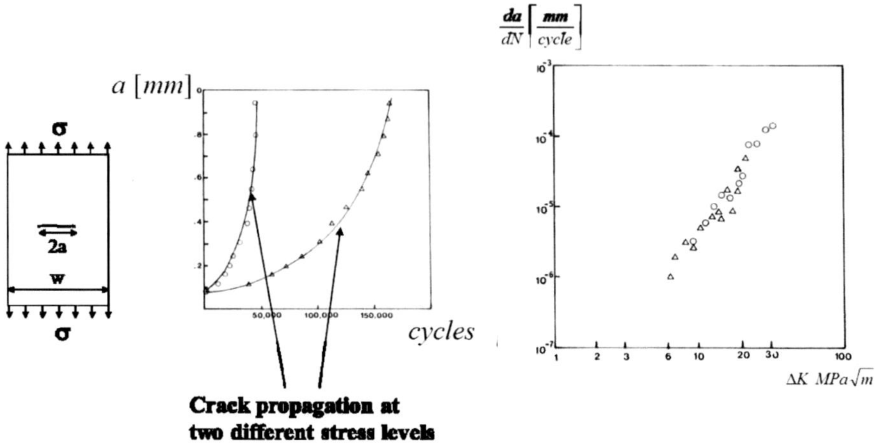

Figure 28. Crack propagation behaviour under two different stress levels

hand diagram on Figure 28. This relationship described as

$$\frac{da}{dN} = C_P \, (\Delta K)^{m_P} \tag{7}$$

has been determined by Paris (18). It has to be mentioned that constants within Eq. (7) are only valid for a specific stress ratio $R = \sigma_{\min}/\sigma_{\max}$ (or strain ratio in case of strain control) and need to be re-determined in case R changes. Furthermore crack propagation rate goes to infinity once it reaches fracture toughness, where stress intensity is expressed in terms of K_c. Forman (19) therefore extended the Paris equation to become

$$\frac{da}{dN} = C_F \frac{\Delta K^{m_F}}{(1 - R)\, K_c - \Delta K} \tag{8}$$

Crack propagation calculations and hence estimations can now be made loading cycle by loading cycle in accordance to the following steps:

1. Assume an initial crack length a_0 that can be reliably detected by means of non-destructive testing or SHM;
2. Determine the geometry factor Y either from handbooks, literature or by numerical analysis;
3. Determine the nominal stress from the maximum or minimum load to be applied within the load cycle considered;

4. Calculate the minimum and maximum stress intensity factors K_{min} and K_{max} as well as resulting ΔK;
5. Calculate crack propagation rate da/dN and hence crack extension Δa for the load cycle considered;
6. Add crack extension Δa to the initial crack length a_0 to determine the new crack length a to be used as the initial crack length in step 1 for the next load cycle;
7. Continue with step 2 accordingly.

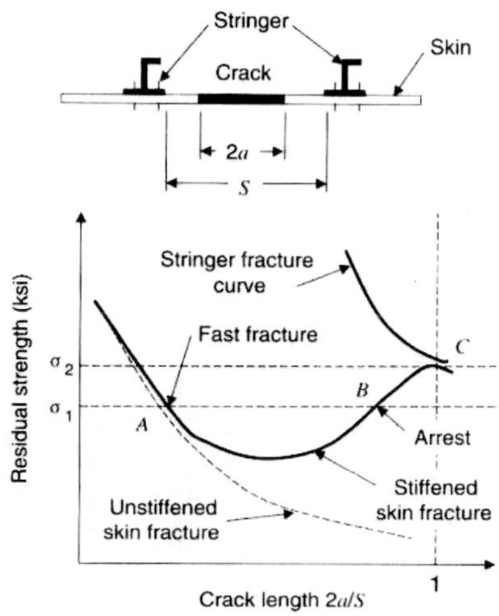

Figure 29. Influence of stiffeners along a damage tolerant structural design (source: (20))

That crack propagation calculation is a necessity within damage tolerance considerations can be seen from the examples provided in Figures 29 and 30 respectively. Residual strength of a cracked structure follows a hyperbolic curve in case strength is plotted over crack length and the crack is considered to propagate in an infinite environment. However if a crack stopper or at least retarder is in the crack progression's way such as a stringer in an aircraft structural component then residual strength of a crack in such a stiffened panel is again reduced such as shown schematically in Figure 29. On the other hand the stiffening stringer has a residual strength curve too which has to be significantly above the one for the panel such that a crack

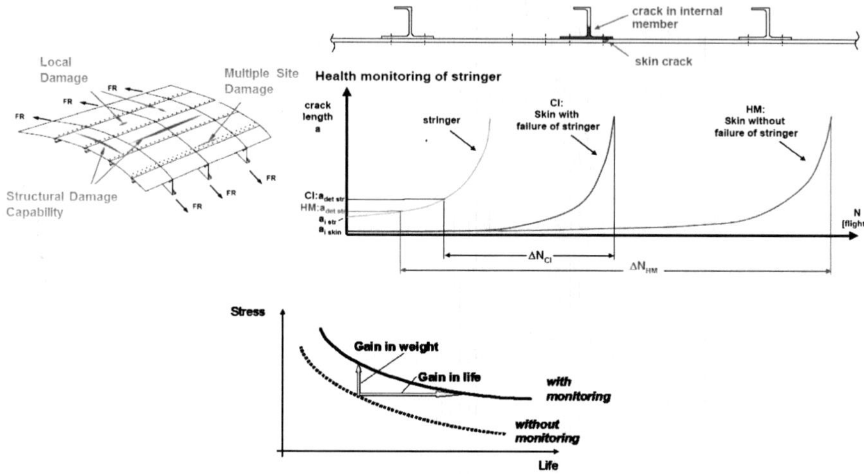

Figure 30. Gain in structural life through health monitoring (HM) of a stringer in an aircraft panel when compared to conventional inspection (CI) (Source: H-J Schmidt, Airbus)

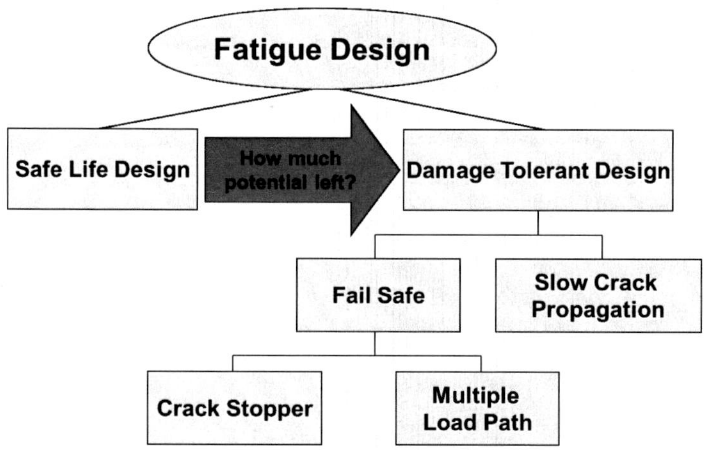

Figure 31. Transition from safe life to damage tolerant design

not having been detected in the panel will not exceed a critical size in case the structure would experience the maximum load along the given design spectrum. Furthermore an initial crack size has to be detected reliably. Only under those circumstances a crack propagation life can be calculated and a resulting inspection period defined. However if a component cannot be inspected, assumptions have to be made with regard to its crack propagation life where crack propagation estimates will have to be on the safe side. Assume an aircraft fuselage component of the type shown in Figure 30 and the condition that frames and stringers might not be inspectable, hence the structure can only be inspected from the aircraft's outer side. Under those conditions frames and stringers have to be considered broken in case crack propagation life and hence the inspection interval has to be based on crack propagation observed on the fuselage's outer surface only. However if stringers and frames were inspectable such as with an SHM system then assumptions for crack propagation could be changed leading to a significant increase in crack propagation life determined or inspection interval defined. If this increase in fatigue life would not be of interest for the case considered then another option would be to increase stresses applied which would then lead to a decrease in structural material to be used and hence result in lighter weight design which again is identical to the effect already observed with damage tolerant design. Implementing SHM is therefore in many cases nothing else than enhancing the idea of damage tolerant design or in other words determining how much potential a structure has in accordance to the damage tolerance principle shown in Figure 31. That damage tolerance has a significant impact on current aircraft structural design can be seen from Figure 32 where areas where crack growth (C.G.) is the relevant design parameter have been clearly marked. Those areas have to be inspected in well-defined intervals where the interval is defined by the metallic material considered, the component's shape, the loading condition including the sequence, an initial detectable crack and a final allowable crack.

Size of the initial detectable crack again depends on a variety of parameters such as the non-destructive testing (NDT) means used as well as the location of the crack to initiate. Figure 33 provides an overview of NDT techniques traditionally used in aeronautics and the possible sensitivities to be achieved. In that context it has to be kept in mind that more than 90% of an aircraft's structure is inspected by visual means and that only a very limited number of critical components is inspected by techniques such as ultrasonics, eddy current or radiography. What is important in that regard is that the probability of detection (POD) of the damage (i.e. crack) considered has to be > 95 % hence the initial crack assumed has to be larger than what the NDT method applied may be virtually able to detect in average.

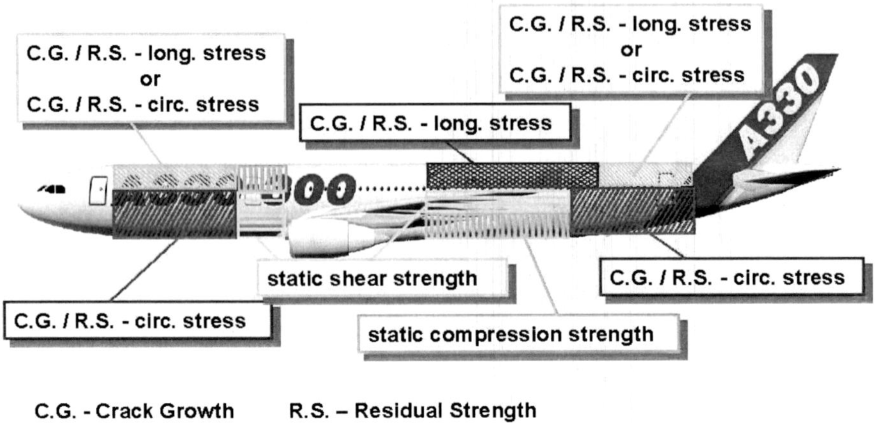

Figure 32. Critical design parameters along an aircraft fuselage (Source H-J Schmidt, Airbus)

NDT METHOD	MINIMUM DETECTABLE CRACK LENGTH	HIGH PROBABILITY DETECTABLE CRACK LENGTH (> 95%)	QUALIFYING REMARKS
X-Ray (XR)	4 mm	10 mm	Dependent on structure configuration, thickness and geometry. Detectable values increase considerably over 12 mm thickness.
Ultrasonic (US)	2 mm	5 mm - Outer fuselage skins 6.5 mm - Inner fuselage skins 2-6 mm – Forgings, spars, extrusions, etc.	Cracks from fasteners holes with fasteners in place Dependent on structure geometry and material type
High frequency Eddy current (HF/EC)	2 mm (surface) 0.5 mm (bore holes)	2.5 mm – Surface 1.0 mm – Bore holes	Surface cracks in accessible areas Bore holes with fasteners removed, using rotating probes
Low frequency Eddy current (LF/EC)	2 mm	4.5 mm – Outer fuselage skins 8 mm – Inner fuselage skins	Cracks from fastener holes with fasteners in place. For thicker structures, the detectable length is approximately equal to the thickness up to 5 mm. Between 5 mm and up to a maximum of 12 mm the detectable length is approximately equal to 2x the thickness. Above 12 mm detection cannot be guaranteed.
Magnetic particle (MP)	2 mm	4 mm Surface	Surface breaking – component removed from aircraft
Liquid penetrant (LP)	2 mm	10 mm Surface	Surface breaking – dependent on surface condition

Figure 33. Detectable crack length for different popular NDT techniques

	CONGESTION RATING		
VIEW RATING	1	2	3
0	0	0	0
1	1	2	3
2	2	3	4
3	3	4	5

	CONGESTION RATING		
SURFACE RATING	1	2	3
1	1	2	3
2	2	3	4

	ACCESS RATING				
SIZE RATING	1	2	3	4	5
1	1	1	2	2	3
2	1	2	2	3	3
3	1	2	3	4	4
4	1	2	3	4	5

	ACCESS RATING			
PRACTI_CALITY RATING	1	2	3	4
1	1	2	3	4
2	2	4	5	6
3	3	5	7	8
4	4	6	9	10
5	5	8	10	11

FEASIBILITY RATING

Figure 34. Rating scheme to determine feasibility ratings in damage tolerant structural design

To determine the initial crack length detectable in a damage tolerant structure a fairly systematic rating procedure has been introduced which is shown in Figure 34. The procedure starts from assessing a congestion rating (the degree of congestion where the component is placed) on a scale between 1 and 3 and a viewing rating (the degree a damage critical location can be viewed) on a scale between 0 and 3 and correlating the two ratings to a new rating called the access rating. This access rating is then further correlated to a size rating (size of the component considered) which then results in a practicality rating. Furthermore the congestion rating considered before is additionally correlated with a surface rating (surface condition of the damage location considered) which results in another access rating. This latter access rating is finally correlated with the practicality rating determined before and results in a feasibility rating, a number which ranges between 1 and 11 and is an indication of the initial crack length to be assumed. An indication of those initial crack lengths versus the feasibility ratings is provided in Figure 35 and shows that those initial crack length can go up to a size of 300 mm easily. The initial crack length determined from Figure 35 is what is also called the basic crack length L_{BAS}.

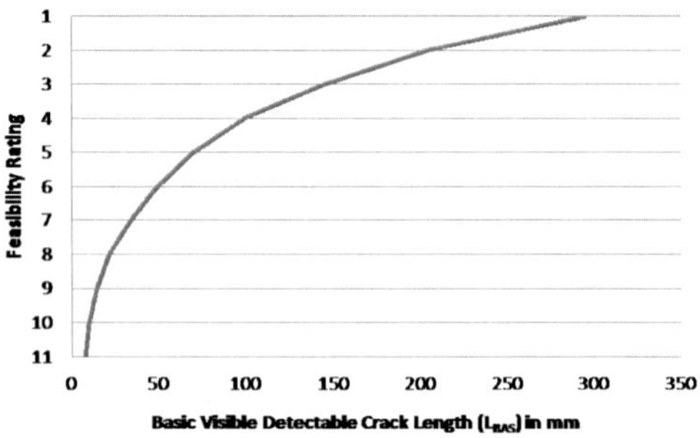

Feasibility Rating	L_{BAS} [mm]
11	8
10	10
9	15
8	22
7	35
6	50
5	70
4	100
3	145
2	205
1	295

Figure 35. Correlation between feasibility ratings and initial crack lengths for damage tolerant design

L_{BAS} may however not be the final crack length to be considered, as can be seen from the example shown in Figure 36. L_{BAS} may therefore be further multiplied by a gauge effect and an edge effect which are based on the following:

Gauge effect

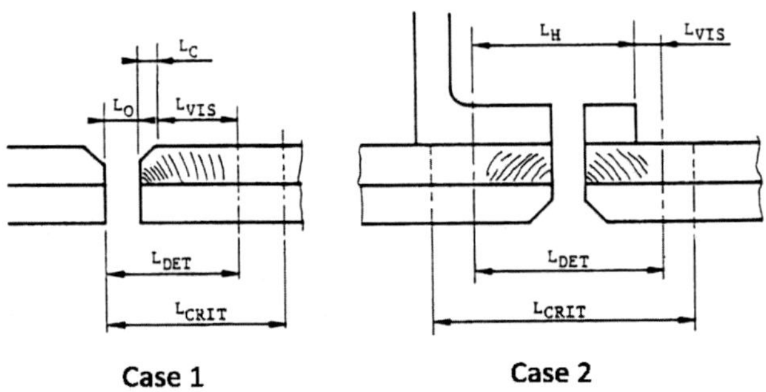

$$L_{DET} = L_0 + L_C + L_{VIS} \qquad L_{DET} = L_H + L_{VIS}$$

Figure 36. Definitions of a detectable crack length L_{DET}

The thicker the material of an item the tighter a crack is held together and therefore the more difficult it is to detect. To cater for this the basic detectable length is adjusted as follows:

1. for material thickness < 5 mm: multiply by 1
2. for material thickness 5 to 10 mm: multiply by 1.25
3. for material thickness > 10 mm: multiply by 1.5

Edge effect

For cracks which originate from or terminate at the edge of a member the detectable length is to be multiplied by 0.5, for non-edge cracks by 1.

Considering the following definitions

L_{BAS} Basic visible length

L_{VIS} Visible detectable length = L_{BAS} × gauge effect × edge effect

L_{CRIT} Critical crack length

L_C Concealed crack length (e.g. under fastener head or other fitting)

L_0 Effective crack length due to hole (as determined by stress office)

L_H Hidden crack length

a detectable crack length L_{DET} can be calculated which is the initial crack length for a crack propagation calculation to follow which allows crack propagation life and hence an inspection interval to be determined. It is amazing to note that such an initial crack length could be up to 500 mm in the most critical circumstances. This demonstrates what potential damage tolerant design might have in designing engineering structures lighter weight.

3.3 SHM, Loads Monitoring and Damage Tolerance

The application of the damage tolerance design principle requires the load sequence applied to be known. Knowledge of the load sequence can be achieved by assuming the load sequence in accordance the different standard load sequences having been generated over the past and where new standard load sequences are still generated for different new applications. Those load sequences have to include a respective safety factor such that they do not generate any critical assumptions. The standardized load sequences provided do usually refer to a single location on the structure only such as a location around the centre of gravity in the case of an aircraft or the attachment of the wing. However would those load sequences also apply to any other location on the aircraft, say to a fitting at the nose landing gear? In that case it might be more advisable to place a sensor at the specific location considered. With SHM this is possible today. A fibre Bragg grating sensor might be easily placed at the location of interest and the real operational load spectrum could be used to evaluate damage accumulated and the location where damage might be due to occur. This way of local loads monitoring would also help to better determine the incident when damage would have to occur as well as potential inspection intervals in terms of damage tolerant design when compared to a global loads monitoring such as around the centre of gravity. However it has to be kept in mind that damage initiation and accumulation is a process driven by a significant amount of statistical parameters. It is therefore useful that loads monitoring based fatigue evaluation procedures can be used to determine the locations prone to fracture around which then a sensor system can be placed that effectively monitors critical damage occurring in accordance to the damage tolerance criteria set. This approach does also guarantee an optimum usage of sensors to be integrated onto the structure to be monitored and provides an optimum strategy with regard to the SHM concept applied.

4 Non-Destructive Evaluation – Some Basic Principles

4.1 Historic Background

Non-destructive testing (NDT) is a science which correlates physical properties to materials' and structural conditions. Historically this correlation is known for thousands of years. Around 8,500 years ago acoustics was determined as a phenomenon for monitoring, followed by magnetism and temperature around 3000 years ago. It is however only during the last 200 years where those phenomena have been converted into what we are considering as science today. In the early 19th century vibration analysis was introduced by individuals such as Fourier, followed by magnetism through Faraday in 1831, radiography by Röntgen in 1895 or dye penetrant around the same time. It is amazing that ultrasonics as a science only dates back to the 1920ies, followed by nuclear magnetic resonance in 1938 and recognised by awarding the Nobel prize to Rabi (in 1944) and Bloch and Purcell (in 1952). It is only in the 1980ies where giant magnetic resistors were discovered with the Nobel prize awarded to Grünberg and Fert in 2007. NDT is therefore a fairly young area of research.

SHM has inherited a variety of the NDT techniques developed with regard to damage monitoring so far. NDT techniques may be categorized into standard and non-standard techniques as to Table 3 below. From those some of them are known to be used within the context of SHM where others are still mainly undiscovered within the SHM context. Table 3 therefore summarises NDT techniques with SHM potentials.

Within SHM partially a different or more specific wording is used for literally the same thing. The NDT techniques with SHM potential determined in Table 3 are therefore compared to some popular SHM terms along Table 4. What turns out is that a variety of specific techniques have been selected less systematically in SHM than this is now established in NDT. Some of the NDT principles can therefore be found in SHM, however there is still much potential for SHM to be explored in NDT where a few examples are explained in more detail in the sub-chapters to follow.

4.2 Magnetism

Magnetism is a physical parameter inherent to any ferro- and even paramagnetic material. It can be measured by inducing a magnetic field into the component to be monitored and sensed by the resulting magnetic response in accordance to the principle shown in Figure 37. This will allow material properties to be monitored up to a specific depth depending on the magnetic field to be introduced as well as effects (i.e. cracks) to be observed on the specimen's surface.

Microstructure of a magnetic material is composed of domains which can be considered as small magnets arbitrarily oriented in the specimen considered in case the specimen has not been considered as a permanent magnet. Those domains are separated by Bloch walls which are nothing

- **Standard**
  - ***Visual***
  - ***Vibrations***
  - ***Ultrasonics***
  - ***Eddy Current***
  - X-Ray
  - ***Thermography***
  - Microscopy
- **Non-Standard**
  - ***Acoustic Emission***
  - ***Magnetics***
  - Nuclear Magnetic Resonance
  - ***Microwave/Radar***
  - ***Optical Coatings***
  - ***Nanoagents/-coatings***

Table 3. NDT techniques with SHM potential (in bold italic)

- **Acoustics**
  - Guided Waves
  - Acoustic Emission
  - Piezoelectric Transducers
- **Eddy Current**
  - EM Foils
- **Modal/Vibration Analysis**
- **Miscellaneous**
  - Thermal
  - Comparative Vacuum Monitoring

- **NDT with SHM potential**
  - ***Visual***
  - ***Ultrasonics***
  - ***Acoustic Emission***
  - ***Vibrations***
  - ***Magnetics/Eddy Current***
  - ***Thermography***
  - ***Optical Coatings***
  - ***Nanoagents/-coatings***

Table 4. NDT techniques associated with SHM

else than a sharp conversion of the electrons' orientation with regard to each of the neighbouring domains' magnetic orientation. Bloch walls are 'pinned' to lattice imperfections within the material would those imperfections be dislocations, precipitates, grain and phase boundaries or any other type of imperfection. If a magnetic material is exposed to a magnetic field the way this is shown in Figure 37 Bloch walls will gradually move and jump from one imperfection to the next leading the domains to gradually orient towards the magnetic field being imposed. This process of enforced reorientation continues with increasing magnetic field until a saturation level is achieved along which all domains are finally oriented in the direction of the enforced magnetisation. Figure 38 shows schematically how the magnetisation process is introduced and three stages of such a magnetisation process in form of some microscopic images can be seen in Figure 39. Once the magnetic field imposed is again reduced Bloch walls and domains do gradually establish again and the process might be inverted in case the magnetic field will be oriented into the opposite direction. This does result in a hysteresis loop like the one shown in Figure 40 which correlates the magnetic field H generated versus the magnetic flux density B generated in the specimen. Further details on magnetisation of materials can be found in textbooks such as [21,22].

Studying phenomena of magnetisation along the hysteresis loop shown in Figure 40 in more detail will elucidate a variety of interesting phenomena. First of all the shape of the hysteresis loop itself is a characteristic where the value at zero magnetic flux is considered as coercivity and the value of

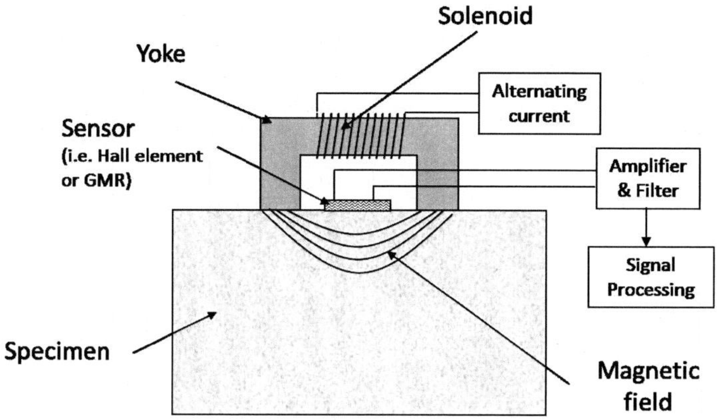

Figure 37. Working principle of magnetics monitoring

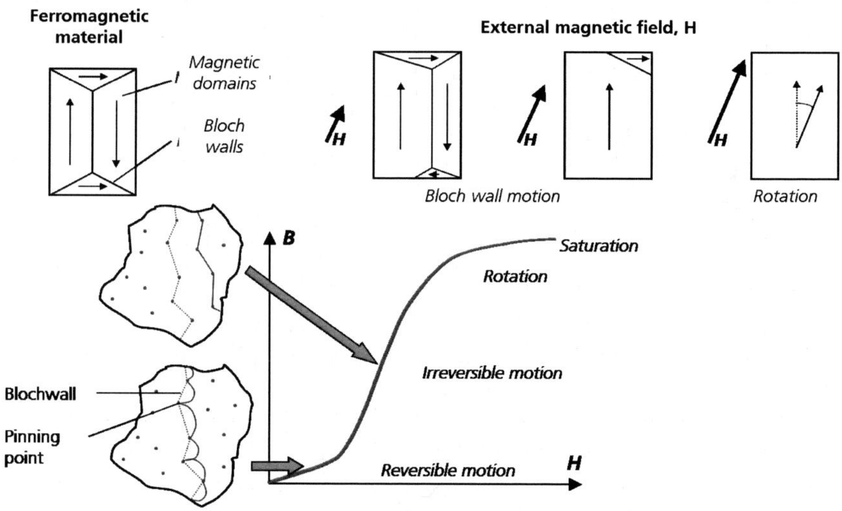

Figure 38. Bloch wall distribution and initial magnetic hysteresis curve in accordance to the four domain model by Cullity (21)

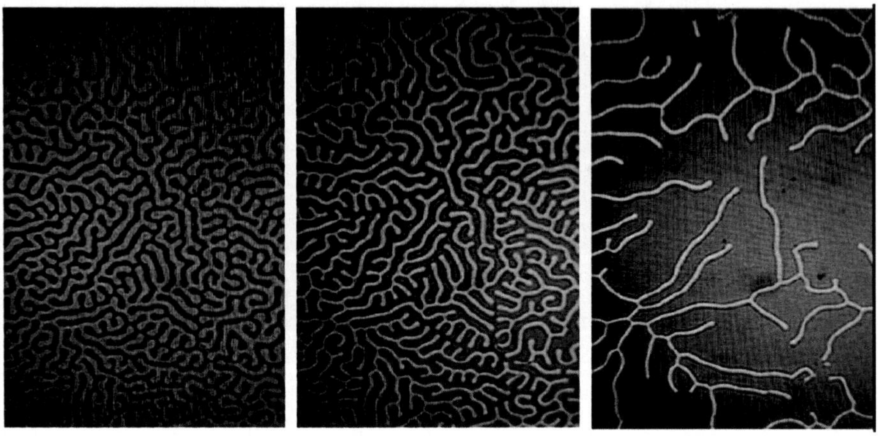

Figure 39. Change of domain structures with increasing magnetisation (from left to right)

zero magnetic field as remanence respectively. It can be further observed that Bloch wall movement at the higher magnetisation scales as well as at rotation processes of electrons around the saturation level are both stress sensitive.

Stress sensitivities may also be observed around the point of coercivity (BW2) in an indirect way. The phenomenon observed in that region is Barkhausen noise for which the principle is shown in Figure 41. Barkhausen noise is generated when Bloch walls move uncontinuously from one imperfection to another and this movement generates pulsed eddy currents to be measured by a sensor which need to be adequately amplified. The resulting output signal is currently characterised by its amplitude and also envelope. When performing a Barkhausen noise test on a specimen being exposed to different loads and hence stresses a change of the maximum amplitude of the Barkhausen noise signal is observed as can be seen from the results shown in Figure 42. Furthermore the field strength where this maximum occurs may change too, which also becomes obvious when looking at the shape of the magnetic hysteresis loop being generated since the maximum of the Barkhausen noise is correlated with coercivity in a large number of cases. Plotting the maximum of the Barkhausen noise signal over the stress applied on the specimen leads to a low scattered non-linear relationship (23). Reasons for this non-linearity are various where magnetostriction may be one with others being subject of on-going and future research.

Since electromagnetic and magneto-elastic quantities of a ferromagnetic

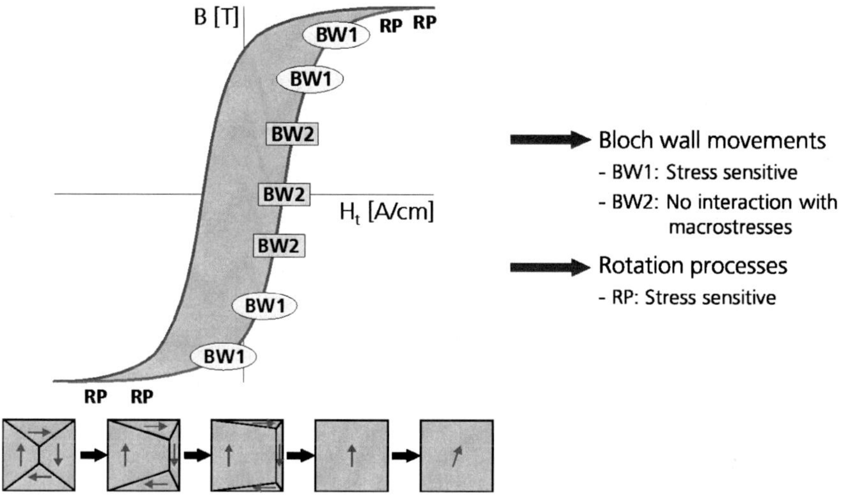

Figure 40. Typical magnetisation hysteresis loop

material are influenced by stress and microstructure and any resulting changes from this, a calibration may be considerable. This may be achieved on either a tensile or bending test. A one dimensional calibration based on magneto-elastic Barkhausen noise measurement is usually sufficient where the amplitude of the magnetic Barkhausen noise is measured as a function of the tension and compression applied to the specimen. This calibration seems to be sufficient as long as the stress in the second principal direction is less than about 25% of the elastic limit of the material. Otherwise biaxial calibrations have to be performed (24) or a calibration procedure based on specific calibration functions as described in (25).

Besides Barkhausen noise and the shape of the hysteresis loop and hence magnetic permeability being the derivative of the hysteresis loop in general there are two further magnetization effects of significance where one of them is the impedance measurement of an induction coil known as a parameter from conventional eddy current testing and the other is higher harmonics analysis. Higher harmonic analysis of the tangential field strength is the magnetic response signal resulting from an external magnetic field excitation alternating at a specific frequency and being measured as a response signal from the component analysed in terms of a fast Fourier transform analysis, resulting in a harmonic distortion factor mainly considering the first odd harmonics. All of those magnetics phenomena are summarised in the chart shown in Figure 43 below. Looking into further details Fraunhofer IZFP

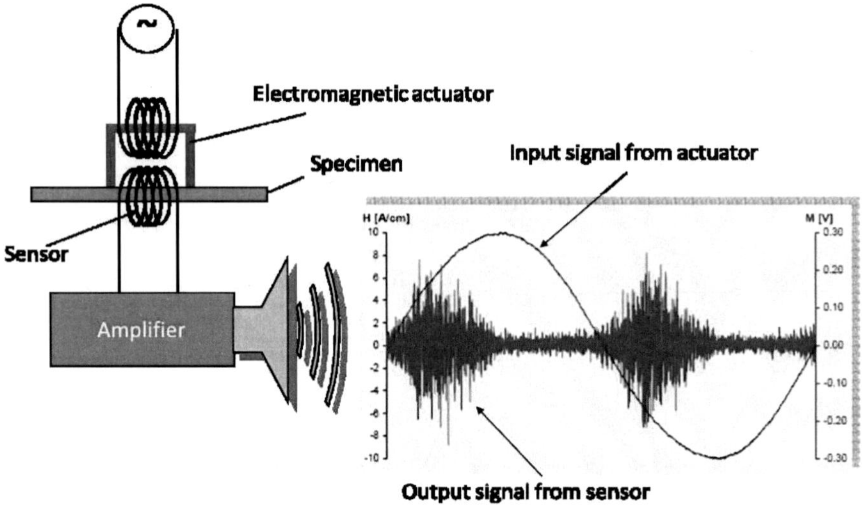

Figure 41. The Barkhausen noise principle

has identified up to 42 parameters in that regard which are selected and tuned in accordance to the specific application to be considered (26).

That micromagnetic techniques can be efficiently used for monitoring ageing structures which has been clearly shown in the case of a steam pipe material (27) that can burst due to ageing phenomena. The steel pipes made of WB36 and operated at temperatures around 350 °C generate copper precipitates over time and results in a significant increase in hardness, brittleness and hence fracture stress intensity. Those copper precipitates are excellent pinpoints for Bloch walls and may be indirectly measurable through the Barkhausen noise being generated as shown as a principle in Figure 44. That the assumption of correlating magnetic with mechanical hardness has indeed become true has been validated experimentally where material of the WB36 steel pipe has been continuously aged and hardness increase measured has been compared to the micromagnetic signals monitored. Results of this comparison are shown in Figure 45 and a good correlation can be observed which recognises micromagnetics already as a promising technique for monitoring ageing phenomena in magnetic materials.

As already explained in the context of Figure 40 micromagnetic techniques are also sensitive to stress measurement. This has made those techniques attractive for stress measurement in deep drawing steels such as

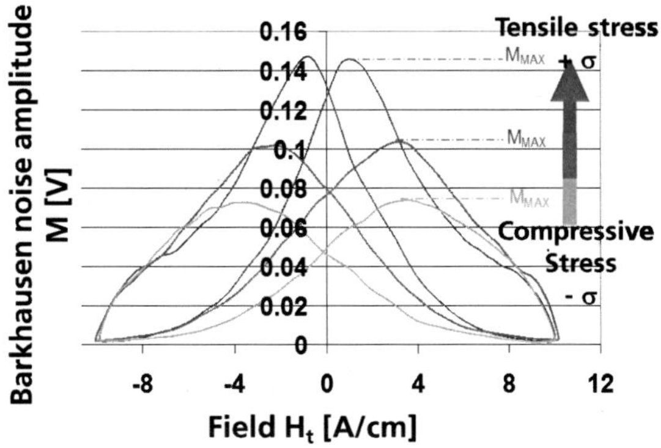

Figure 42. Barkhausen noise sensitivity to applied stress under uniaxial loading

used for automotive structures. Figure 46 shows a comparison of residual stress measurement on formed sheet metal for automotive application and demonstrates the significant advantage micromagnetics has when compared to radiography.

Sensing with micromagnetic techniques is traditionally done with a coil or possibly even a Hall sensor. A significantly emerging alternative is however the use of giant magnetic resistors (GMR), which have become a standard in all kinds of electronic data communication and specifically data storage devices such as harddisks. As schematically shown in Figure 47 a GMR is a composite system consisting of at least two ferroelectric layers switched in opposite directions and separated by an electrically conductive paramagnetic layer. For an antiparallel configuration of the electron spin orientation the electrical resistance is high and hence any impedance measured when placing the GMR sensor in an electromagnetic field will become high as well. Figure 47 shows on the right hand side the size of such a sensor (in that case of four sensors due to a Wheatstone bridge) which demonstrates that such a sensor could be implemented in damage critical notches of fairly small size.

Monitoring structural components with GMR sensors has provided some interesting results where some have been published in (28). One of those examples is shown in Figure 48 and demonstrates that electrical impedance measured by a GMR sensor system can be well correlated to plastic strain

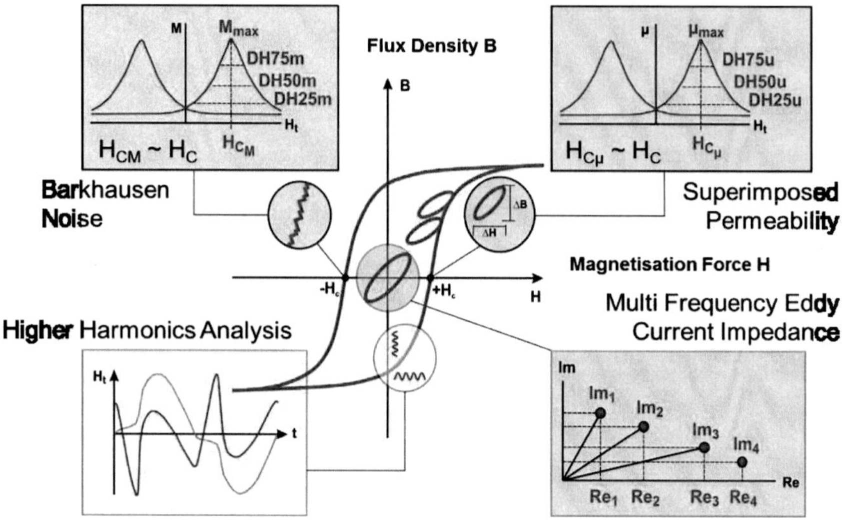

Figure 43. Magnetic hysteresis curve and magnetic phenomena considered

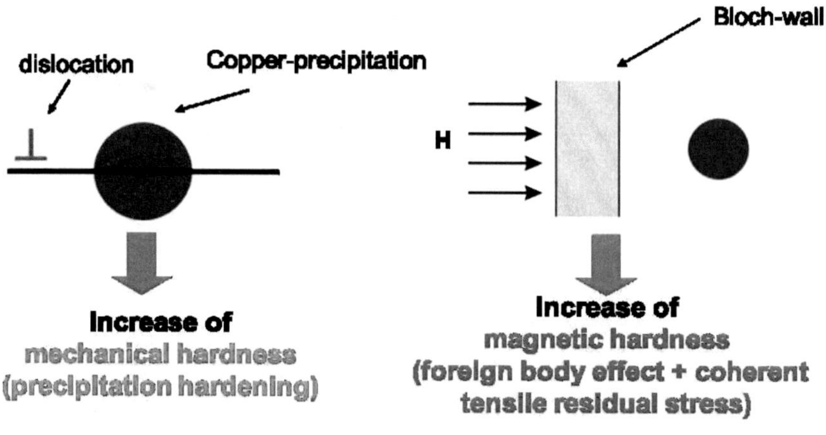

Figure 44. Analogy between dislocation movement due to mechanical loads and Bloch-wall movement under magnetization

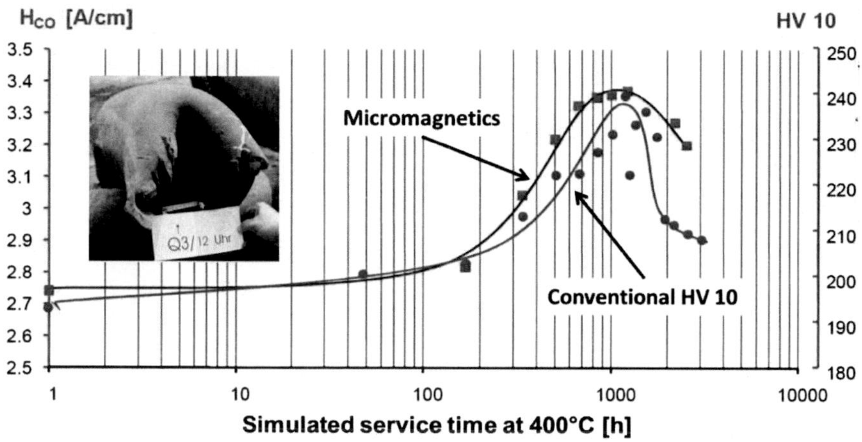

Figure 45. Analogy between mechanical and magnetic hardness

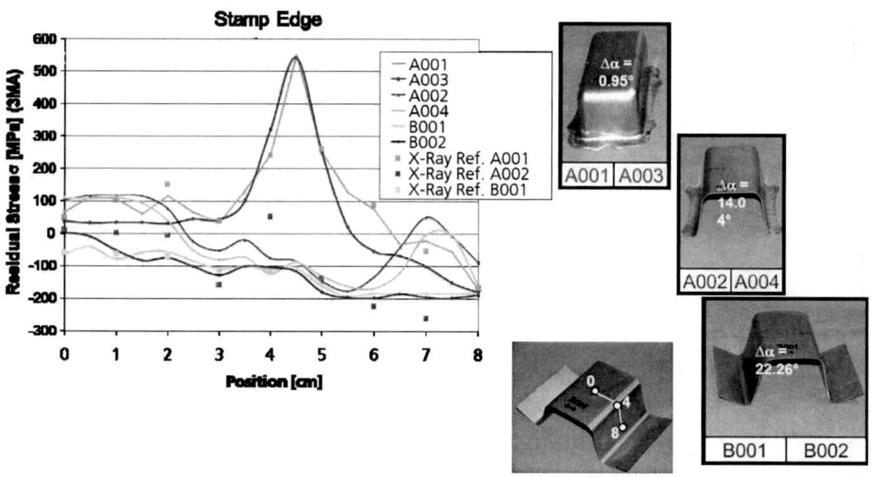

Figure 46. Measurement of residual stresses in formed steel using micromagnetics and compared to conventional X-ray analysis

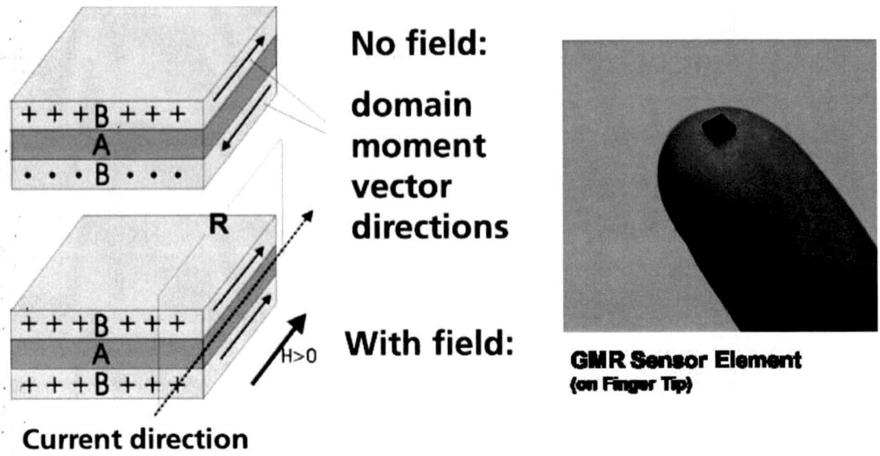

Figure 47. Giant magneto resistor (GMR) sensing device

being generated along a fatigue test under random loading.

All together micromagnetics shows a highly promising but far from being exhaustively explored area for monitoring ferro- and paramagnetic materials and structures. Since the technique is also sensitive to microstructural effects it can be used to determine material properties at the early stage of damage incubation too, much earlier than any crack monitoring technique does. An interesting correlation shown in Figure 49 below can therefore be principally observed on a microstructural level between micromagnetic values on the one side and mechanical values on the other.

4.3 Electromagnetic Ultrasound (EMUS)

The principle of electromagnetism shown in Figure 37 above can be also applied in a wider sense, such as generating an ultrasonic wave the way this is shown as an example in Figure 50. In this case the dynamic magnetostriction is generating an ultrasonic wave in addition to the magnetic properties described before, which can provide additional information of a component's condition, specifically when a larger volume of the component considered is intended to be monitored. This is what is also called electromagnetic ultrasound (EMUS). An interesting and also highly successful application is the examination of railway wheel rims as shown in Figure 51 below. In that case a Rayleigh (surface) acoustic wave is sent around the rim with echoes and through-transmitted signals being recorded. Any turn of the

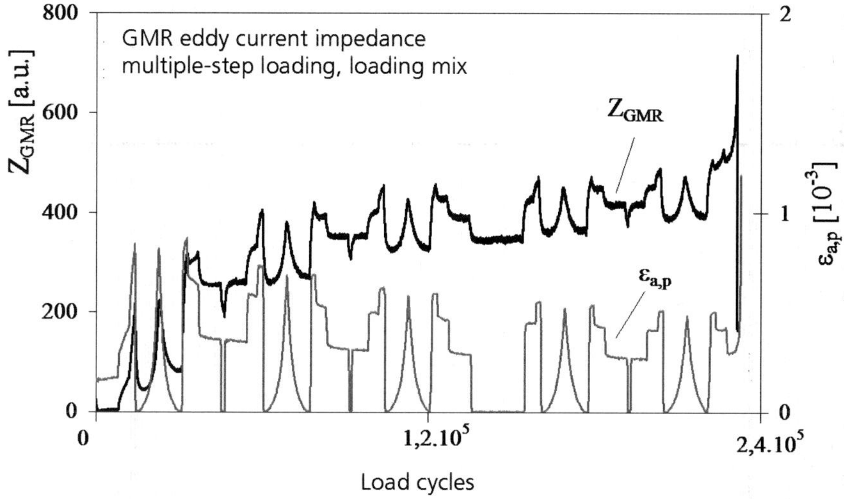

Figure 48. Plastic Strain Characterisation with GMR Sensors for AISI 321 Steel under Service Loading

signal around the wheel can be recognised through the time of the signal's flight while any imperfections on the rim are recorded by echoes observed in between. The EMUS actuation and sensing unit is placed within the rail and allows trains to roll over at speeds up to 15 km/h while the inspection process is performed. EMUS has also been successfully applied in the context of inspecting welded sheet metal such as tailored blanks, where guided waves are generated being sent through the sheet metal and being reflected by the weld in accordance to a defined condition. A major advantage of EMUS is its ability to generate shear horizontal waves which piezoelectric actuators are less able to do and the avoidance of a coupling medium which allows EMUS to be also applied at elevated temperatures.

4.4 Eddy Current

Eddy current is another electromagnetic technique applied. Instead of a yoke with a coil operating as a solenoid a simple coil is used to generate an alternating magnetic field through an alternating current. This will generate eddy currents in any electrically conductive material which can be measured by the electrical impedance. Figure 52 shows the principle

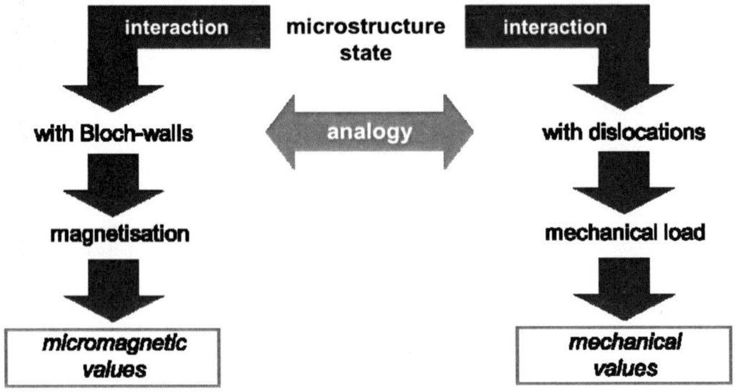

Figure 49. Correlation between micromagnetic and mechanical values

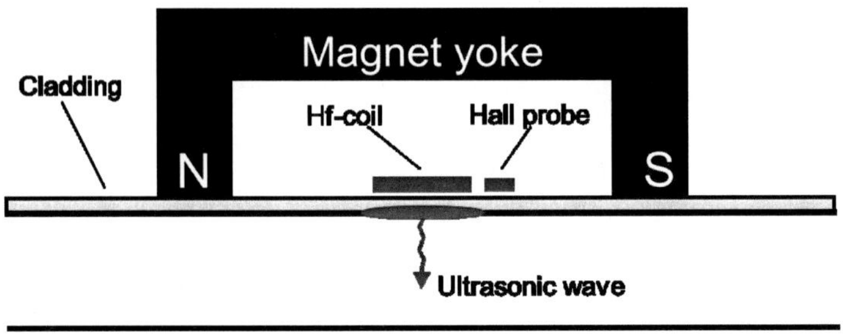

Figure 50. EMUS for monitoring a material's condition under a cladding

of eddy current inspection. Eddy current testing can be performed in a wide range of frequencies depending on the size of damage to be detected and the location of the damage below a surface. Principally the higher the frequency becomes the smaller a defect can be detected and the lower the damage below the surface can be. Eddy current testing may also be used for monitoring the thickness of coatings.

Eddy current testing has been traditionally used for the inspection of metallic materials. However this may also be used for other electrically conductive materials such as carbon fibre reinforced composites (CFRP). Recently a scanner has been developed shown in Figure 53 which allows for the inspection of CFRP materials up to a defined thickness using high

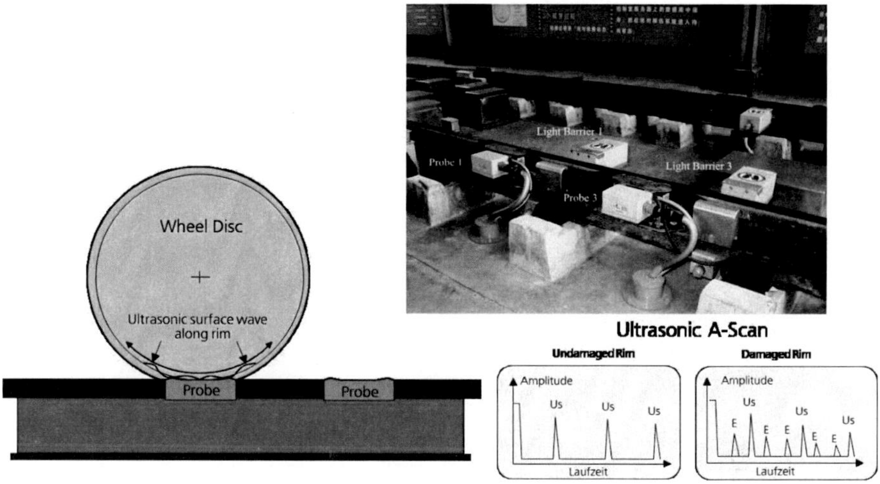

Figure 51. In-motion EMUS testing on railway wheel rims

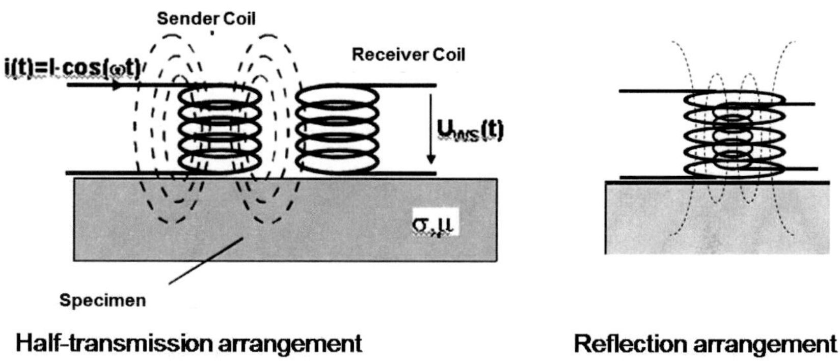

Figure 52. Principle of eddy current testing

frequency eddy current (29). The sensor being used is a very specific matrix sensor which allows for monitoring techniques to be used similar to what is known in the context of phased array techniques in acoustics.

Eddy current has been also applied in the context of SHM as eddy current foils where coil typed structures have been placed on Kapton foils which again could be placed on or between lap joints such as for corrosion detection (30).

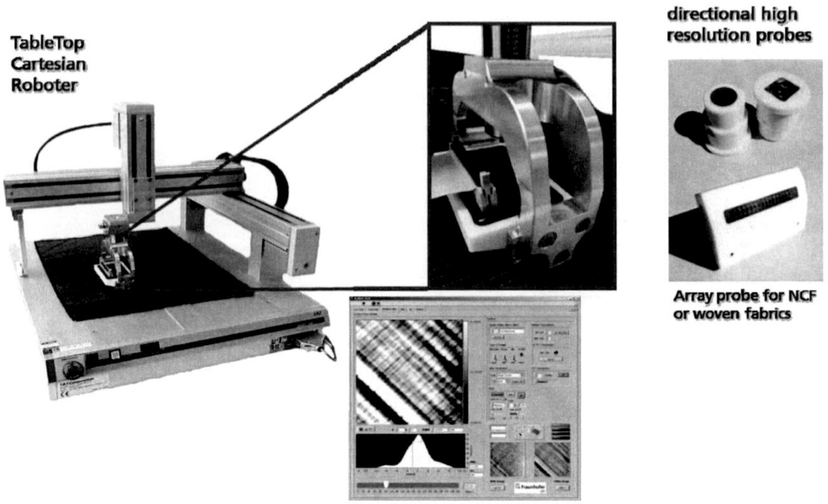

Figure 53. High frequency eddy current based scanner for CFRP testing (Courtesy of Fraunhofer IZFP)

4.5 Ultrasonics

Acoustics and hence ultrasonics using piezoelectric transducers is the physical principle possibly used most in NDT. Current emphasis is specifically on the use of phased arrays combined with advanced data management. A method in that regard principally shown in Figure 54 is sampling phased array where a signal is sequentially sent by one of the transducers and the echo is recorded by all the others. All time domain data generated is stored in a data base which is then taken as a basis to determine an optimum spatial and contrast resolution by calculating the image recorded mainly through phase shifting and the Synthetic Aperture Focusing Technique (SAFT). Images obtained in near to real time are 3D volumetric pictures of high resolution as shown as an example in Figure 55. This approach may also be used in the context of SHM where transducers are sparsely distributed around a structure and the resolution of images obtained may be enhanced.

4.6 Other Conventional NDT Techniques and Emerging Technologies

There is a variety of other NDT techniques widely considered in structural inspection. One of them is thermography where a heat impulse such

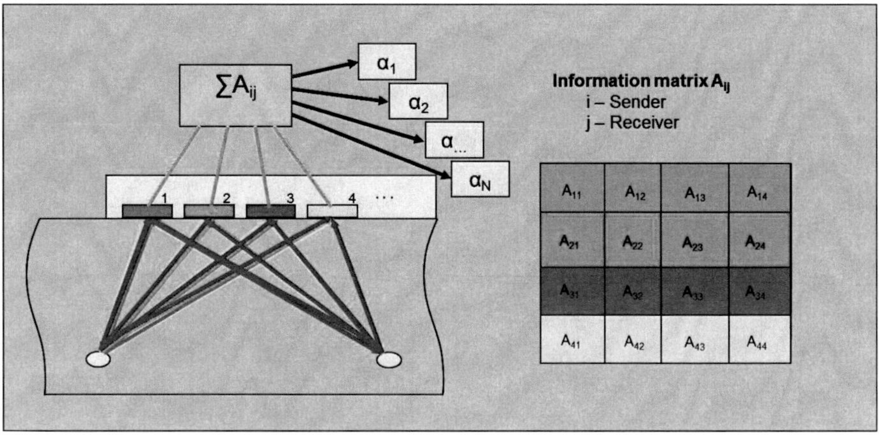

Figure 54. Data sampling procedure using the sampling phased array principle

as generated through a halogen lamp is sent into the structure to be monitored and the temporal change of the reflected temperature signal is integrally recorded by a camera system such as an IR camera (Figure 56). Any changes in the material's condition due to defects may be observed through a change in density, heat conductivity and specific heat capacity observed by a change in amplitude, phase or even both being measured. Heat actuation may be also generated through vibration such as a sonotrode or some multi-use sparse piezoelectric transducers within the structure being considered. Another alternative explored is local heating by a laser. The advantage of thermography is its integral, non-contact and non-invasive way of monitoring which makes it attractive for any high speed monitoring. However resolution may still have to be enhanced where current research still demonstrates significant margin for improvement.

Another NDT-technique being highly popular for quality assurance purposes is radiography as shown as a principle in Figure 57. The object to be inspected is exposed to x-ray radiation while the projection image is recorded by a digital image recorder. Since the object is turned around on a manipulator data acquisition is performed for a variety of different angles. Through data processing and use of a computing tomography algorithm a 3D image is generated which allows the size and volume of a damage to be determined quantitatively. X-ray sources are currently more and more placed on robotic systems which provide further inspection flexibility when the image recorder can be moved around accordingly such as shown with

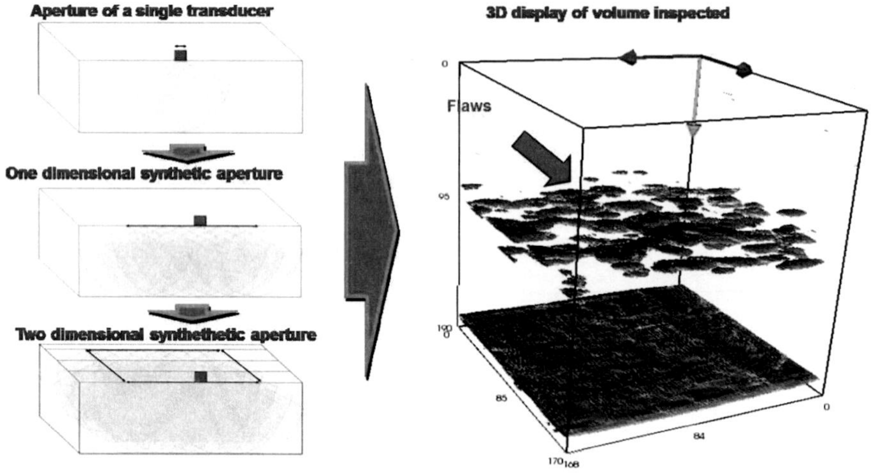

Figure 55. Quantitative 3D imaging using the sampling phased array approach

laminography in Figure 58. However due to being invasive radiology can be used as a validation method for SHM only.

Beyond all this there is a multitude of other optical techniques to be explored would those be based on planaroptical microresonators, optical coatings or any type of nanoparticles influencing fluorescence, all of those components to be adapted or integrated into the structural material considered. Doping structural materials with damage sensitive elements is a current trend in materials research ongoing with significant expectations regarding the future.

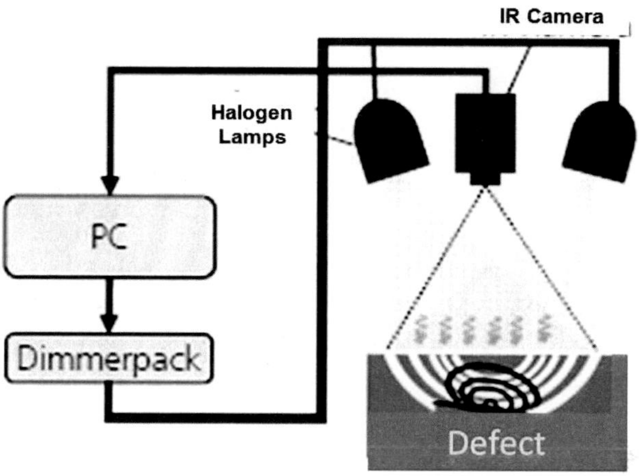

Figure 56. Principle of lock-in thermography testing

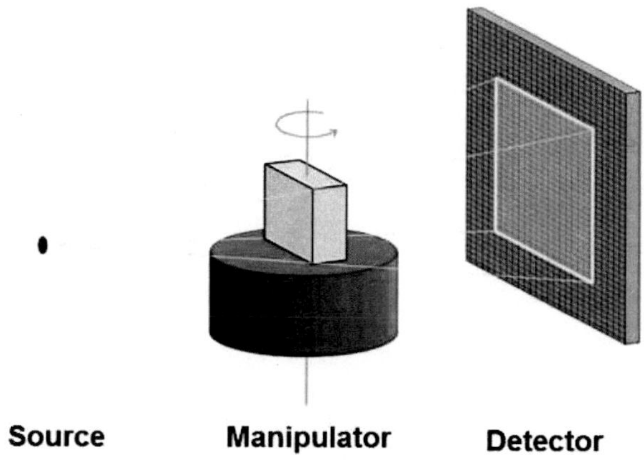

Figure 57. Principle of x-ray testing

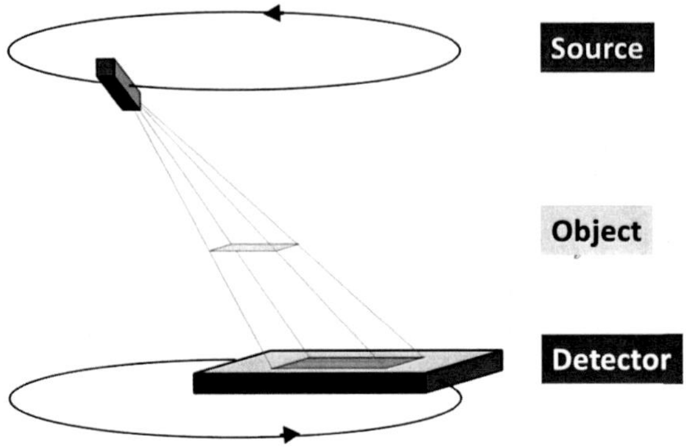

Figure 58. Laminography as a new way of x-ray inspection

5 Non-Destructive Evaluation in Composite Materials

The increasing request for glass and specifically carbon fibre reinforced plastic (CFRP) materials not just in aviation but recently also in the automotive sector has triggered a lot of thought for further improvement of NDT techniques to be used. Current testing equipment is rather bulky, costly and specifically time consuming. Techniques used so far in aviation are ultrasonic testing (UT) and radiography based. Furthermore damage mechanisms in composite materials are much less known when compared to metals. Hence a safety factor has to be imposed along the design of those CFRP-structures as long as damage mechanisms are not sufficiently known and a damage tolerance principle cannot be applied. The objective of NDT for CFRP structures is in enhancement of flaw detectability and quantification, reducing inspection time and cost.

When looking at state-of-the-art NDT testing for CFRP structures, ultrasonic based phased array testing is the procedure most commonly used. X-ray testing, mainly on a computer tomography (CT) basis, is fairly exact but a costly and time consuming NDT procedure which is therefore considered for verification purposes only. There are other NDT techniques considered as well, however they have not been approved in many fields of application (i.e. aviation) due to lack of certification so far. To better understand the current state-of-the-art in CFRP-related NDT technology a benchmark test has been performed. Driven by leading edge technology

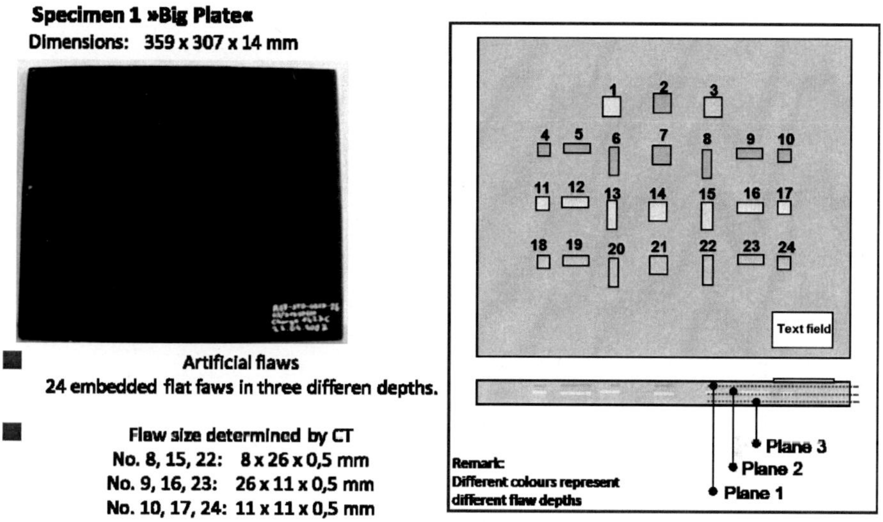

Figure 59. Plate with flaws of different size and at three different depths (Specimen 1)

such as sampling phased array in the case of UT and 3D imaging algorithms in the case of radiography, those two traditional techniques have been compared to emerging NDT techniques in the CFRP field including thermography and eddy current testing. The components considered for validation have been mainly stepped flat panels with integrated artificial flaws such as shown in Figures 5.1 to 5.4 respectively.

Results obtained in detecting the various flaws are summarised in Table 5. None of the specimens were a challenge for the two 'classical' NDT techniques being phased array UT and x-ray CT where 100% of the flaws were detected with even excellent resolution. Specifically UT benefitted in resolution from using sampling phased array. Flaws in the knee point of specimen 4 could be detected after a sensor head was used that was in accordance to the specimen's geometry, being an indication of UT limitations with regard to increased complexity of a component's geometry. Limitations with x-ray CT has to be seen in the weak distinction between material densities and also the high effort required in general would it be time, cost and safety. However space is also a limitation if a conventional CT-equipment is considered to be used.

The major reason why high frequency eddy current (HFEC) and ther-

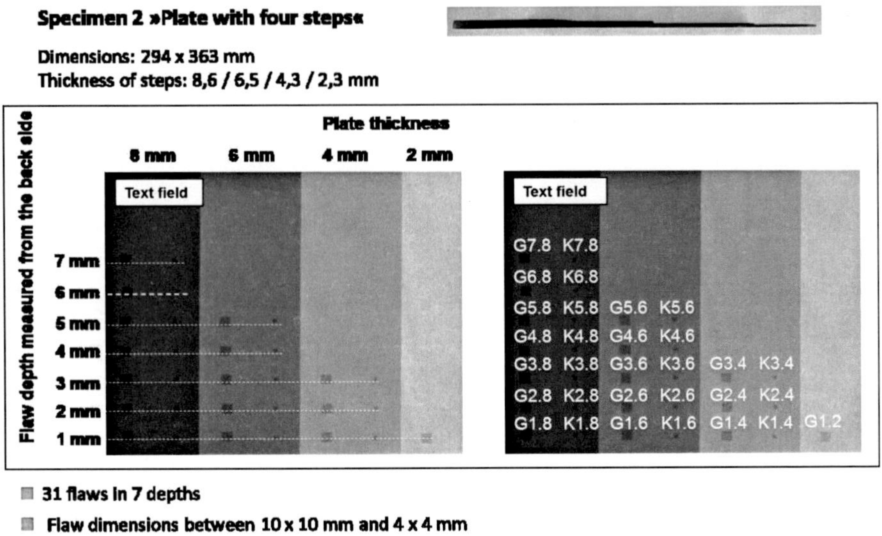

Figure 60. Four stepped plate with same sized flaws at different depth (Specimen 2)

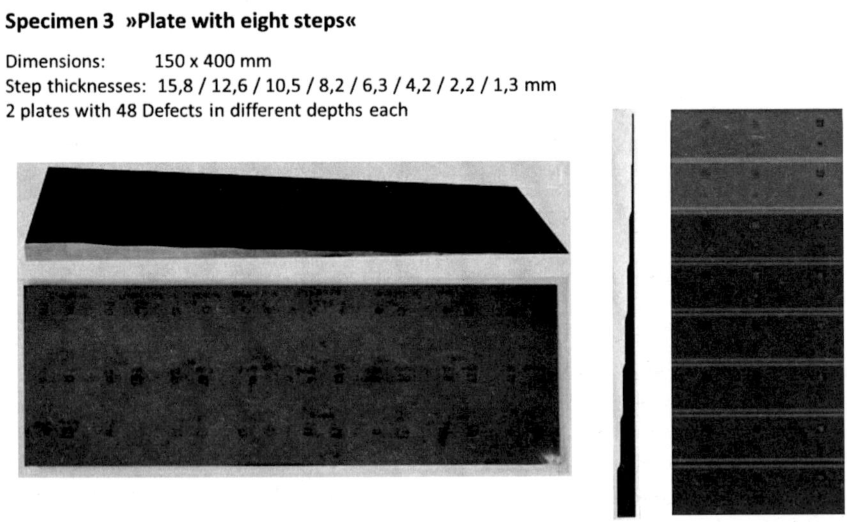

Figure 61. Eight stepped plate with same sized flaws at different depths (Specimen 3)

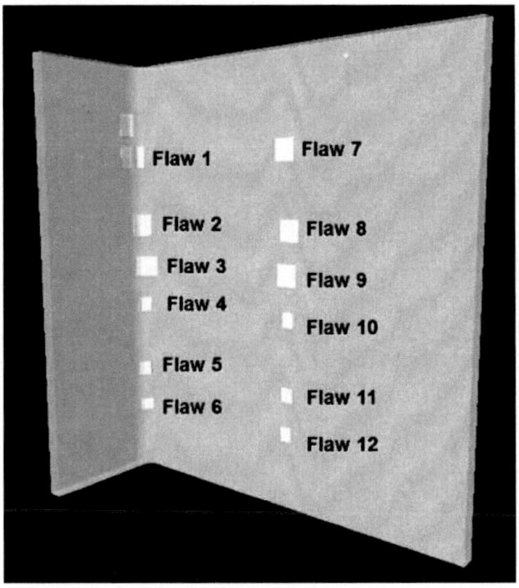

Dimensions: L-shaped 40 x111 mm, thickness 4.2 mm

Figure 62. L-shaped clip type specimen with different flaws at three different depths (Specimen 4)

mography have only been able to detect a little more than half of the flaws results from the fact that their resolution in depth is very much limited. So far this is limited to about 7 mm in depth. Results can be improved if the specimen is allowed to be inspected from two sides. Further limitations mentioned in Table 5 currently still prevail. HFEC and thermography have other significant advantages when compared to UT and CT where non-contact sensing in case of both and large space monitoring in the specific case of thermography have to be mentioned. A summary of various advantages and disadvantages of the different NDT-techniques considered is provided in Table 6.

Since inspection time is a critical parameter in CFRP component manufacturing a slightly more detailed look has been placed on this parameter. A 30 x 30 cm CFRP panel has therefore been selected as a reference being slightly similar to the larger specimens 1 and 2 discussed above. Although that comparison could not be made in all details a time target was at least set such that inspection time would be achieved below 30 seconds. Results from that comparison can be seen in Table 7. Current state-of-the-art technology shows that inspecting a component of the respective size is a matter of minutes. UT can so far only benefit from phased array testing and high performance signal processing and hence computing and has significant po-

NDT method	Flaws detected %	(Physical) limitations
UT / SPA	100 % (163/163)	- Complex geometry
CT	100 % (163/163)	- Weak distinctions in density
EC	52.8 % (86/163)	- Weak distinctions in conductivity especially in depth direction - Only flaws near the surface
Thermography	54.8 % (86/157)	- Weak distinctions in density, heat capacity and heat conductivity - Only near surface flaws detectable; - Rule of thumb: lateral dimensions > depth

Table 5. Results of NDT-technique benchmark on CFRP specimens

NDT method	Advantages	Disadvantages
UT / SPA	3D volume capture High flaw detectability Quantitative flaw evaluation	Coupling is required – complex mechanics Significant equipment costs for complex geometries
CT	3D volume capture High flaw detectability High spacial resolution Quantitative flaw evaluation	Limitations on object size Significant equipment costs
EC	Fast, flexible, coupling free Low cost equipment High potential for automation	Limited sensitivity Limited depth range Only qualitative detection, flaw depth/size can only be estimated
Thermography	Simple testing procedure (no contact, less scanning) Fastest possible testing for large components and also complex geometries	Limited sensitivity Limited depth range Only qualitative detection, flaw depth/size can only be estimated

Table 6. Advantages and disadvantages of NDT-techniques benchmarked

NDT method	Pure inspection time without arrangement	
CT CT Laminography	from 5 min to 30 min < 30 s	(surface 30 cm x 30 cm)
UT / SPA	< 30 s	(surface 30 cm x 30 cm)
Lock-In thermography	ca. 100 s – 400 s	(surface 30 cm x 30 cm)
Pulsed thermography	< 30 s	(surface 30 cm x 30 cm)
EC	< 30 s	(surface 30 cm x 30 cm)

Table 7. Benchmark related to inspection time of a laboratory based component

tential to meet the time target set. With CT laminography will be the method of choice helping to speed up the process while for thermography pulsed thermography may become the preferred method to choose. HFEC will easily meet the target specifically when matrix sensors will be used and the variety of signal processing options will still have been explored to their full extent.

In general it can be concluded that there is still a huge potential in enhancing NDT-based CFRP inspection which will allow inspection effort to be reduced by factors and times possibly too. These enhancements will certainly have a significant impact on further techniques to be applied with SHM and it is advisable to observe developments made in NDT and vice versa.

6 Approaches to Structural Health Monitoring

Although there is a lot of options available to get SHM established there are still a variety of open questions to be answered to get SHM implemented in the first place. Frequently asked questions in that regard may include: What does SHM cost? How to establish SHM? When to establish SHM? When will SHM pay off? This chapter tries to give some answers.

SHM is not something completely new. In aviation SHM was recognised after the serious Comet accidents in the early 1950ies although nobody used the expression of SHM at that time. Acceleration sensors were implemented on fighter airplanes and load spectra were continuously tracked, which then resulted in a variety of standard load sequences to be used for aircraft qualification [7,31]. Today a modern fighter aircraft like Eurofighter Typhoon has a 16 strain gauge operational loads monitoring system which allows real load sequences to be tracked and damage accumulation to be determined. Similar approaches are made with bridges and other load critical structures and components. With the development of optical fibre Bragg sensing a completely new option of sensing has been provided that allows complexity of sensing to be reduced by orders of magnitude. One hundred conventional strain gauges with two wires each can now be easily replaced by a single fibre, giving rise to a multitude of interesting new applications.

A major rule in implementing SHM must be: *Maximise a SHM system's efficiency by minimising the number of sensors to the most efficient ones.* Sensors are only a part of an SHM system, which contains much more. A major element of an SHM system is simulation, which can be large computational machines not doing anything else than simulating the performance of a structure under various conditions. This simulation does require an input from sensors and the question is: *Which input will have which impact on an SHM system's performance output?* This is nothing else than an optimisation task which as a consequence asks for minimising the sensor's input to maximise a structure's performance output. In the case of SHM this means to start where structural design starts as well and this is with strength analysis of materials and structures. A strength analysis requires as an input material properties, the structure's geometric shape and operational loads. Material properties can be considered to be given, specifically when materials have been qualified through a quality assurance process. Geometric shape can also be considered to be given, where no change is to be expected over the structure's life. The only major unknown at this stage is operational loads, where assumptions are made prior to a structure's realisation and where safety factors have to be implemented to cover any unexpected uncertainties. Reducing those uncertainties is where sensors have to be implemented first and where major benefits can be expected due to the high safety (or better penalty factors) imposed initially. Determination of the type and sensors' best locations therefore has to come from the structural simulation process itself.

Once an optimum of loads monitoring sensors has been determined, placed on the structure and operating reliably safety factors can be reduced resulting in either lighter weight design and/or longer endurance, depending on what the operational target of the structure considered would be. A next factor of uncertainty within the structure's operational life is damage possibly combined with a change in the structural material's condition and the probabilistic nature of damage to occur. Again simulation might be a tool of major benefit although this time the tools might be more towards sim-

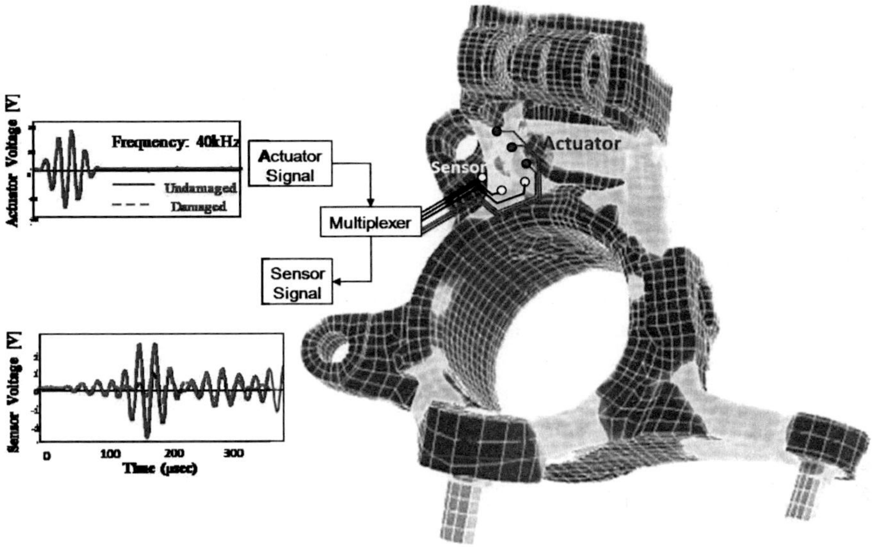

Figure 63. Fatigue life and damage evaluation to determine 'hot spots' for damage monitoring

ulation of materials properties on a molecular and possibly even atomistic level which might generate ideas for sensing other than those traditionally looked at in terms of strain monitoring. With all that information available from the load and material condition monitoring sensors as well as from the simulation algorithms applied locations of likely damage occurrence can be determined even on a prognostic level which will again allow an SHM-based damage monitoring system to be determined on a targeted and hence optimised level. An example for such a concept is shown in Figure 63 where a map indicating the locations ('hot spots') of highest damage accumulation is obtained through prognosis and hence simulation. It is for those 'hot spots' where simulation has then to be performed again, this time on the basis of simulating physical waves travelling around the hot spot and the allowable damage to occur such that the damage tolerated can be reliably detected with the NDT-based transducer configuration to be determined through simulation and successively then realised in hardware.

Although Figure 63 might imply that actuators and sensors or hence transducers might be the key elements of SHM this is by far not the case. What is essential with SHM is the SHM system. An SHM system as shown

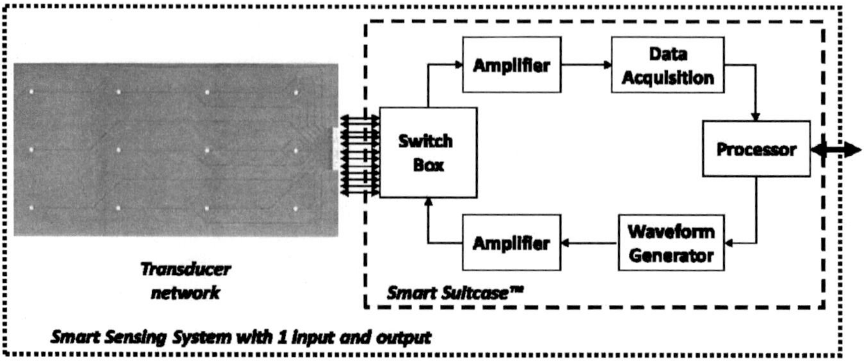

Figure 64. Principle of an SHM sensing system for damage monitoring

schematically in Figure 64 consists of a defined number of transducers which are managed by a signal generator, amplifiers, a multiplexer, a data acquisition unit and a processor. The most important requirement with this SHM system is that the single input and single output interface of the SHM system provides a reliability of information defined by the SHM system user requirements. If the rate of false alarms is set to 10^{-7} by the SHM system's user, then it is the obligation of the SHM system provider to get a system realised that will meet that requirement irrespective of how many transducers the system will need as long as the SHM system meets any other requirements set such as weight and cost.

Damage monitoring in aviation has been very much triggered by the Aloha Airlines accident which happened in April 1988. This accident made the public very much aware of a phenomenon called multi-site damage (MSD) which is when a structure suddenly fails like a perforated paper, because small cracks not fully detectable by means of NDT have been mainly generated at each of the rivets along a rivet line. As a consequence all aircraft types with an age of 15 years and more were analysed in terms of their damage critical components where additional care is required for those components with regard to damage tolerance which has resulted in enhanced inspection effort. This additional inspection effort is an interesting opportunity for SHM to be applied and it was therefore explored how well SHM might identify MSD along rivet lines. A substantial overview regarding all of this has been provided in (32) with a major conclusion, that cracks of 5 mm in length and above could be well determined with SHM when compared to conventional NDT. This was confirmed in a follow-on study (33) where the reasons for limited monitoring sensitivity with SHM were ana-

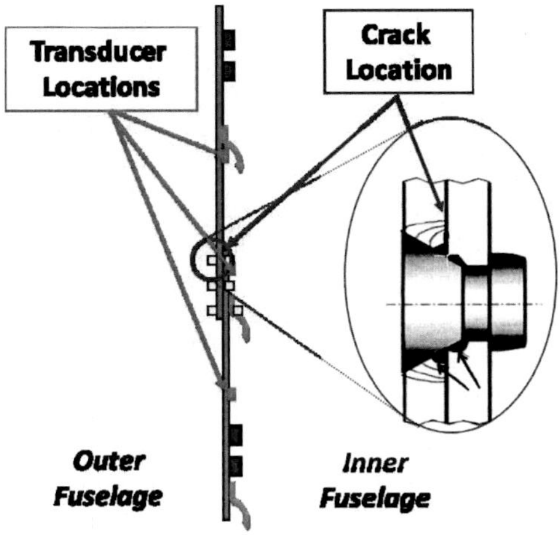

Figure 65. Acoustics based SHM on a riveted lap-joint panel

lytically explored and could be explained. A first explanation can be given when looking at the cross-section of a lapped joint of an aircraft and how an acoustic signal would have to travel as shown schematically in Figure 65. For pragmatic and possibly also aerodynamic reasons a monitoring unit might only be placed on the inside of the aircraft structure considered. The transducer layers were placed at decently separate locations mainly for the reason of being able to generate a fairly homogeneous guided wave. However this wave had to cross the lap and hence an interface which is a major source for attenuating the guided wave signal. Furthermore the area around the rivet is far from being of a simple shape. When a crack is therefore due to generate as indicated in Figure 65 it will start in an area where the interaction with the guided wave is by no means easily described nor even very much influential on the guided wave as long as the crack hasn't at least achieved a decent size.

Even when not considering the geometric complexity of a rivet hole the scattered signal resulting from a pure longitudinal Lamb wave generated is by far not as strong as one would believe. In (33) some simulation results shown in Figure 66 have been reported where a longitudinal wave has been sent through a plate with several holes and one of the holes had a crack of increasing length.

The top row of Figure 66 shows on the right hand side the intensity of the scattered wave distribution behind the hole for a quarter section

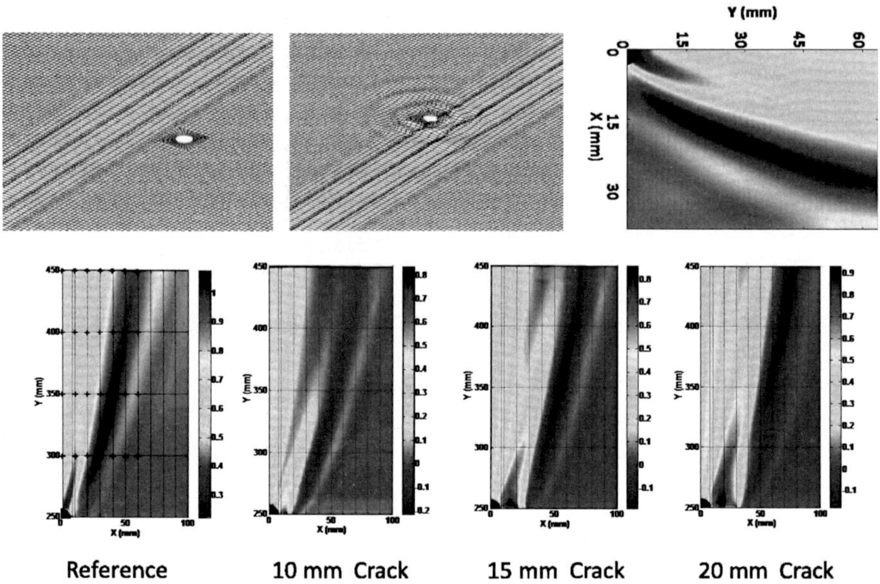

Figure 66. Influence of a hole in an infinite pate on the scattered wave behind the hole (top) and influence of a crack starting from the hole on the differential signal of the scattered wave (bottom)

only. It can be seen that the maximum distribution follows a hyperbolic distribution that fades away around 15 to 20 cm behind the hole. This signal shown also on the bottom left of Figure 66 has been taken as a reference for determining the difference to the signal determined for a crack emanating from this hole, shown on the bottom row of diagrams in Figure 66. It can be observed that the crack must have a substantial size to be detectable at a fairly still low distance to the hole and the crack itself. A rule of thumb can be set up saying: A crack length of 2.5 times the diameter of a hole (rivet) has to be monitored no further away than 12.5 times the size of the hole (rivet) diameter. This shows that damage monitoring is a 'hot spot affair' in all regards and that state-of-the-art SHM is currently possibly not able to detect damage of a size smaller than traditional NDT can do today.

However operators of damage tolerant structures might not show a very keen interest in getting damages detected at a size smaller than what has been considered to be tolerable. Within the world of structural integrity

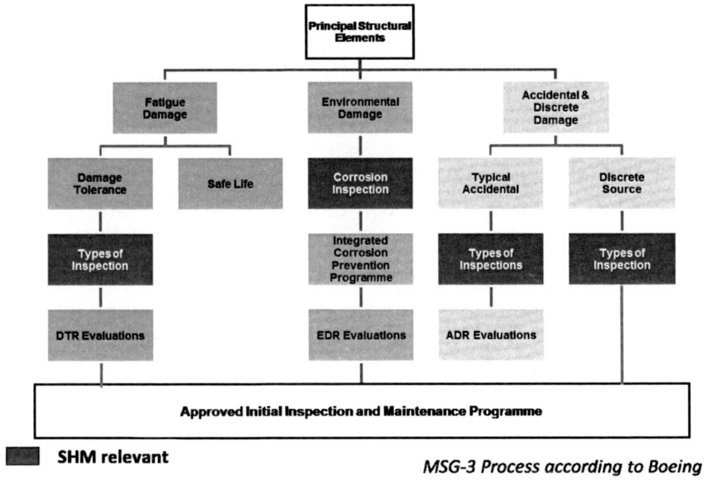

Figure 67. Options for SHM to go into the aircraft maintenance process

where fatigue, environmental and accidental damage play a major role there is a significant margin for SHM to interact as can be seen from the structure shown in Figure 67 under which the international aircraft maintenance steering group (MSG) works.

That SHM can become a core economic element with high asset structures can be easily shown for a structure such as an aircraft. Every operational hour lost on a commercial aircraft does generate cost of a few thousand Euros per hour each. Hence comparing this cost with the hourly rate of an inspector or technician does easily allow a number of twenty or more inspectors and technicians to work. Figures get even extreme when it comes to compare the cost of SHM equipment versus aircraft operability lost. As a conclusion: Operability of high asset value infrastructure is core.

In a larger study performed to analyse aircraft maintenance processes and to determine the potential of SHM within aircraft maintenance a major civil aircraft type flown worldwide had been considered (34). The maintenance process along a D-check, the major inspection an aircraft faces during its operational life, was selected and modelled with discrete event simulation to identify time critical structural items and estimate the potential benefit SHM could bring. Modelling, essentially involved mapping the workflow of steps needed to fully perform the interval and analysing the steps to flag time critical items. Generation of information required a large number of different types of documents to be studied. This started from Maintenance

Planning Documents (MPD) issued by the manufacturer. Since times for the different maintenance actions provided by the MPD are rather optimistic job cards had to be included which better reflect the local specifics of the maintainer and/or aircraft operator.

The maintenance process was modelled using a simulation tool named Arena (35) which is a discrete event simulation tool to model intervals. Arena has the ability to incorporate and analyse stochastic times easily. An estimate of the individual step times and a probability distribution for modelling is sufficient. The tool works on the basis of a graphical interface with processes to be constructed in a flow chart fashion. However initially it is important to plan the model before construction. Measuring the aircraft downtime is a key performance criterion in understanding where SHM could be used. The entity moving through the process would therefore be the aircraft. As the aircraft moves along the steps, the duration time would alter. Having the structure of the model such that the aircraft is the entity with maintenance actions as blocks does ensure that time critical items can be traced.

Each item in the D-Check was segregated into stages set by planners, and modelled in parallel as process blocks within a stage. The five stages in a D-Check can be summarised as:

- Stage 1: Removal
- Stage 2: Inspection
- Stage 3: Defect rectification and rebuild
- Stage 4: Paint
- Stage 5 : Final Checks

Figure 68 shows the structure of the D-check as modelled in an Arena model for greater clarity.

With the model established simulations could now be performed in accordance to the procedure generally described in Figure 69. This procedure can be split into six steps of which steps 1 to 3 have been already described for the D-check example above. For any structural item appearing on the critical path along step 3 an SHM solution has to be determined which is then fed back into the maintenance process (step 4) and its impact analysed by comparing the maintenance duration determined with the one achievable without SHM. This may be further varied in case alternative SHM technologies are available (step 6). Further SHM potentials might turn up in case the critical path along the maintenance process changes as a consequence of an SHM implementation and further components with SHM potential might turn up.

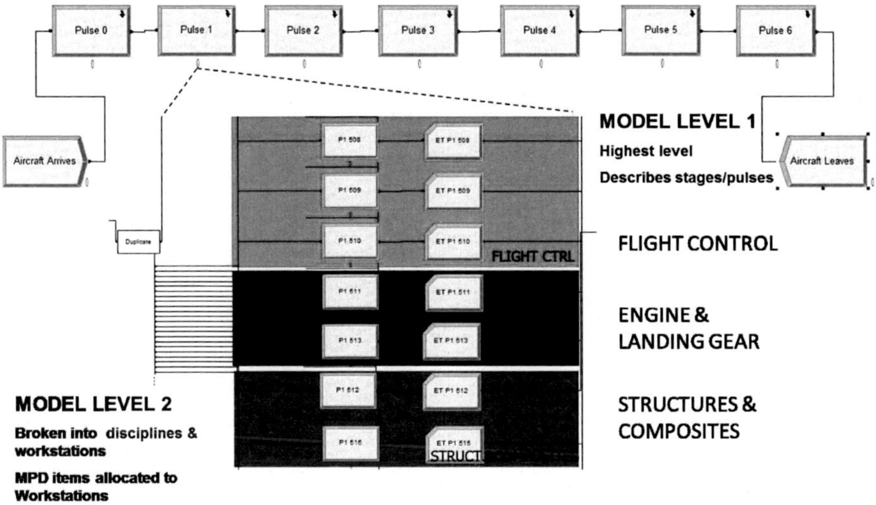

Figure 68. Structure of the D-check model. Level 1 shows the stages. A section of stage 1 has been expanded to show level 2. Here items are broken in parallel to work stations.

Maintenance work along the D-check mentioned above was organised in terms of so called process blocks. These process blocks have been the result of an optimisation of the maintenance process. Figure 70 shows the histogram for all process blocks that were considered critical. The number of hits represents the number of times a process block became critical in 100 replications, which is a result of some uncertainties specifically applied to the simulation model. The critical blocks were therefore altered due to uncertainties elicited from maintenance planners. The process blocks where SHM could be implemented are highlighted. At those process blocks there were a number of SHM solutions inapplicable (non inspections, larger visual checks) since items to be inspected required identical access. Therefore, this access would be necessary even if a SHM alternative was made available. For this reason, time savings were mainly marginal.

Figure 71 shows that the average saving between the existing process and the improved process with SHM at hot spot locations resulted in average savings of nearly 3 hours per interval. The average total saving was less than the minimum saving, as the highlighted critical process blocks were not critical 100% of the time when running the replications. Considering

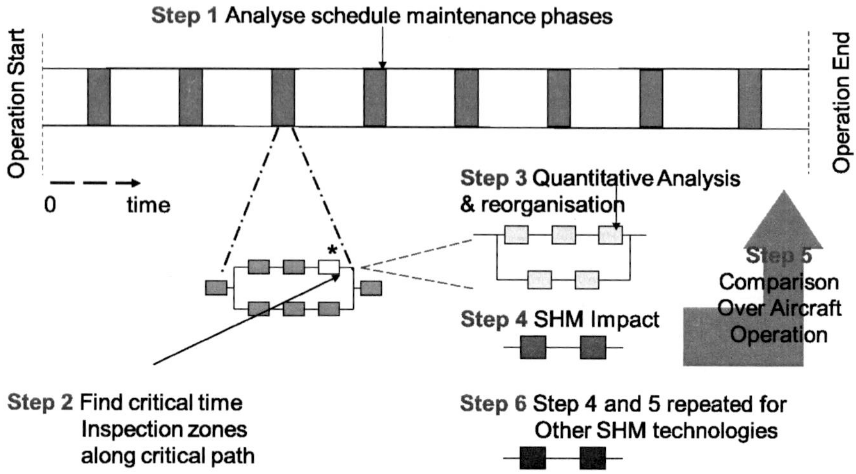

Figure 69. Procedure for analysing operational life cycle SHM potentials along the maintenance process of en engineering system

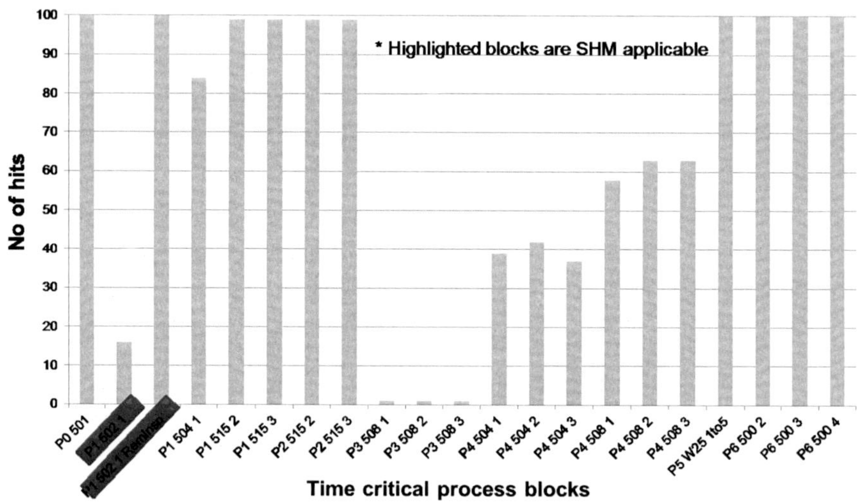

Figure 70. Number of times a process block is critical after 100 replications of D-Check simulation considering uncertainties. Highlighted blocks are where SHM is applicable

	AV. tot time hrs	Min hrs	Max hrs
Existing	413.54	371.78	460.83
SHM hot spots	410.61	367.18	451.39
SHM global	387.34	341.85	424.75
Savings	26.2	29.93	36.08

Figure 71. Comparison of the two simulation durations: one modelling the existing process and the other replicating if identified hot spot regions were eliminated.

replacement for only a handful of hot spot areas in a D-Check would yield very small savings only. The gain sought in operational capability would not even improve significantly if all structural components were equipped with SHM which is the scenario denoted as 'SHM global' in Figure 71 below. Even here the difference might not exceed 36 hrs. which is a marginal number when keeping the uncertainty and the D-check as a possibly singular event in an aircraft's life in mind. What those figures provide is the perfect match between engineering design and maintenance or in other words, aircraft structures today are designed for being well maintained. Should maintenance principles therefore change then those would most likely also have a significant implication on design. Implementing SHM into aircraft designed today or in the past is therefore most likely not to provide the expected benefits.

A major question now emerging as a consequence of the conclusion made above is now: Can SHM have a benefit for current aircraft at all? The answer to that question is 'yes' and the potentials have to be seen in something called 'drop out'. A drop-out is a damaging situation which occurs unexpectedly over an aircraft's operational life. An example is repairs, which are often not predictable and do require specific care over the aircraft's remaining life. Other drop outs are damages which occur while an aircraft is ageing and which could not be predicted before. Drop outs are most likely to define their own inspection intervals and as such those intervals are most likely not to match with the optimised maintenance plans mentioned above. Hence inspections will have to be performed with regards to those drop outs only and this may result in a fairly expensive inspection action. An example where such an SHM impact could be validated has been the gantries on the Airbus A300. Gantries as shown in Figure 72 are the major longitudinal

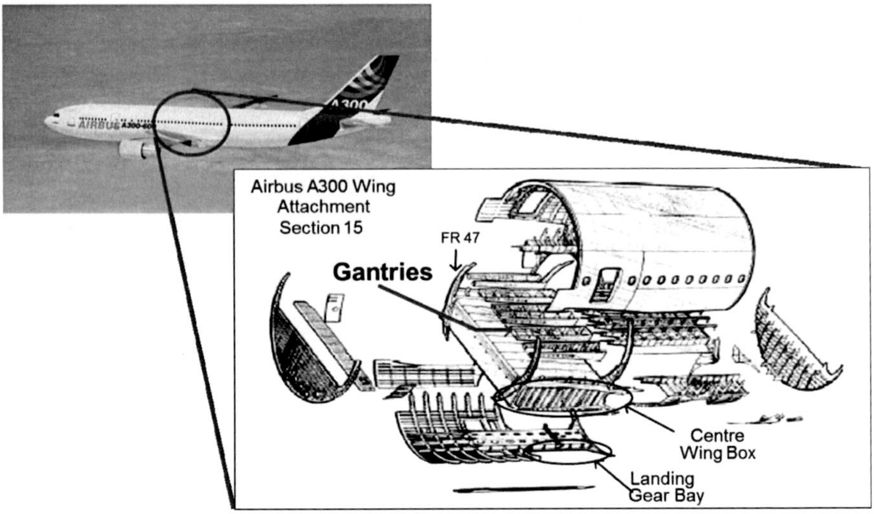

Figure 72. Drop out item gantries and its location within the aircraft

beams over the wing box of an aircraft. They are exposed to major bending loads of the aircraft fuselage as well as cabin pressure loads, which are downloaded from bolted diaphragm panels into the flanges of the gantries. The complexity of stress states in that region may result in cracks to emerge which need to be detected through inspection and be removed at intervals most likely not to be in line with the conventional maintenance process. Getting those locations inspected also requires a substantial amount of dis- and re-assembly of various components in the near-field.

In the case of gantries' inspection the process was simulated with Arena initially by varying the numbers of inspectors and workers being involved in the inspection process. Results obtained are shown in Figure 73 and it can be seen by increasing the number of workers and inspectors that the duration for inspecting the gantries will decrease, however not below a threshold value of around 120 hrs of maintenance. This result has been compared to a solution where an SHM system has been implemented around the locations prone to cracking as shown in Figure 74. This solution would only require a plug to be contacted for inspection in the aircraft's floor and would avoid any disassembly of seats, carpets, floors and possibly more. Inspection time would be reduced to 20 hours for an aircraft's remaining life cycle resulting in around 100 hours per aircraft of operability gained or

Model	Number of MRB Workers	Number of Inspectors	Total Number of Workers	Total Maintenance Time for whole life-cycle (hrs)
1	3	2	5	264.47
2	6	4	10	171.33
3	9	6	15	145.90
4	12	8	20	128.99
5	15	10	25	121.75

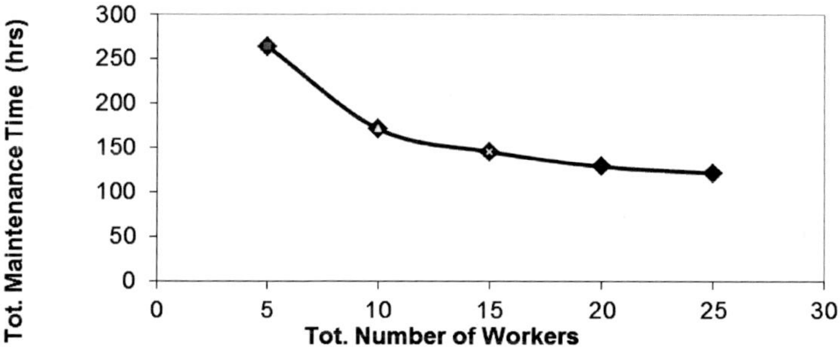

Figure 73. Total life cycle maintenance time for gantries' inspection depending on labour involved

a significant six digit Euro volume of operational cost.

SHM has its short term benefits therefore specifically in automating unscheduled inspection processes. This requires flexible solutions being application tailored. Implementing SHM at a product's life cycle onset requires the product's design principles to be modified such that an overall economic benefit can be seen. This approach is definitely only possible with products being designed from scratch but hardly from those existing today.

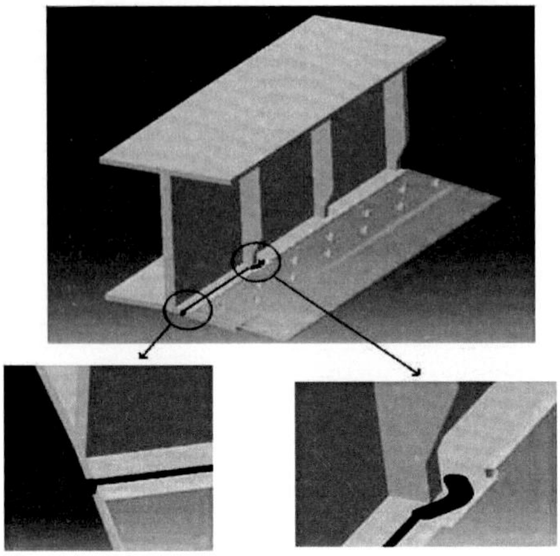

Figure 74. SHM solution considered for gantries' inspection

7 Emerging SHM Applications and Concluding Remarks

There has been a large amount of research and development done in the area of SHM. However not too much has been shown so far that has matured to a true commercial SHM product being applied. This may result from the fact that most of the areas being addressed for SHM would this be aeronautics, mechanical and/or civil engineering are disciplines being well established in engineering. Design principles would therefore have to be changed if SHM was to be implemented at the onset of a product's life cycle.

An area where design principles are much less established and where consequently technology pull plays a much bigger role is wind energy generation. Wind energy rotor blades are critical components in many regards and monitoring their loads, condition and hence damage is significant. In view of their size and limited access they are ideal for SHM applications. Along some longer standing development an SHM system is on its way to be realised (36) that is a part of a wind energy rotor blade design process and can be integrated and updated in accordance to the rotor blade's operational needs. Starting from numerical simulation locations for sensors have been determined as shown in Figure 75 below. With a hardware acoustic system which is able to work on a wireless basis acoustic signals can be generated and/or recorded which allow the relatively large allowable dam-

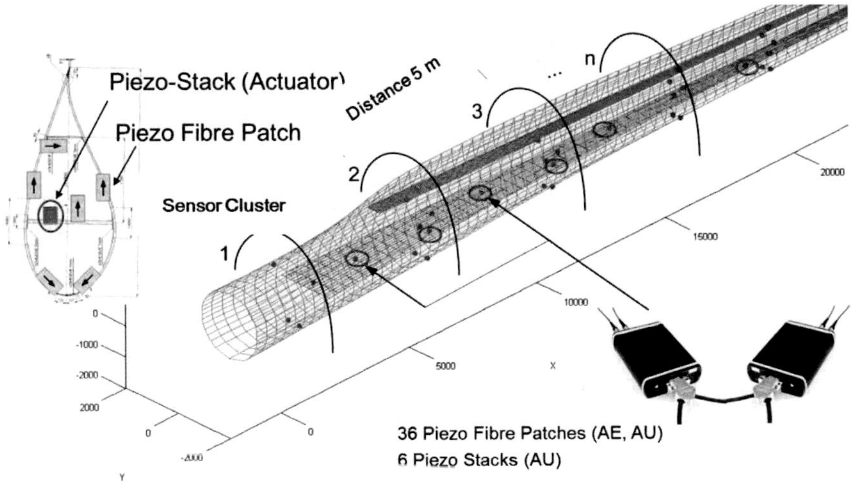

Figure 75. Numeric simulation for SHM system configuration (36)

ages to be observed in a rotor blade. Techniques considered are acoustics based including acoustic emission and acousto-ultrasonics. An additional option is provided through a fibre optic Bragg grating sensing system which allows modal shapes of the rotor blade to be monitored as an additional monitoring technique. Figure 76 shows the correlation coefficient of defined acousto-ultrasonic paths where the actual signal has been correlated with the signal from the pristine stage and a lack in correlation is considered as the damage indicator. Transducers placed in the rotor blade have also been used as sensors for acoustic emission monitoring only where damages in the rotor blade have been clearly identified as can be seen from some of the results shown in Figure 77.

Implementing SHM into wind energy systems in general is a most rewarding application for SHM technology demonstration. Components to be inspected are comparatively large and difficult to inspect. So far there are no certified inspection techniques and procedures available which give SHM a chance to become one of them. Allowable damages are relatively large when compared to aircraft and shutdown times relatively short, which is another advantage for getting SHM implemented. However there are also some challenges still to be met where lightning strikes is one of the major ones besides energy harvesting, environmental ruggedness and longevity of the SHM system which all can be nicely demonstrated on the wind energy turbine blades themselves. Challenges have also to be seen in the sensors'

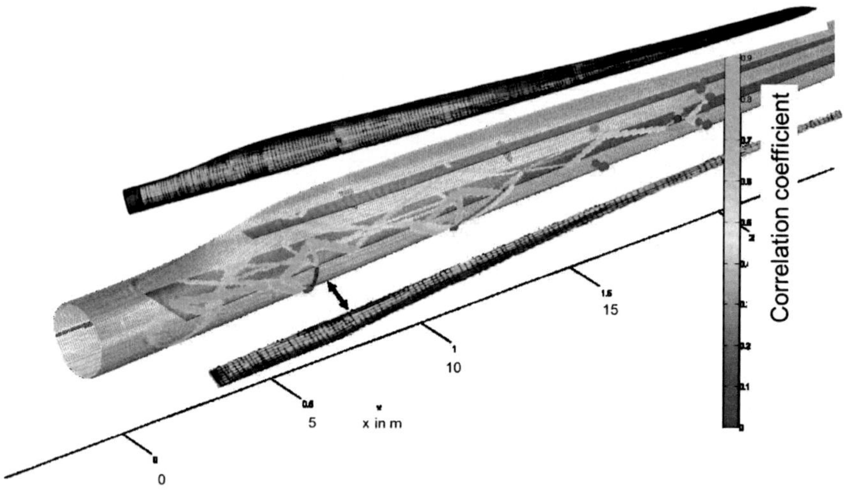

Figure 76. Acousto-ultrasonic signal correlation coefficient as a means for damage detection (36)

damage detection performance. Sensor algorithms do have to distinguish between real damage and other effects such as resulting from environmental, system failure or other external influences. SHM systems do also have to provide very high probability of damage detection capabilities which also results from an SHM system's high system integrity with basically no false positives being provided. Validation of the SHM system has also to be performed under the various environmental conditions the system has to operate in resulting in adequate SHM systems to be finally provided on the market.

Other major issues include sensor durability and a possible option for a SHM system repair. Slow or long-term damage progression resulting from corrosion, fatigue cracking, etc. has to be considered as well. A sensor/SHM system must therefore be robust to survive well beyond any of those damaging events. Repair as well as replacement of the SHM system has to be low-cost with the SHM system still being able to meet original detection requirements. SHM systems must not interfere with structural repairs or do have to be configured such that they do circumvent those or possibly even get those included.

The reasons why not too much SHM has been applied in aeronautics are

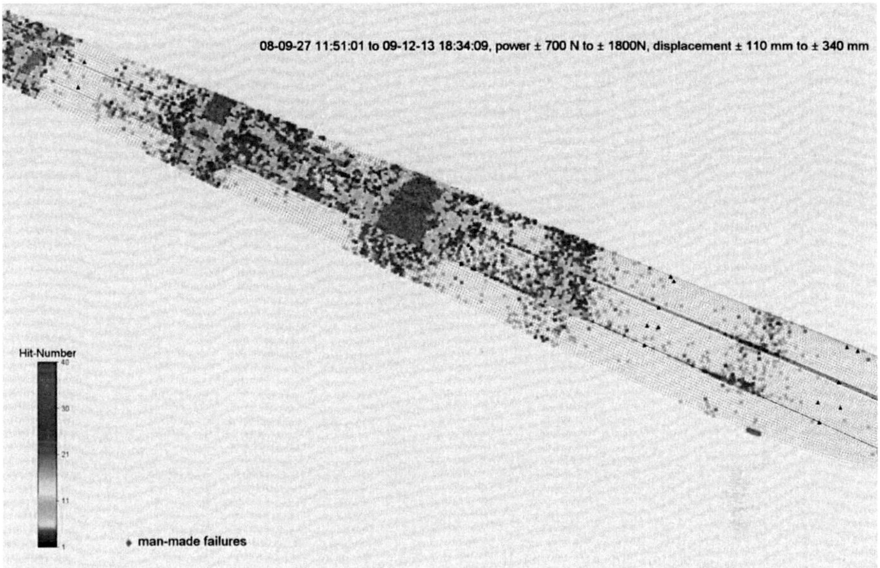

Figure 77. Localisation of acoustic emissions in wind energy rotor blade (36)

various and do include significantly the insufficient technology readiness of most of the SHM techniques proposed. Another reason is also that SHM potentials are still hardly explored and hence mainly unknown specifically from a quantitative point of view. Aircraft manufacturers mainly consider SHM within the context of new aircraft, where potentials can only be seen in very dedicated applications such as collision monitoring around cargo doors. Current and hence short term potentials are however more with ageing aircraft which is a different segment of business in which aircraft maintenance companies are more active. However those companies are reluctant to adopt SHM as a future technology since it automates current inspection processes reducing the value of maintenance business ongoing. Aircraft operators are also not aware of SHM and the SHM potentials which can only be determined once the operational conditions of their aircraft are known. However aircraft operators might be reluctant to release this information. Furthermore they are reluctant in getting more sensors on board their aircraft since they already feel sufficiently penalised with all the other sensor systems being on board ranging from avionics to wellbeing and entertainment. Airworthiness authorities could be better integrated into the SHM development process at present although they are finally the institutions declaring an SHM system to be airworthy. Airworthiness certification may therefore become another stumbling stone along the way to get SHM into operation. Then there is the wide field of data integration. The amount of data an SHM system generates is huge and will have to be included in a data management system. In aviation data management is already a major issue within the context of avionics, flight performance and systems such as engines, landing gear, fuel and others. Data formats at any of the interfaces have already been fixed and likelihood is more than low that those will be changed or adapted with regard to SHM. Hence data generated by SHM will have to be adapted to the formats being already set in aviation so far. Finally there is too little appreciation from the 'SHM-community' with respect to activities happening in neighbouring fields. There are engine condition monitoring systems operating with success now and the same can be mentioned for the health and usage monitoring systems (HUMS) in helicopters or any type of built-in test equipment (BITE) applies in avionics.

With all this critique said there is however a variety of things that can be done straight away to get SHM realised already today. It starts with implementing loads monitoring at discrete locations and feeding the information recorded back into the fatigue life evaluation process. In a time where most of our engineering systems are designed digitally there exists a digital model that may allow for predicting damage critical locations based on the operational loads the system is going through. In a damage tolerant structure there are many components which allow for fairly large damages going far into the multi-centimetre range such as broken stringers. These damages are quite easy to detect with the SHM techniques being available provided the techniques do withstand all other environmental requirements being set. Maintenance processes can be easily simulated today which allow the potentials of SHM to be determined on a completely numerical basis before investing into SHM hardware to be realised. And last but not least one can automate the inspection process as much as possible because a human generating in nearly 100% of the cases the information 'no failure found' (NFF) gets easily bored. It would be much more beneficial having an automated inspection system filtering out most of the NFF cases and providing a residual of borderline cases which the human brain might be better challenged to tackle with since a human's capabilities in correlation are still beyond any machine as long as being challenged adequately.

Bibliography

[1] Günther G, 1993, DASA (now Cassidian) (private communication)

[2] Boller C, F-K Chang and Y Fujino (Ed.s), 2009: *Encyclopedia of Structural Health Monitoring*; 5 Volumes; John Wiley & Sons

[3] Armitag S R and D M Holford, 1998: Future Fatigue Monitoring Systems; NATO RTO AVT Proc. 7 (RTO MP-7), Paper 2

[4] Haibach E, 2006: *Betriebsfestigkeit;* Springer-Verlag, 3rd Edition (in German)

[5] Matsuishi M and T Endo, 1968: Fatigue of metals subjected to varying stress – Fatigue lives under random loading; Preliminary Proc. of JSME Kyushu District Meeting, pp. 37-40 (in Japanese)

[6] Potter J M and R T Watanabe (Ed.s), 1989: *Development of Fatigue Loading Spectra,* ASTM STP 1006

[7] de Jonge J B, D Schütz, H Lowak and J Schijve, 1973: *A standardized load sequence for flight simulation tests on transport aircraft wing structures;* Nat. Aerospace Lab. (NLR) technical report TR 72018 and Fraunhofer LBF report FB-106

[8] Fischer R, M Hück, H-G Köbler and W Schütz, 1970: *Eine dem stationären Gaußprozess verwandte Beanspruchungs-Zeit-Funktion für Betriebsfestigkeitsversuche;* Fortschr.-Ber. VDI-Z Reihe 5, Nr. 30 (in German)

[9] Peterson R E, 1974: *Stress Concentration Factors,* John Wiley & Sons, New York

[10] MIL-Handbook 5J: http://femci.gsfc.nasa.gov/links.html (last visited 03/12)

[11] Palmgren A, 1924: Die Lebensdauer von Kugellagern, VDI-Z 58, pp. 339-341 (in German)

[12] Miner M A, 1945: Cumulative damage in fatigue, J. Appl. Mech., 12, pp. 159-164

[13] Schmidt H-J, B Schmidt-Brandecker and G Tober, 1999 : Design of Modern Aircraft Structure and the Role of NDI, Proc. of 7th ECNDT, Vol. 4, Nr. 6 (also available on [41])

[14] Tada H, P C Paris and G R Irwin, 1973: *The stress analysis of cracks handbook*; Del Research

[15] Sih G C, 1973: *Handbook of stress intensity factors*, Lehigh University Press

[16] Rooke D P and D.J. Cartwright, 1976: *Compendium of stress intensity factors*, Her Majesty's Stationery Office

[17] Murakami Y, *Stress Intensity Factors Handbook*, Japan Soc. Mater. Sci., Tokyo/Japan

[18] Paris P C and F Erdogan, 1963: A critical analysis of crack propagation laws; Trans. ASME, Series D, Vo. 85, pp 528-535

[19] Forman R G, V E Kearney and R M Engle, 1967: Numerical analysis of crack propagation in cyclic-loaded structures; J. Basic Engrg., Trans. ASME, Vol. D89, pp. 459-464
[20] Grandt Jr. A F, 2004: *Fundamentals of Structural Integrity*, John Wiley & Sons
[21] Cullity B D, 1972: *Introduction to magnetic materials*, Addison Wesley, London
[22] Jiles D C, 1988: Review of magnetic methods for non-destructive evaluation, NDT International, vol. 21, 311
[23] Altpeter I, G Dobmann, M Kröning, M Rabung and S Szielasko, 2009: Micro-magnetic evaluation of micro residual stresses of the IInd and IIIrd order, NDT & E International, Vol. 42, Issue 4, pp 283-290
[24] Tiitto S, 1996: Magnetic Methods. Handbook of measurement of residual stresses; J Lu (Ed.), Soc. for Experimental Mechanics; The Fairmont Press Inc Lilburn, pp 179-223
[25] Cikalova U, B Bendjus and J Schreiber, 2009: Bewertung des Spannungszustandes und der Materialschädigung von Komponenten industrieller Anlagen, Materials Testing, Vol. 51/10, pp 678-685
[26] Altpeter I, R Becker, G Dobmann, R Kern, W Theiner, A Yashan, 2002: Robust Solutions of Inverse Problems in Eletromagnetic Non-Destructive Evaluation, Inverse Problems, **18**, pp. 1907-1921
[27] Altpeter, I, G Dobmann, K-H Katerbau, M Schick, P Binkele, P Kizler and S Schmauder, 2001: Copper Precipitates in 15 NiCuMoNb 5 (WB 36) Steel: Material Properties and Microstructure, Atomistic Simulation, and Micromagnetic NDE Techniques, Nuclear Engineering and Design 206, pp 337-350
[28] Yashan A, 2009: *Über die Wirbelstromprüfung und magnetische Streuflussprüfung mittels GMR-Sensoren*, Doctoral Thesis, Saarland University, Saarbrücken/Germany (in German)
[29] Schulze M, Heuer H, Küttner M, Meyendorf N, 2010: High-resolution eddy current sensor system for quality assessment of carbon fiber materials; Microsystem Technologies, Springer, Volume: 16 Issue: 5, pp 791-797
[30] Goldfine N, V Zilberstein, D Schlicker and D Grundy, 2009: Eddy-current in situ sensors for SHM, in [12], pp. 1051 - 1064
[31] Schütz W, 1989: *Standardized stress-time histories: An overview.* Development of Fatigue Load Spectra, ASTM STP 1006, pp. 3-16
[32] Boller C and W J Staszewski, 2003: *Aircraft Structural Health and Usage Monitoring;* in [38], pp. 29-73
[33] Boller C and M R Mofakhami, 2008: *From Structural Mechanics to Inspection Processes: Getting Structural Health Monitoring into Application for Riveted Metallic Structures*; in B Dattaguru, S Gopalakrishnan and V K Aatre(Ed.s.): IUTAM Symposium on Multi-Functional

Material Structures and Systems; IUTAM Bookseries Vol 19, Springer, pp. 175-184

[34] Kapoor H, C Braun and C Boller, 2010: Modelling and optimisation of maintenance intervals to realise Structural Health Monitoring applications on aircraft; Proc. of 5th European Workshop on Structural Health Monitoring

[35] Kelton D., Sadowski P., and Sadowski A., 2002: *Simulation with Arena*, McGraw-Hill

[36] Frankenstein B, L Schubert, B Weihnacht, E Schulze, C Ebert, H Friedmann and T Granert, 2009: Development of condition monitoring systems for rotor blades of windmills; Prof. of 7th Internat. Workshop on SHM, DESTech Publ., pp. 1595-1602

Further Reading:

[37] Niu M C Y, 2002: Airframe Structural Design, Hong Kong Conmilit Press Ltd., 2nd Edition

[38] Staszewski W J, C Boller and G R Tomlinson, 2003: Health Monitoring of Aerospace Structures; J. Wiley & Sons

[39] Broek D, 1989: The Practical Use of Fracture Mechanics; Kluwer Academic Publ., Dordrecht, Netherlands

[40] Dowling N, 1998: Mechanical Behaviour of Materials, Prentice Hall

[41] Non-destructive testing: http://www.ndt-ed.org/index_flash.htm or http://www.ndt.net (last visited 03/12)

[42] Bray D and R K Stanley, 1989: Nondestructive Evaluation; McGraw Hill

[43] Aircraft accidental reports: http://www.ntsb.gov (last visited 03/12)

[44] Crack propagation software AFGROW http://www.afgrow.net (last visited 03/12)

[45] The Damage Tolerance Design Handbook http://www.afgrow.net/applications/DTDHandbook/pdfs.aspx (last visited 03/12)

[46] Balageas D, C-P Fritzen and A Güemes, 2006: *Structural Health Monitoring*, ISTE Ltd.

Structural Health Monitoring of Aircraft Structures

B. Rocha[1], C. Silva[2], C. Keulen[3], M. Yildiz[4], and A. Suleman[3*]

[1] Instituto Superior Tecnico, Lisbon, Portugal
[2] Portuguese Air Force Academy, Sintra, Portugal
[3] University of Victoria, Victoria, BC, Canada
[4] Sabanci University, Istanbul, Turkey

1 Introduction

Aircraft structures operate in harsh conditions sustaining high loads, fatigue cycles and extreme temperature variations. Therefore, inspections to assess the structural condition is of the utmost importance for safe and efficient operation of aircraft. Also, to achieve lighter structures, damages are allowed to exist in aircraft during operation as long as they are within predetermined and safe limits. Thus, aircraft structures are designed according to a damage tolerance philosophy.

Non-destructive testing and evaluation techniques developed in the last few decades and currently in application to assess the health of aircraft structures in operation, fail to detect localized damage, i.e. they are able to detect damage uniquely in limited areas in the vicinity of the inspection region. To obtain a full inspection, repetitive execution of these techniques in different areas of the structure requiring direct access is necessary. This involves complicated, time consuming and consequently expensive operations; particularly disassembly and re-assembly procedures that force aircraft to be grounded. When an aircraft is grounded and not in operation it is not profitable and represents an added cost for the operator (maintenance, hangar time, leasing, etc). Thus, structural inspections and maintenance are performed on a scheduled basis with increasing pressure from aircraft fleet operators to reduce inspection and maintenance time and extend intervals in between consecutive interventions. This increases the risk of damage

* Principal Investigator, IDMEC-IST, Lisbon, Portugal

existence and growth between inspections. Damage might also exist and grow in regions not inspected. All of these issues point to a need for a reliable, persistent and integrated real time and global structural integrity evaluation system.

In the early years of aviation (1920-1940), structures were designed according to the Infinite Life concept, where they would be operated below their fatigue stress limit. Obviously the application of such a concept in design resulted in very heavy structures thus limiting aircraft performance. It is noted that increasing aircraft weight must be counterbalanced by increasing generated lift, leading to an increase in either wing area, or wing loading, and consequently increasing structural weight. Increasing lift generation also increases drag generation that in turn must be counterbalanced by increasing thrust. This requires the use of powerful and likley larger and heavier powerplants, with increased fuel consumption, leading again to the increase of overall weight. Moreover, higher structural weight for a certain lift generated decreases the available payload weight (both passenger and cargo capacity), maneuvering capabilities and aircraft performance such as endurance and range, since available fuel weight is reduced.

To overcome these chellenges a Safe Life structural design and operation approach was then adopted. In this concept, a finite service life was established, within which the probability of fatigue cracks to initiate and develop was extremely remote. The derivation process to determine structural life span was therefore based on the crack initiation stage of the fatigue failure process. This concept was applied in structural design during several years. However, it did not account for flaws generated in manufacture or other forms of damage that reduce structural life, such as corrosion or accidental damage. Damage models and stress analysis were also often inaccurate and inadequate. These inaccuracies have led to multiple accidents, like those of the five DeHavilland COMET, between May 1952 and January 1954. In the last accident, the aircraft had only about 1000 flights, while its flight simulation tests predicted an operational life of 3060 flights. Additionally, in the Safe Life concept, the detection of defects in one component was automatically a cause for its removal from service. In the early seventies, the development of inspection technology, in particular the ability to detect smaller flaws, led to an increasing number of parts being rejected. At that time, advances in the discipline of fracture mechanics enabled the prediction of whether a crack of a given size would induce the failure of a component under a particular load if particular material properties and fracture resistance were known. Models were developed to predict the growth rate of cracks subject to cyclic loads (fatigue). It became possible to have structures in service with existing defects as long as their existence and dimensions were known and smaller than calculated maximum damage dimensions determined to lead to structural collapse.

A Fail Safe structural design and operation concept was then introduced in which a structure is designed to be capable of retaining required residual strength for a period of unrepaired service after sustaining a failure or partial failure of a primary structural element. Structures designed according to this concept present multiple load paths and use crack stoppers, so that stress levels after crack initiation are kept low providing controlled and slow crack growth rates. Simultaneously, structural design is performed to assure a high probability of crack detection before strength is reduced below limit load capabilities. The Fail Safe structural design and life concept evolved to the currently used Damage Tolerance philosophy. In this concept, worst case scenario assumptions of initial manufacturing defects are considered. Also details on sensitivity of inspection techniques (minimum detectable damage dimensions and other limitations imposed by the structures geometry, potential locations and characteristics of damages to detect, their orientation, etc), to be applied for in service crack detection, are accounted for. A service life before next inspection and structural repair is then predicted, based on quantitative calculations of crack growth rates and predicted flight loads severity. A regime of repeated inspections is defined to ensure a high probability of crack detection prior to catastrophic failure, i.e., a scheduled based inspection and maintenance is established (DOD, 1987).

2 Structural Health Monitoring Architectures

Two main transducer architecture strategies based on Lamb-wave propagation were implemented and evaluated, namely piezo networks and phased arrays. Optimum solutions with a minimum number of transducer are sought. The design variables range from sensor type, position, distances between sensors, and damage search area. In both architectures, wave emission and sensing based methods rely on the assumption that a discontinuity such as damage will produce an extra reflection, such as an echo. Figure 1 shows the echo received by a sensor after having produced a pulse. Based on the propagation velocity and time of flight it is possible to determine the distance between the echo generator and the sensor. With this information, one can determine a region where the damage may be present.

In order to detect the damage location with a network of sensors, a minimum of three transducers are required. If each transducer discloses a region of possible damage there will be a single point where the three regions intersect. This is the location of the damage. This technique is called triangulation. For phased arrays, the wave front propagation direction is

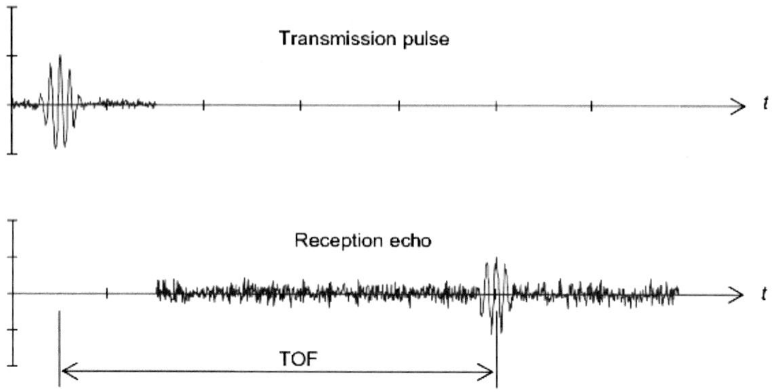

Figure 1. ToF echo detection

predetermined when the array is activated. Knowing the direction and the echo time of travel, it is possible to determine the damage location. The algorithm necessary to carry out this task depends on the approach selected. Network and phased array techniques rely on different algorithms but each can be based on the following principles. One principle involves executing a simple wave sweep. In this case, the damage presence results in the generation of unexpected reflections. The echo and permanent reflections created by the boundaries, other sensors, other wave types and discontinuities of the specimen being tested are sensed simultaneously. The second principle requires a comparison between the response from a damaged structure to an undamaged structure. The undamaged response, acting as a baseline is subtracted from the damaged response; the differences between the two responses indicate a change from the undamaged state. This approach allows for the isolation of the damage reflection eliminating the permanent reflections. Despite the higher complexity associated with the second approach due to the requirement of storing the undamaged response at all times, it does ease the damage detection. On one hand, the problem of having a superposition of permanent reflections along with damage is ruled out. On the other hand, the specimen deterioration, for example by aging, can result in various non-damage related patterns.

2.1 Sensors Network Architecture

The sensors network consists of an architecture where the piezo sensors are distributed in a pre-determined form on the thin structure. Two different damage detection techniques can be implemented for the network approach. A direct technique that uses all of the transducers as both actuators and sensors, and a second, more complex method that uses one transducer as both sensor and actuator and the remaining as only sensors. Figure 2 shows both methods, with dots representing the transducer location.

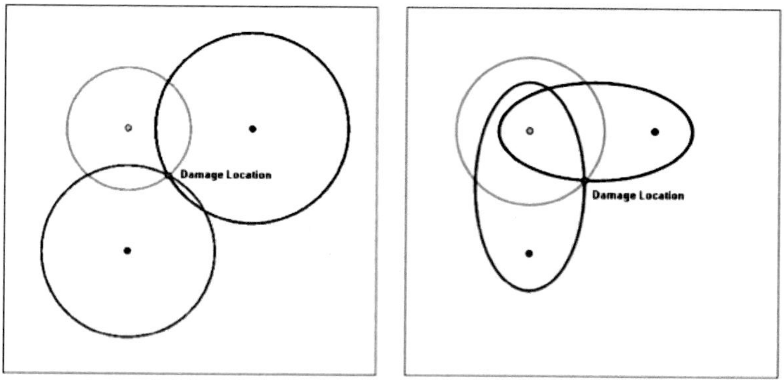

Figure 2. Regions of possible damage and damage location for a three sensor network

Using the first technique, assuming that the propagation velocity is equal in every direction, three circles indicating the location of an echo can be determined (shown on the left of Figure 2). The circles result from the fact that echoes and generated waves travel on the same path but in opposite directions. With the second technique, one circle and two ellipses are calculated (shown on the right). The circle is applied around the actuator transducer. The response gathered by the sensor transducers will receive an echo created by the damage. The time at which this echo is sensed will result from the summation of two different distances. The first will correspond to the distance travelled by the generated wave until it reaches the damage. The second will be the one travelled by the echo between the damage and the sensor. The geometric figure that matches this characteristic is an ellipse since the sum of the distances from any point of the ellipse to the actuator and sensor remains constant. For the first case presented, knowing the wave propagation speed and the time spent between its emission and echo detection, the travelled distance can be calculated. This principle looks at each piezo as both actuator and sensor. The algorithm consists of these steps:

1. An actuation frequency is selected and the wave propagation speed V

in the material is determined by simulation of experimentation.
2. Separate tests are performed using each actuator individually.
3. The sensed response gathered by only the actuator transducer is recorded.
4. Step 2 is repeated on the damaged structure.
5. Step 3 is repeated. At this point responses for both undamaged and damage profiles are obtained.
6. For each pair of data, the time t_i, at which the echo occurred (i is the transducer number) is determined.
7. Knowing the travel speed and the time the radial distance to the damage (r_i) can be determined. The travelled distance will correspond to two times the radius as in (1):

$$2r_i = Vt_i \tag{1}$$

8. For each transducer a circle using the transducer centre position (x_i, y_i) and the distance found on Step 7 as radius r_i is defined:

$$(x - x_i)^2 + (y - y_i)^2 = r_i^2 \tag{2}$$

9. The system of three equations for the unknown coordinates (x, y) can now be solved. They correspond to the damage position.

This approach does not take advantage of the presence of the other two transducers during Step 2. In order to use them, another assumption must be considered. Assuming that the damage acts as a source of reflections, besides the one that is sent to the actuator, others are sent in the remaining directions. Based on this and knowing the wave's propagation speed and the time spent between its emission and reflection detection, the travelled distance can be calculated. This principle looks at each transducer as a sensor. The algorithm follows the next steps for each test using just one actuation, the second technique described above:

1. An actuation frequency is selected and the wave propagation speed (V) in the material is determined by simulation or experimentation.
2. A single test is performed using just one transducer as an actuator.
3. The sensed response gathered by all transducers is recorded.
4. Step 2 is repeated on the damaged structure.
5. Step 3 is repeated. At this point responses for both undamaged and damage profiles are obtained.
6. For each pair of data, the time, t_{ij} at which the echo occurred (i is the actuator number, j is the sensor number) is determined.

7. Knowing the travel speed and the time, the distance travelled by the wave, d_i can be determined. This distance is the sum of the two, the first is the trip between the actuator and damage and the second is the trip between the damage and sensor:

$$d_{ij} = Vt_{ij} \tag{3}$$

8. For each transducer an ellipse can be defined. The actuator/sensor transducer positions are the two foci. The centre (h, k), semi-major axis a, distance between the two foci c, and semi-minor axis b can be calculated as follows (Important notes: x axis is parallel to the major axes and one ellipse will be in fact a circle, which corresponds to the actuator PZT).

$$a_{ij} = \frac{d_{ij}}{2} \tag{4}$$

$$c_{ij} = \frac{\sqrt{(x_i - x_j)^2} + \sqrt{(y_i - y_j)^2}}{2} \tag{5}$$

$$b_{ij} = \sqrt{a_{ij}^2 - c_{ij}^2} \tag{6}$$

$$h'_{ij} = x_i + \frac{(x_j - x_i)}{2} \tag{7}$$

$$k'_{ij} = y_i + \frac{(y_j - y_i)}{2} \tag{8}$$

$$\frac{(x' - h'_{ij})^2}{a_{ij}^2} + \frac{(y' - k'_{ij})^2}{b_{ij}^2} = 1 \tag{9}$$

9. A common coordinate system is selected and the coordinates x', y', h' and k' are rotated in order to get x, y, h and k
10. The system of three equations (two ellipses and one circle) for the unknown coordinates (x, y) can now be solved. They correspond to the damage position.

Despite the increased complexity the latter algorithm provides extra data per test run.

2.2 Phased Arrays Architecture

To implement phased arrays, a number of PZTs must be actuated in a predetermined and synchronous way. The lag between each individual actuation determines the travel direction of the generated wave front. These lags are predetermined by the actuation system, which also sets the actuation frequency and consequently, wave propagation velocity (V). Nevertheless, these lags are calculated using the acquired data. The orientation is determined by the lags. Consider that the array reference position is (x_A, y_A). Figure 3 describes the algorithm. The red line shows the admissible damage locations, blue lines represent the wave front positioning at two different time instances, orange lines represent the wave front boundaries, (d_{WF} is the distance travelled by the wave front, d_R is the distance travelled by the echo produced by the damage, d_E is the length of the line that completes the right triangle created by d_{WF} and d_R.

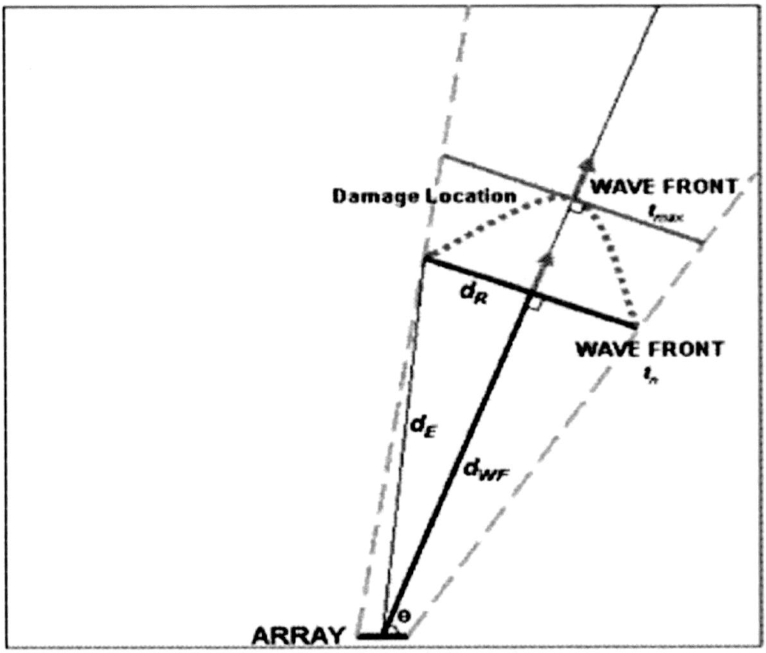

Figure 3. Damage location for a linear phased array

For the previously defined orientation, the algorithm follows these steps:

1. One test is run using the phased array.
2. The sensed response gathered by all array transducers is stored.
3. Step 1 is repeated on the damaged structure.
4. For each transducer, the undamaged data is subtracted from the damaged data and the time at which the echo occurred, t_i (i = transducer number) is determined. This time corresponds to the time required

for the wave front to reach the defect added to the time spent by the reflection to reach the transducer.

5. Since the travel velocity (V_i) and the time (t_i) is known, the total distance (d_i) resulting from the one travelled by the wave front (d_{WF}) added to the echo one (d_E) can be determined for each transducer:

$$d_i = Vt_i \tag{10}$$

$$d_i = d_{WF} + d_E = Vt_i \tag{11}$$

6. Note that the line established by the damage and the array may not correspond to the wave front propagation line direction. In order to determine the damage position, an iterative calculation must be made.
7. The wave front travelled distance at each time.

$$d_{WF} = Vt_{WF}, \;...\; t_{WF} = 0 \;...\; t_{max} \tag{12}$$

Where t_{max} corresponds to the condition where the line, established by the damage and the array orientation equals the wave front propagation direction. In this particular situation:

$$d_{WF} = d_E \tag{13}$$

8. For generic conditions, first the echo travelled distance, resulting from the subtraction between the total distance and travelled wave front distance has to be determined iteratively:

$$d_E = d_i - d_{WF} \tag{14}$$

9. For each time, the coordinates (x_{WF}, y_{WF}) corresponding to point at which the wave front is intersected by the line having the wave front propagation orientation is calculated:

$$x_{WF} = x_A + d_{WF}cos(\theta) \tag{15}$$

$$y_{WF} = y_A + d_{WF}sin(\theta) \tag{16}$$

10. Considering the right triangle constituted by d_{WF}, d_E and d_R one can calculate d_R. This distance is necessary to determine the two possible damage locations at each time instance:

$$d_R = \sqrt{d_{WF}^2 + d_E^2} \tag{17}$$

11. With this distance calculated, one can determine the two locations. Consider a left and a right point, with respect to (x_{WF}, y_{WF}):

$$x_{LEFT} = x_{WF} - d_R cos(\theta - \frac{\pi}{2}) \tag{18}$$

$$y_{LEFT} = y_{WF} - d_R sin(\theta - \frac{\pi}{2}) \tag{19}$$

$$x_{RIGHT} = x_{WF} + d_R cos(\theta - \frac{\pi}{2}) \tag{20}$$

$$y_{RIGHT} = y_{WF} + d_R sin(\theta - \frac{\pi}{2}) \tag{21}$$

The algorithm determines the probable damage location region for a given wave front orientation and echo detection. There will be as many regions as the number of transducers constituting the array. The network and phased array architectures presented are valid for constant thickness specimens made of isotropic materials. Only for these particular cases, wave propagation velocity can be considered constant. For anisotropic materials or thickness changing specimens, the velocity will not remain constant.

3 Piezoelectric Networks

In SHM experiments, with the transducer network setup defined, for each (and all) reference (considered undamaged) condition and for the actual (potential damaged) condition, each transducer in the network is actuated at a time, each actuation corresponding to one scan. The maximum amplitude of the actuation signal was established by the maximum amplitude enabled by the signal generation equipment ($\pm 15V$, for the NI PXI-5421). The frequency of actuation applied was of $340kHz$, according to what was established using the dispersion curves calculated for the aluminum plate (and its thickness of $2mm$); and for the PZT disc diameter selected to be $8mm$ - corresponding to an enhancement of actuation and sensing of the S_0 wave mode with a wavelength of $16mm$ (double of PZT diameter), which in turn corresponds to the (actuation) Lamb wave frequency selected, according to the dispersion curves calculated.

Sensor signals from the different transducers of the network are then recorded for each actuation (scan, or inspection). The comparisons between the actual, and potential damaged state, to a reference condition are then performed by subtracting the corresponding sensor signals for those two conditions. Specifically, the sensor signals obtained for those two conditions, for the same sensor, and when the same transducer is actuated, are subtracted. This is performed for all sensors in the network and for all scans, i.e., for all actuations, meaning for all transducers being actuated, of course, one at a time. With this process, the original excited waves and boundary reflection corresponding sensor signals are removed from the resulting subtracted signal. Signals corresponding to reflections generated by the other transducers in the network and potentially by geometric features and other characteristics of the host material and component being inspected, such as material non uniformity, or structural reinforcements, also disappear in the resulting signal. Furthermore, if the reference state considered had already existing (and known) damages, the reflections generated by those damages are also eliminated from the resulting subtracted signal. Only the signals corresponding to the actual state (to new damages with relation to the reference condition), damage generated wave reflections exist in the resulting signal (and noise).

Damage generated reflection waves can then be identified (damage detection) and their ToF determined. The ToF corresponds to the time of travel of the original, excited wave, from actuator to damage position, plus the time of travel of the damage echo to the sensor which signal is being considered. Knowing the wave propagation velocity (constant for all radial directions in an isotropic material) and the ToF, the total travelled distance can be determined. If the same PZT is used as actuator and sensor, possible damage positions define a circle centered in the considered transducer, with a radius of half the determined total travelled distance. If a pair of PZTs is used, one as actuator and the other as sensor, then possible damage positions define an ellipse with its focus points being the transducer pair considered since the sum of the distances of any point in the ellipse to its two focus points is a constant, in this case, equal to the total travelled distance. Damage position can then be determined as the intersection of the different formed circles and ellipses for all scans performed.

Maximum noise levels, in initial experiments, were verified to be, on average, around 1/1000 of the maximum amplitude of the actuation signal and the low amplitude of damage echoes was around 1/250 of the maximum amplitude of the actuation signal. This low SNR, introduce considerable difficulties in the implementation of the SHM method. Such problems are diminished by the fact that for each inspection/health condition of the component, with the set transducer network, three actuators are used in different positions, at a time, for each scan with the sensed signals from all transducers in the network being recorded for each scan. For each inspection, three scans are then performed, obtaining three sets of one circle and two ellipses to intersect. Although, more computationally demanding, random noise influence is decreased and accuracy of results is enhanced.

To decrease the influence of noise and potential small errors in actuation (in obtaining the same trigger time and amplitude for all actuations), the origin of time for all sensor signals, for each actuation, is tuned according to the actuator trigger time, for each scan. Sensor signal amplitudes are also normalized with relation to the actual verified maximum signal amplitude in the actuator PZT signal (maximum amplitude of actuation). Furthermore, since the SHM method is developed around the defined frequency of actuation (the same as the interesting frequency of generated, propagating and reflected waves), bandpass filters are applied to sensor signals (second order Butterworth bandpass filters) and FFTs/inverse FFTs to decrease the influence of off tone random noise (and potential other actuations in other frequencies).

Even with the application of all previous referred techniques, false reflections (positives) still exist in the resulting subtracted sensor signals due to noise. The true damage reflection is then contained in the sensor signal in between different false reflections and being masked by them. The duration of each scan is very small, inferior to $1ms$, already accounting for a time interval between consecutive scans to enable the damping of propagating reflected Lamb waves (second and third reflections from the plate boundaries, etc). Furthermore, memory requirements to record data from tests are in the order of what is usually available in a personal computer. To further decrease the influence of random noise, each scan (or each actuation, for each PZT in the network, for both reference and actual states) is repeated fifty times, i.e., for each scan fifty sensor signals are obtained for each sensor. Statistical operations are then applied to the fifty sensor signals (for the same sensor, actuating PZT and condition), such as averaging and determination of average maximums and minimums (within 90% of absolute maximums and minimums), for each and all times. A signal band is then obtained. To identify the corresponding signal of potential damage reflections, sensor signal bands (for the same sensor and actuating PZT), for both the actual and reference conditions are compared/subtracted. When, the sensor signal band corresponding to the actual state leaves (in more than 50% of its width) the sensor signal band corresponding to the reference state, a signal corresponding to a potential damage echo is identified. The process is then continue according to what was explained before (i.e., determination of ToF, etc).

To decrease the influence of the possible existence of false positives, all detected potential reflections are considered. Sets of circles and ellipses are determined for all potential reflections and intersections are performed, enhancing the indication of the location of the actual damage and decreasing (and eliminating) the indication of false (random) damage positions.

In this process, the computational time required for the determination of the equations of the circles and ellipses and their intersections is considerable. Furthermore, some potential imprecision in the determination of such equations must be accounted for. This results in a possible space imprecision in the determination of the points contained in circles and ellipses, derived from errors in sensor signal time and consequent imprecision in determined wave propagation velocities and in the determination of ToFs for the corresponding signals for different damage reflected waves. If these errors are of enough magnitude, the different circles and ellipses might not even generate any feasible intersection.

Considering these aspects and to decrease their potential negative effects in the code created to implement the SHM method the plate is meshed numerically. A quadrangular, single layer mesh is applied with a spacing of $5mm$. An array of 300 x 300 (for the $1.5m$ x $1.5m$ plate) is created, corresponding each of its positions to one point in the mesh. In a first approach, circle and ellipse equations are determined and a region of influence of 5mm is created around each circle and ellipse. For each circle and ellipse, each point in the mesh is checked upon if it falls in the region of influence of such circle or ellipse. If this condition is verified a unity value is introduced in the array in the corresponding position. For the subsequently considered circles and ellipses, the unity values are added to the values already existing in the array in the corresponding positions superimposed results. The point in the mesh with higher final value (damage index) will indicate the most probable damage position. In the end, all the values in the resultant, final matrix are divided by the highest value, becoming then a percentage proportional to the highest (equal to one in the position corresponding to the highest value) and lowest probability of damage existing in such region. With this method the determination of damage location is performed with a precision of $5mm$, i.e., it will be contained in a circle region with $5mm$ of diameter, around the determined point in the mesh (related with the mesh spacing).

3.1 Damage Detection: Metallic Plates

The transducer network SHM system was experimentally tested in a $1.5m$ x $1.5m$ x $2mm$ aluminum plate, subject to different boundary conditions, ranging from totally supported (to increase Lamb waves propagation damping), to simply supported and riveted in the boundaries. Experiments were conducted initially in a laboratory setting and afterwards in an aircraft maintenance hangar, with no surrounding noise control and limited temperature control. Experiments were performed with the introduction (cumulatively) of surface and through the thickness circular holes and cuts (with different orientations), with a maximum dimension of $1mm$, to simulate damage (Figure 4). Damages were inflicted in different positions in the entire plate as can be observed in Figure 5.

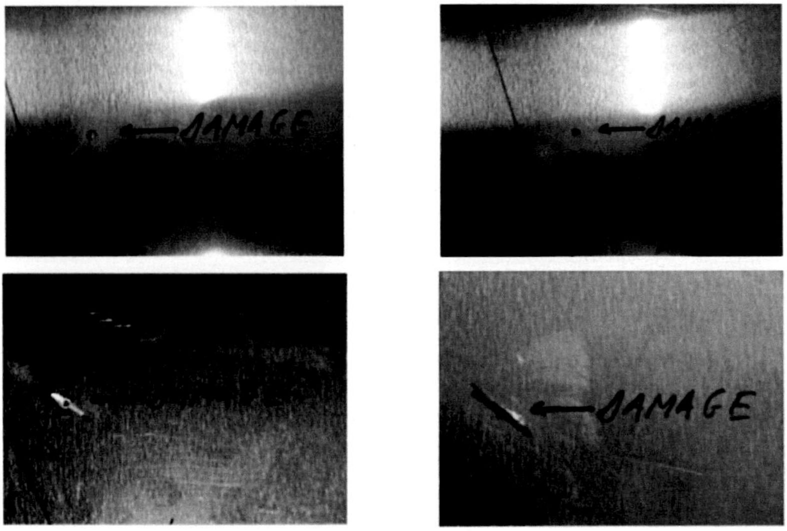

Figure 4. Inflicted damage types

The software for automated inspection was developed in LABVIEW ®, with embedded MATLAB ®codes for signal processing and to implement the damage detection algorithms. The software developed also enables the execution of an initial frequency sweep execution of scans with different frequencies of actuation. Using the position of transducers as an input in the network, in the plate, and plate dimensions, the software calculates the distances in between transducers, and to plate boundaries. With the user input, through the determination of ToF of different direct waves originally excited waves travelling between the actuator and sensors, and boundary reflected waves, during the frequency sweep, it calculates the propagation velocity of S_0 and A_0 waves for the different frequencies. The software determines then the experimental dispersion curves for the usable frequency range within the frequency sweep. The software also accepts as inputs the material/mechanical properties, such as Young modulus, Poisson coefficient and density for an isotropic material, and a thickness for the plate. If requested by the user the software can calculate the theoretical dispersion curves and can compare these with the experimentally obtained curves.

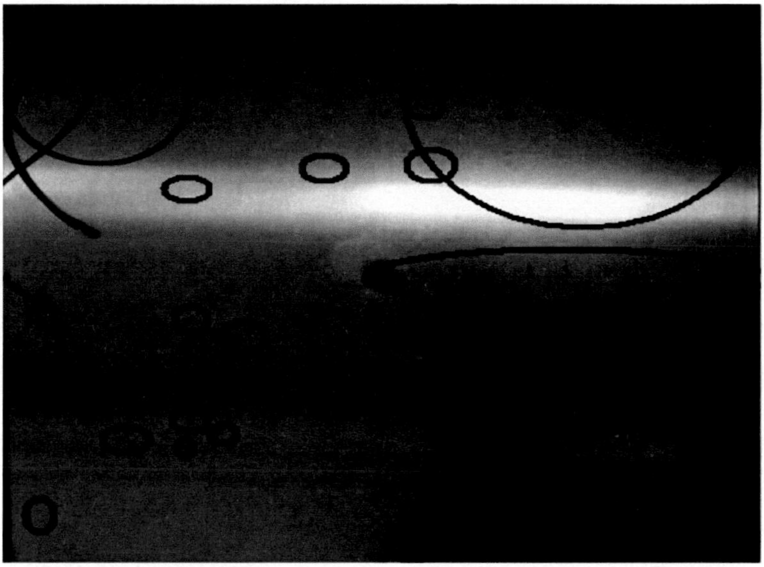

Figure 5. Aluminum plate with all, introduced damages (positions) successfully identified

Based on the experimentally and theoretically determined dispersion curves, it calculates also the wavelengths for the different frequencies. Based on the input for the PZT diameters, it indicates the frequencies at which the actuation and sensing of the S_0 and A_0 modes are enhanced (in which their wavelength is double of the PZT diameter introduced). This indication is backed by an analysis that is enabled to the user by the code, of the evolution of the amplitude of the waves during the frequency sweep. After the calculation of the different wavelengths, the software also presents a tuning curve for the usable frequency range within the frequency sweep. This allows the user to select the optimum (excitation) frequency, and for the determination of wave propagation velocities, to implement subsequently the SHM method.

In the following Figures 6, 7 and 8 the plots presented by the software to the user are depicted. These plots show the damage location algorithm running, i.e., the points corresponding to the circles and ellipses determined for each detected possible damage reflection in corresponding sensor signals (determined through the differences between actual and reference states) for each set of fifty scans when each PZT in the three transducer network is being actuated. Determined intersection points are also presented. Damage will most likely exist in an area that shows the largest density of intersections (of every color).

To note that in the following figures the plate dimensions are presented non-dimensionalized to 1, and the three transducers in the network are positioned in the following manner: PZT1 positioned in the center of the plate; PZT2 positioned in the middle of the plate boundary presented at the bottom of the plots; and PZT3 positioned in the middle of the plate boundary presented at the left in the plots. In the different plots, it is presented the PZT actuated and the points corresponding to the determined circles (for the actuator PZT) and the different ellipses (determined for the two different pairs of actuator-sensor) are plotted in different colors. Each set of ellipses, determined for each pair actuator-sensor (for all detected potential damage reflections, with each ellipse representing the possible damage positions for each one of those possible different damages), is presented with the same color.

Figure 6 was obtained when a surface cut (with a depth of half of the plate thickness, i.e., of $1mm$), with $1mm$ in length, oriented at $45°$ with relation to the plate boundaries was imposed in the aluminum plate in a position corresponding to the bottom-center position in the figures.

A surface circular hole with $1mm$ of diameter was then drilled in the aluminum plate, in the same position where the previously referred cut was imposed Figure 7.

This circular hole was drilled afterwards through the thickness of the plate. Figure 8 presents the software generated plots depicting the execution of the damage location algorithm (determined points for circles and ellipses, intersection points and most probable location for damage, for all scans), for such introduced damage.

For brevity, only the previous results are presented, however, it must be noted that all applied damages (with $1mm$ of maximum dimension and applied cumulatively to the aluminum plate) were successfully detected and located with the developed automatic SHM system. This was achieved for surface imposed damages, for cuts with different orientations with relation to the transducers in the network and with the considerable (reflection) wave scattering introduced by the round shape/interface of circular holes. Nonetheless, damages introduced behind already existing/imposed defects, in the aluminum plate, far from the implementation region of the transducer network were more difficult to detect and accurately locate. Such difficulties were also verified for cuts, and multiple damages, oriented according to a radial direction, with the centre in one of the transducers in the network, and particularly when aligned with a direction defined by two of the transducers

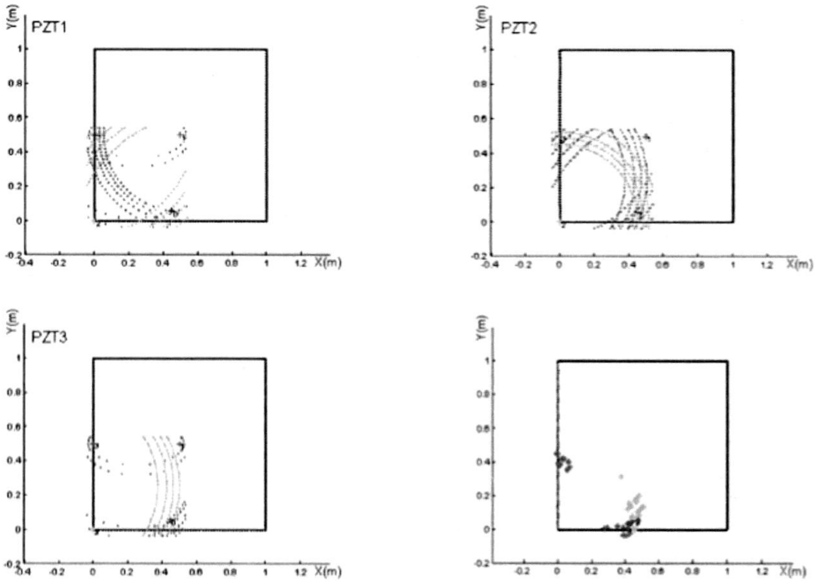

Figure 6. Software generated plots, presented to the user, of the execution of damage location algorithm for a $1mm$ cut imposed in the aluminum plate

in the network (and even more difficult when introduced in regions farther from the area of implementation of the network).

The capability of the transducer network to detect the different introduced damages, and one of its advantage, is related with the fact that by being a network of transducers in considerably different positions in the plate, the actuation origin is changed considerably when the different transducers are actuated. Sensing capabilities of the global network are also enhanced due to the same fact, improving the accuracy in the implementation of the triangulation methods for damage position determination. For instance, for this configuration, it is impossible for cuts, or multiple damages to be aligned with all transducers in the network. The capability of this system to detect damages introduced cumulatively is related with the possibility, enabled by the system, to perform comparisons of the actual state sensor signals with any previous recorded sensor signals, corresponding to different health conditions, including the ones in which previous damage already existed. This enables the capability of detecting new damages, by comparison with the sensor signals obtained for the last inspection (the last state, just

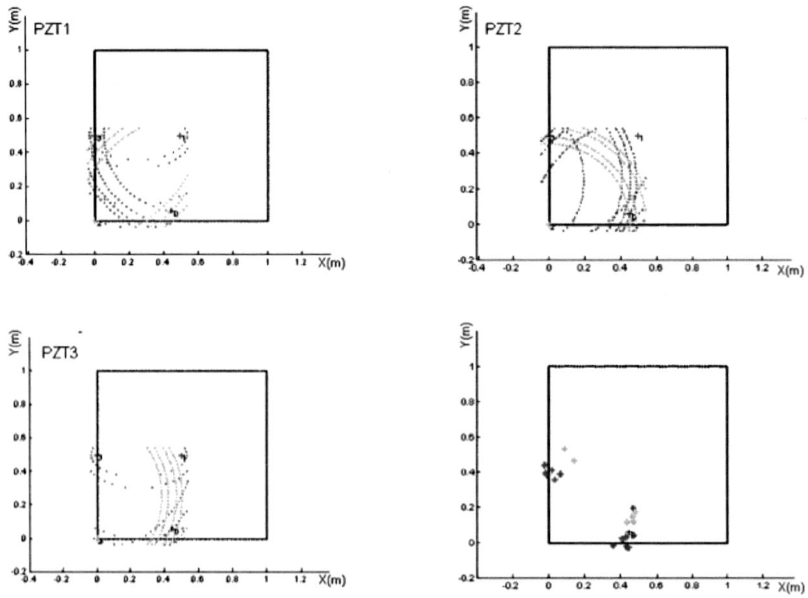

Figure 7. Software generated plots for a 1mm surface circular hole

before the actual inspections, or determined condition). This might also enable future damage growth monitoring. The successful detections, and damage location determination, were achieved although the small actuation capability of this system, based in the single actuation of single transducers at a time, and then small excited waves amplitude and consequently much smaller reflection waves amplitude this is, in fact, the main disadvantage presented by a transducer network.

Particularly for riveted boundaries, i.e., for when experiments were performed with the aluminum plate boundaries riveted to a support, some attention was dedicated to inspect such boundaries (conditions). The plate was riveted at its boundaries with two (staggered) rivet lines, spaced 30mm apart. The same spacing was used in between rivets in the same line and aerospace grade cherry-pop rivets, 2mm in diameter, were used. Inspections were executed to detect and locate loosen rivets, cuts introduced in between rivets and initiating in rivet holes. Any of these introduced types of defects were only successfully detected and (approximately) located when rivets in the interior rivet line (with relation to the plate center) were considered. The inner rivet line scatters the propagating waves with no waves with sig-

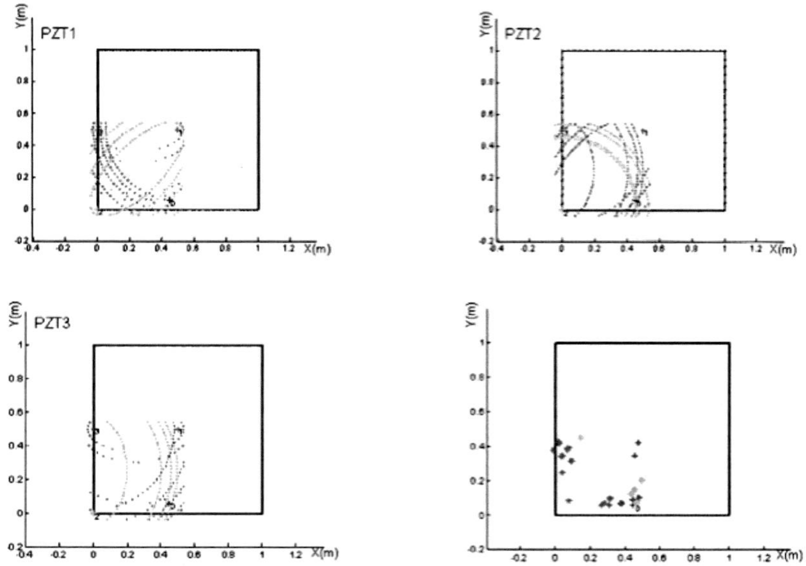

Figure 8. Software generated plots for a 1mm through the plate thickness, drilled circular hole

nificant amplitude propagating beyond those even more true if the already much smaller amplitude of damage reflections are considered. Regarding cuts initiating in rivet holes, these were successfully detected, as long as they are not behind the rivets with relation to the network, i.e., when they were oriented to the interior of the plate, towards the network location. Specifically, cuts aligned with the network were only detected when their length was over $2mm$, due to a minimum required time separation between the detection of their reflection in the corresponding sensor signal and the contiguous rivet hole generated reflection. The minimum required separation time is dependent on the maximum sampling frequency of acquisition of the sensor signals by the oscilloscope (NI digitizer), in this case $60MS/s$, and the wave propagation speed - and also on the minimum number of sampled points for the software reliably determine the existence of such reflection (considering the interference in terms of amplitude between the reflections from the cut and contiguous rivet hole).

At a later stage, an aluminum L stringer was riveted to the plate in the opposed quadrant, with relation to the quadrant of implementation of the network, and perpendicular to a radial direction with center in the center

PZT. The stringer had a height and base width of $30mm$, a thickness of $2mm$ and a length of $500mm$. It was riveted with a single line of rivets, spaced $50mm$. Aerospace grade cherry-pop rivets, with $2mm$ of diameter, were used (as for the riveted plate boundaries). Similar experiments were performed, with similar final results, as the ones regarding the riveted boundaries, described before, i.e., concerning loosened rivets, cuts introduced in between rivets and initiating in rivet holes. The main objective of the introduction of this riveted stringer was to assess the capability of the transducer network based SHM system to detect damage beyond the introduced riveted stringer (and riveted line). All experiments executed proved to be unsuccessful, with no damage imposed beyond the stringer, with relation to the transducer network (application region), being detected. This is explained by the small amplitude of actuated waves and consequently of (damage) reflections, due to the single actuation of single transducers, at a time, in this configuration. The rivet line and stringer completely scatter the excited waves propagation pattern, with no waves with significant amplitude propagating beyond those even more true if the already much smaller amplitude of damage reflections are considered.

3.2 Development of a PCB Board for System Automation

During the experiments executed and presented before, the cables connecting the actuation board and the transducers had to be constantly changed, depending on the actuating PZT since the signal generator used only has one output channel. Furthermore, it was verified that the signals corresponding to reflections, in the actuator PZT sensor signal (i.e., when this transducer was being used as a sensor, just after being actuated), presented amplitudes considerably inferior to the ones verified in the signals of the remaining sensors in the network. This fact created increased difficulties for the damage detection and location algorithm, i.e., for the SHM system, in terms of the detection of such reflections, due to their small SNR, being masked by noise.

Those smaller amplitudes of the signal of the actuator PZT, when used as a sensor after actuation, are related to a mismatch impedance between the actuation and acquisition equipment. The actuation equipment has an impedance to ground that can be configured for 50Ω or 70Ω. This small impedance is required to enable the generation and output of signals with high slew rates, at high frequencies, with small sampling times and high amplitudes. The impedance to ground of the digitizer is $1M\Omega$. A high impedance is desirable to improve the accuracy of data acquisition (based in the measurement of voltage differences, and swings, in between the terminals of such resistance). The ground for the entire system is the same. When a PZT is connected to both equipments (such as the actuator PZT, since the PZTs used as sensors are only connected to the data acquisition equipment) and is being mechanically excited, generating charge (current and voltage) due to piezoelectric effect (PZTs being used as sensors), the current generated will flow to ground through the path with smaller impedance, in this case through the signal generation equipment. This means that for the data acquisition equipment, charge (again current and voltage) is escaping to ground through the actuation equipment. The voltage difference to be measured at the terminals of the resistance in the data acquisition equipment is then smaller.

To solve these two problems, a Printed Circuit Board (PCB) was developed. This board consists of a Micro Control Unit (MCU), an analog signal amplifier circuit (OpAmp) and two Integrated Circuits (ICs) with four switches each. This board is then positioned in between the single channel output of the signal generator and the different actuation channels for the different PZTs in the network. The developed circuit and final PCB, with components mounted and ready to use is presented in Figure 9. To test the developed PCB, a new aluminum plate (with the same dimensions as the previous one) was used. The same transducers and algorithms were applied and similar experiments as the ones presented before were executed. The main difference, besides the use of the PCB, was the introduction of a fourth PZT in the network, positioned in the corner of the aluminum plate (in the quadrant of application of the network). The introduction of this fourth transducer was done to improve the accuracy of the previous network and as a preparation for the application of such PZT network SHM system to composites. It must be referred that after the application of the developed PCB, the amplitude of the signals corresponding to wave reflections, in the actuator PZT sensor signal, were increased by more than 6 times.

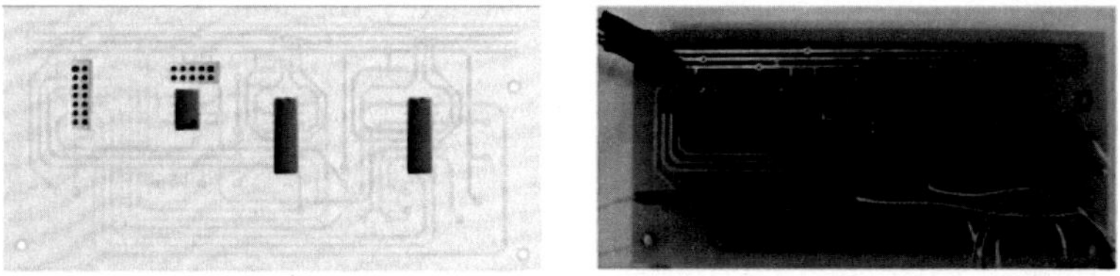

Figure 9. PCB for application in the PZT network SHM system

3.3 Damage Detection: Composite Plates

Composites present different mechanical properties in different directions (anisotropy) and consequently excited Lamb waves present different velocities (wavelengths and dispersion curves), dependent on the direction considered. For the application of the developed SHM system based on a PZT network into composites, damage location algorithms had to be adapted accordingly. In this case, the circles and ellipses determined for damage location degenerate in shapes established according to the different wave propagation velocities in different directions (Figure 10).

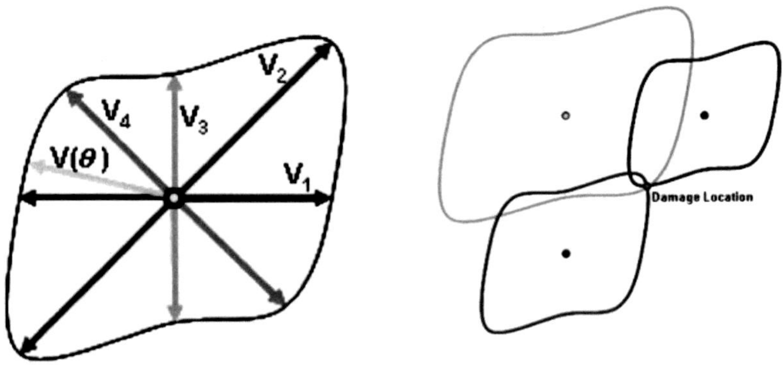

Figure 10. Wave velocity distribution estimate for composite materials

In the initial frequency scans performed by the code, wave propagation velocities (wavelengths and dispersion curves) are now determined for different radial directions in the plate (Figure 11). This is performed with the interaction of the user and analyzing the sensor signals corresponding to the direct excited waves, in between the different transducers in the network, and the plate boundary reflections (using their ToF). The application of a four transducer network becomes of importance to improve the determination of the wave propagation velocities in more radial directions. In terms of the direct waves, the network allows for the determination of such propagation velocities in the directions connecting pairs of transducers. In terms of plate boundary reflections, wave propagation pattern must be tentatively reconstructed, with the determination of the waves propagation velocities in additional directions.

After the determination of dispersion curves (wave propagation velocities and wavelengths for different frequencies) in the different directions, tuning curves and using the method described, are also determined for the different

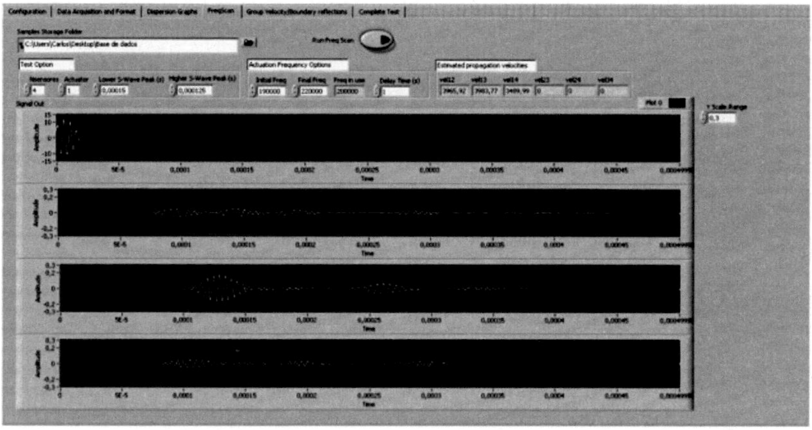

Figure 11. Frequency scan module

directions. These tuning curves will point the frequencies that should be avoided and the ones that tune the SHM system, enhancing the actuation and sensing of the S_0 waves and decreasing the actuation and sensing of A_0 waves, for the different directions. The selection of the (actuation) frequency to base the method upon is performed by the user, with the help of the code. For the diameter of the PZTs being used and for each direction being considered, the frequency that corresponds to a S_0 wavelength double of the PZT diameter being used is determined (to enhance actuation and sensing of S_0 waves). An average frequency should be selected, taking into account also the tuning curves obtained. The results in terms of the selected and analyzed frequencies can also be confirmed by the observation of the waves corresponding signal amplitude, during the frequency scan. As explained before, if the selected frequency corresponds (or is close) to a wavelength that is double of the PZT diameter used, such wave actuation and sensing will be enhanced, and then the waves corresponding signal amplitude (in sensor signals) should be increased. Frequencies corresponding to the highest amplitudes of the S_0 waves signals in sensor signals, during the frequency scans - and preferably lower amplitudes of A_0 waves, are desirable.

It must be noted that the PZT diameter selected a priori may not be the optimum, or even adequate, for this SHM method. To guarantee that, a previous study, with relation to the selection of PZTs and application of the SHM system, should be performed, regarding the composite panels to be inspected. Analytical and numerical methods can be developed to determine the properties of the material and Lamb wave propagation velocities in the different directions. Assumptions and simulation errors must be diminished at all costs, however, these will introduce potentially significant errors in such determinations, and then in the SHM method. The plates can also be characterized experimentally, with the material mechanical properties (and then Lamb wave propagation velocities) being determined in the different directions through tensile tests, etc. Samples of the material must be used, with potential differences in properties, due to (errors or small changes in) their manufacture, and must be tested in different directions. Lasers can also be used to excite and acquire Lamb waves in the component to apply the SHM system on and potentially dispersion curves - propagation velocities and wavelengths for different (excitation) frequencies - will be obtained in different directions. Tuning curves could be determined and together with the dispersion curves an optimum PZT diameter and (actuation) frequency (averaged for the different directions) could be established.

With the determined propagation velocities of interest, in the different directions, a spline is fitted to approximate the propagation velocities in all directions. The plate is then numerically meshed by the code, in the same way as presented before (with the same mesh, spacing, number of nodes, etc). For each node in the mesh, the radial directions connecting such node and the transducers in the network are determined and consequently wave propagation velocities are obtained in those directions. Knowing also the distances between the considered node to the different transducers, the ToF of a wave travelling between those points is calculated for each node and related with each transducer in the network. With this information, sixteen matrices are set. In each matrix, a node in the mesh corresponds to one of its positions. Each matrix is related to each possible scan to be executed, i.e., with each specific PZT in the network being actuated and with the sensor signal being retrieved from each of the four different transducers, i.e., with each actuator-sensor transducer pair. The different matrices are then populated with the corresponding combination of ToFs previously calculated. For the matrix related with the actuation of a certain PZT and the sensor signal being obtained from a specific transducer in the network, i.e., for a specific pair of actuator-sensor transducers, the matrix position corresponding to a certain node in the set numerical mesh is populated with the sum of the ToFs previously calculated for waves travelling between the actuator and point in the mesh, and between that point and sensor. This corresponds to the total ToF, i.e., the sum of the ToFs of the excited wave, until it reaches the point in the plate considered, and a reflection wave departing such point (as if a damage existed in such position), until it reaches the sensor considered.

With this information, when damage detection and location algorithms are executed, comparing the sensor signals corresponding to the actual and reference states for a determined scan, i.e., for a determined actuator and sensor pair; and after the identification of potential signals corresponding to damage reflected waves and the determination of their ToF, this value (ToF) is compared with the values (ToFs) in each position of the matrix corresponding to such actuator-sensor pair, i.e., to such scan. In this comparison it is verified if the first ToF referred falls in between certain positions of the matrix in consideration, with these boundary positions being considered as potential damage location candidates. The remaining process explained before, for the developed SHM system is implemented (formation of damage indexes, probability, intersections, scan repetition, filtering, etc).

After adapting the developed SHM system for its application to composite material plates, accordingly to what was explained before, experiments were performed in a $0.48m$ x $0.48m$ CFRP material plate. This plate has a thickness of $1.6mm$, resulting from ten layers of carbon fiber weave in the following laminate lay-up: $[(0/90)/\pm45/(0/90)/\pm45/(0/90)]S$. By possessing an uneven ratio between $(0/90°)$ and ±45 ° layers, the plate presents a higher stiffness (or Young modulus) along the $0/90°$ directions, being expected to obtain higher wave propagation velocities along the $0/90°$ directions, with relation to the $\pm45°$ (and all other) directions.

A four transducer network SHM system was implemented into the composite material panel, using similar PZTs as before, in the following configuration, presented in Figure 12.

From the initial frequency scans performed for this case, by the SHM system, a frequency of actuation (and to base the SHM method on) of $210kHz$ was established. For this frequency, the S_0 Lamb wave (mode) propagation velocities, determined with the help of the program, in different directions connecting the different transducers in the network (and principal directions in the plate), i.e. for the direct excited waves, are presented in Figure 13.

Observing the calculated S_0 wave propagation velocity by the code, it can be verified that the propagation velocities obtained for each PZT pair, when one is used as an actuator and the other as a sensor and then their function is alternated, are similar. Propagation velocities obtained for the $0/90°$ directions are similar and higher, as expected, for the reasons explained before, than the velocities obtained for $\pm45°$. Regarding the velocities obtained for the $135°$ direction, these are similar for both pairs of PZTs (1 and 4, 2 and 3). However, significant different values of such propagation velocities were obtained for the $45°$ direction, depending on the pair of transducers being used (pair PZT1 and 2 with relation to PZTs 3 and

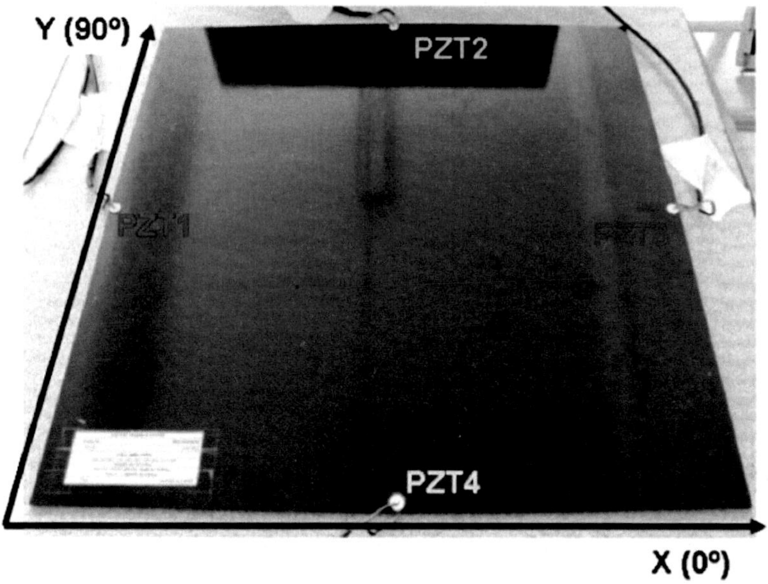

Figure 12. Composite panel, transducer network and experimental setup

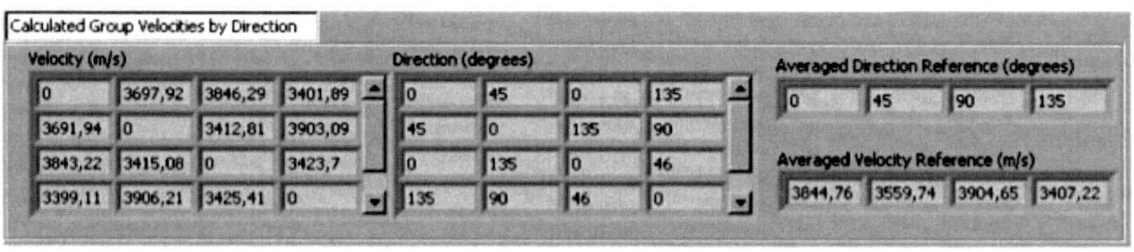

Figure 13. Propagation velocities obtained from the code for different directions

4). Furthermore, since a $\pm 45°$ weave was used, the velocities along 45° and 135° directions should be similar, this is verified when considering the pairs of PZTs 3 and 4, with relation to PZTs pairs 1 and 4, and 2 and 3, but not when comparing all previous pairs with the pair 1 and 2. Unexpectedly, the propagation velocity results from the pair of PZTs 1 and 2 are about 7% higher than those obtained with the remaining pairs referred before. This may result from errors in the manufacture of the composite material plate - manually manufactured. The fact that the weaves were manually laid and no particular control on resin flow was possible may explain the differences in properties at 45°, depending on the area of the plate considered. The velocities for each direction were averaged, and then the velocities of propagation for 45° were also averaged. The considerable differences verified at such direction (due to manufacture imprecision), depending on the location in the plate, and the fact that an average velocity must be used, will introduce errors in the damage detection and location algorithms (although the use of an average velocity tends to decrease existing errors).

Fitting a cubic spline to the average propagation velocity values calculated for the different directions, a velocity distribution can then be estimated for all directions in the entire plate. This is shown in Figure 14.

With this velocity distribution, damage detection and location experiments were then performed. The steps described previously, for the execution of the SHM method, were followed, with the reference or undamaged condition of the plate being characterized (through the execution of the different initial scans). Afterwards, a through the thickness hole of $1.5mm$ in diameter was drilled in the plate at a location ($0.17m$, $0.17m$), considering as a reference the coordinate system indicated in Figure 12, i.e., with its origin in the bottom left corner of the plate. Scans were then executed and the actual damaged state was characterized with the damage detection and location algorithm (described before) being executed afterwards. The execution of the damage detection and location algorithm can be seen in Figure 15, as presented to the user by the developed software. Notice the degenerated circles and ellipses as a result of the dependence of propagation velocity on direction.

The final plot/results (software output) are presented in Figure 16, for this case.

The observation of the final plot by itself is not revealing in an easier and direct manner of the existence and location of damage. However, the observation of the plot together with (and particularly of) the resulting table reveals that the system was capable of detecting and determining three positions in the plate with a higher probability of damage existence. Even though such probability is not considerably high and is just slightly

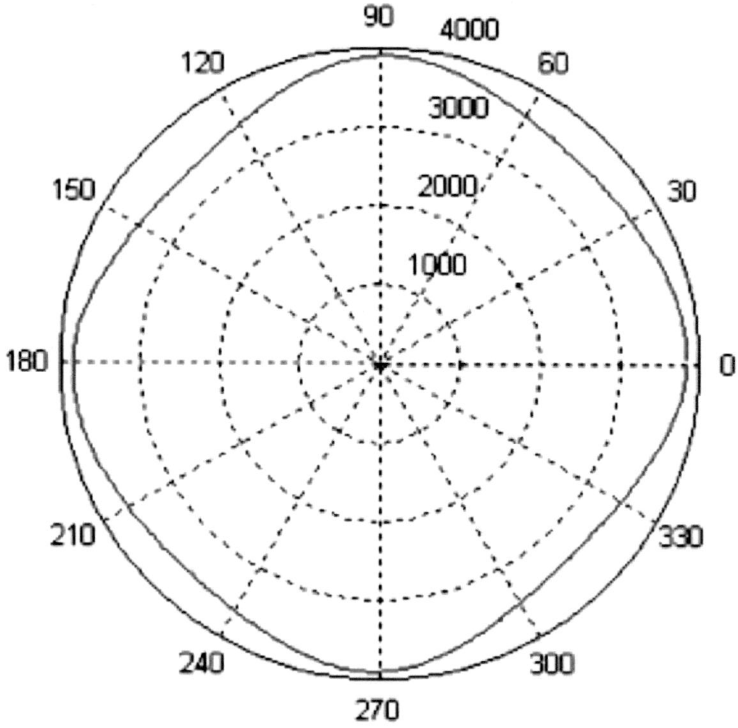

Figure 14. S_0 wave propagation velocity radial distribution

higher than the one determined for the next possible seven candidate points for damage existence; and the system also determined two false positives; the imposed damage was also detected and located.

With the results obtained, noise threshold levels were tuned within the software and newer results are plotted in Figure 17. As it can be observed, also false positives were determined, however the introduced damage was detected and successfully located, obtaining a stronger indication.

3.4 Initial Acoustic Emission Experiments

The idea to execute initial experiments for the detection of Acoustic Emissions (AE), with the implemented SHM system, stemmed from the fact that this system must already be deployed, i.e., a set of PZT transducers is

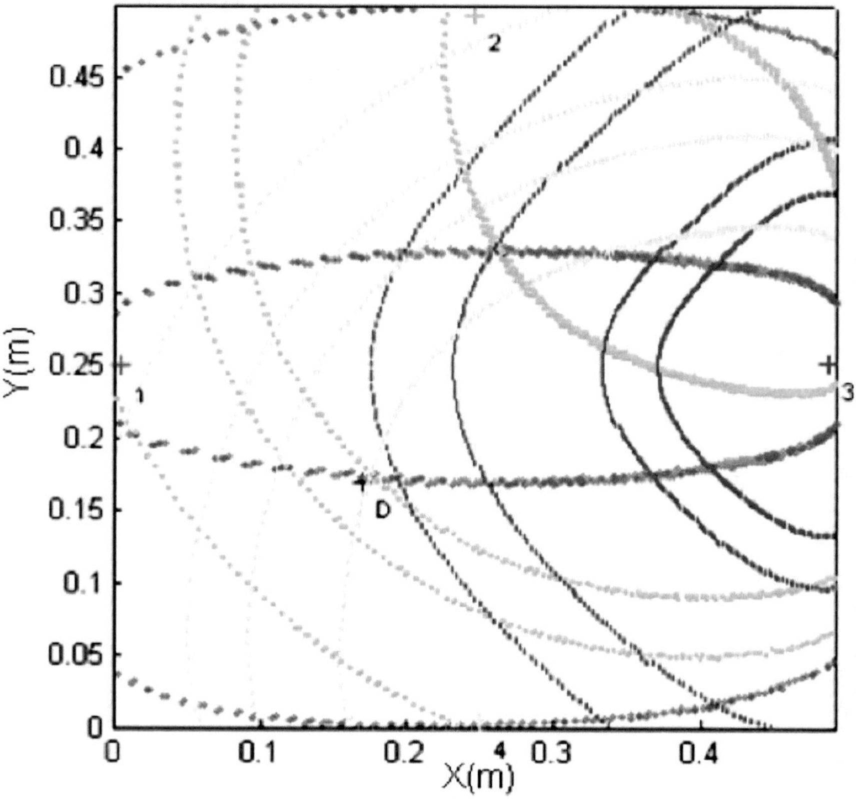

Figure 15. Damage detection and location algorithm

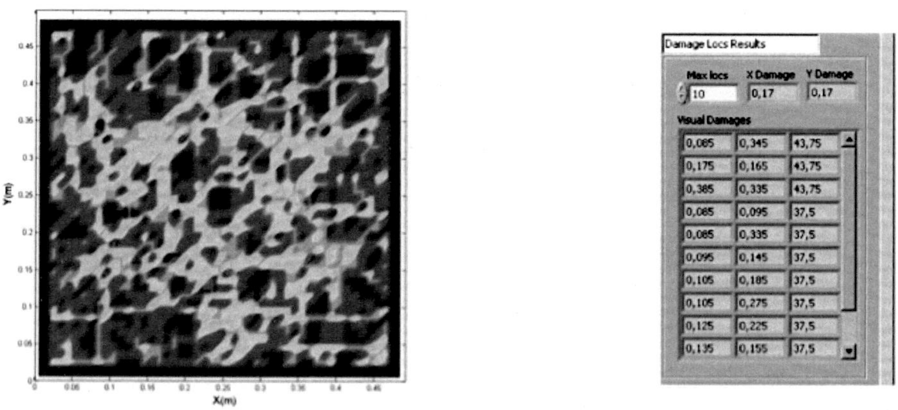

Figure 16. Software output

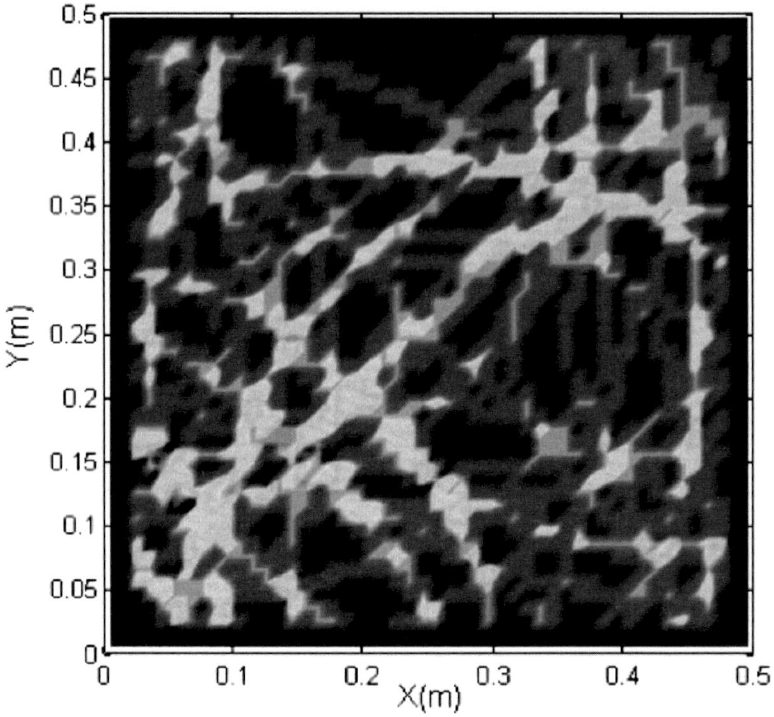

Figure 17. Results obtained after noise threshold levels were tuned

already positioned in the component to inspect and available with the SHM system being executed only at certain times. There is then the possibility to use such system and transducers continuously, in between the execution times.

With the developed SHM system based on a four transducer network, initial AE experiments were performed in an aluminum plate identical to the ones used before, but with the system being used in its passive mode, i.e., no actuation was applied, with the actuation system being disconnected from the network transducers (just connected to the data acquisition system). The software developed was adapted to acquire transducers signals continuously, and just starting to record data from the four sensors, when triggered by detecting the first corresponding signal to a wave in one of the sensor signals. After such detection occurs, the code verifies if it can detect signals corresponding to such wave in the other three sensor signals. If it doesn't, it neglects such data and goes back to the initial state. If signals corresponding to a wave are also detected in the other three sensor signals, data is conserved and a developed damage detection and location algorithm is started.

After the software detects the initial triggering wave corresponding signal, it saves data during $500\mu s$. This value was determined with the following considerations:

- knowing that the system is searching for a wave that was generated by an impact in the plate (possibly also simulating the growth of an existing damage) and not a reflection;
- considering the diagonal of the plate used ($1.5m$ x $1.5m$) as the highest distance between a possible AE origin and any transducer of the network (in this case the corner PZT);
- the SHM system was developed to detect (and then it will only try to detect, or better was optimized to detect) λ_0 waves with a frequency around $340kHz$, i.e., the frequency that, through the previously determined analytically and experimentally obtained dispersion curves for the aluminum plate (as explained before), corresponds to a wavelength double of the PZTs diameter being used, optimizing its sensing (and also according to the tuning curves calculated and frequency scans). The λ_0 waves in the aluminum plate with the referred frequency of $340kHz$ have a propagation velocity of $5390m/s$, as determined before. It must be remembered that the developed system applies passband filters to the acquired data, around such frequency. It can be concluded that the system will search for AE only as Lamb waves, in its first symmetric mode (S_0), with a frequency around $340kHz$;
- the determined value exceeds the calculated value according to the distance and velocity previously referred, for safety.

For the detection of such waves (and to trigger the system), a noise threshold was defined. In the location method to be presented next, multiple detected potential waves are considered for each sensor signal, as before, including false positives due to noise.

After detecting such waves, and since (there is no dedicated actuation) the original time of actuation, i.e., the time at which the AE was generated in its origin point is unknown, it is not possible to use that as a time reference or origin. The initial time at which the system was triggered (by the first wave detected) is then considered the origin of times. Also, due to the fact that the time at which the AE was generated in its point of origin is unknown, the location method must be based in time differences between the different detected waves in the different sensor signals, i.e., relative, or differences in their ToF. With the propagation velocity considered, these ToF differences can be translated in differences of distance travelled by the AE between the different sensors in the network. Based on this information, hyperbolas can be determined for each pair of sensors and for all (combinations of the different) detected potential waves combinations are formed with the data from one wave detected by one of the sensors of the pair and the data from other wave detected by the other sensor in the pair. Such hyperbolas represent the locations in the plate that the distances to the pair of transducers considered are related by the determined differences (traveled by AE to the different sensors).

The location algorithm, previously based on circles and ellipses, is now adapted to the referred hyperbolas. The remaining algorithm, with the meshing of the plate and search algorithm is used to obtain the intersections of such hyperbolas and then determine each candidate position for damage existence and the corresponding probability.

The generation of AE was simulated by dropping a steel sphere, of $10mm$ in radius from a height of $15mm$ (from its center) with relation to the plates surface, in different positions in the plate. Due to the small amplitude of generated AE, these were detected and their origin successfully located in 50% of the experiments performed.

4 Phased Arrays

Earlier, experiments were performed with the implementation of a PZT network for SHM applied to aluminum and composite material panels. During that study the small amplitude of generated and consequently (boundaries and damages) reflected Lamb waves was verified. Particularly, damage generated wave reflections were of extremely small amplitude and prone to noise interference making them more difficult to detect (in corresponding sensor signals). Aggravating this fact, damage reflection wave amplitudes decrease considerably when smaller damages are considered, introducing paramount difficulties when trying to decrease the detectable minimum damage dimensions. Furthermore, it was verified that with the small amplitude of the excited waves and subsequent reflections (generated by a single transducer), damage beyond structural reinforcements and riveted lines (with relation to the system location) were not detected. Such features scatter the excited waves propagation pattern (and energy), creating reflections, and with no waves with significant amplitude propagating beyond those, for the small wave amplitudes (energy) involved.

This is related with the single actuation of a single transducer (at a time) that occurs in a transducer network. A possible way to increase the amplitude of emitted waves would be the use of actuation signal amplification with the use of amplifiers. This solution introduces, however, two main difficulties: inclusion and amplification of noise with relatively small improvements in SNR and the fact that the actuation signal applied in the referred experiments (in the range of ± 18V, permitted by the maximum output swing of the signal generator) was already close to the maximum voltage and charge permitted by the PZT transducers. Previous experiments resulted in the depolarization and damage of the small transducers when the actuation signal was amplified to a range of ± 21V.

The solution to increase amplitude levels of actuation and reflection waves (specifically of damage, when smaller sizes have to be considered in order to reduce detectable damage dimensions) was then to use multiple actuation. The objective is then to promote constructive interference in predetermined positions of the plate between the waves excited by several single actuators (beam forming, or generation of a wavefront). Since actuators will have to be in different positions on the panels a phased actuation is necessary. The phased actuation implies that small time delays will be introduced in between the actuation of neighbouring transducers. The time delays to introduce are related with the relative positions at which the transducers are applied on the plate (configuration of the array), the distance between them (considering a linear array, its pitch), the propagation velocity of waves to excite and the propagation direction of the generated wavefront. By varying the time delays the wavefront propagation direction can be modified (steered). Inspection can then be focused into predetermined regions of the component. By modifying time delays in between different scans the component can be entirely inspected.

In the PZT network experiments, a National Instruments (NI) PXI-5421, $100MS/s$ Arbitrary Waveform Generator with 16 Bit resolution, $8MB$ of internal memory and a single output actuation channel was used for signal generation and actuation. Due to the high propagation velocities of Lamb waves to excite, in order to achieve successful constructive interference, time delays to be introduced will be extremely small. This means that even when phased actuation is considered, at certain times multiple transducers will be actuating simultaneously. The implication of this fact is that a single actuation channel simply demultiplexed cannot be applied. Delay circuits would then have to be considered, with the introduction of noise and imprecision in the establishment of time delays. The conclusion is that multiple actuation channels are required. As stated before, besides presenting extremely high cost of acquisition, existing systems with multiple actuation channels commercially available, or under research, suffer either from:

- Not offering enough precision in time definition for the correct establishment of smaller time delays required for the generation of fast propagating S_0 wavefronts. Some of the commercial available systems are capable only to establish the correct actuation signal to generate S_0 waves in a single channel at a time, while phased array systems under research generate slower propagating A_0 wavefronts, relaxing the requirement of high definition in time, since to generate slower wavefronts, the time delays to introduce are higher;

- or not even being capable of the generation of the correct actuation signal for each individual actuator, in terms of time or (higher) frequency definition or at least not offering the capability of the user to define such actuation signal (black/closed box systems), either in terms of waveform or frequency selection, restricting its application to certain conditions, materials and components.

The obvious conclusion is that a lower cost, dedicated multiple channel actuation system had to be developed. The system must be capable of a high time definition to precisely establish the required small time delays in between consecutive channels/transducers to be actuated for the correct generation of the fast propagating S_0 wavefronts and the correct actuation signal waveform and (high) frequency for the generation of the single, fast S_0 waves by each transducer.

For the reasons stated previously, PZT transducers were selected to be applied in the phased array SHM system. This selection is based on the fact that a single PZT transducer can be used as an actuator and as a sensor. Furthermore, PZT transducers can be used in techniques involving high frequency generation and sensing. As the basis for a future development of arrays with different configurations (cross, star, circular shapes, etc) it was decided to implement a linear array to excite fast propagating S_0 wavefronts (the selection of this wave mode was based in the reasons stated before). A linear array presents advantages in terms of a simpler development. At the same time, different array configurations can be seen as a combination of multiple linear arrays, while solving the disadvantages in terms of inspections introduced by a single linear array (also referred before).

To simplify the implementation and testing of this phased array SHM system in development. The initial application was to aluminum plates (isotropic materials, with homogeneous properties and then constant wave propagation velocities in all directions). However, in its development it was also considered its future application to panels made of composite materials (orthotropic), presenting different mechanical properties in different directions and then with wave propagation velocities depending on the direction considered. This fact became important even for the experiments executed in aluminum panels, since, due to their manufacturing processes mechanical properties are not exactly the same in all directions. Small relative differences in properties exist, for instance due to the lamination direction in aluminum plates. These small relative differences are translated to considerable absolute values when the high values of Young modulus and wave propagation velocities are considered. In the previous experiments with the PZT network SHM system, differences around $200m/s$ (however less than 0.5%) were verified in the propagation velocity of the S_0 wave in aluminum plates in different directions.

The development of the phased array SHM system accounted for its future implementation in real structural panel components in the sense that it was considered, in its development, its experimentation in representative plates with more stringent boundary conditions. For instance, experiments with the plate totally supported (increasing wave propagation damping), with simply supported and riveted boundaries were considered.

Furthermore, experiments were considered to be performed in environmental conditions close to reality. Some experiments were considered to be executed in an aircraft maintenance hangar, without any surrounding sound/noise control and with limited temperature control.

4.1 Dispersion Curves

As referred before, the experiments involving phased arrays were performed in aluminum plates identical to the ones used before, for the implemented transducer networks same material, thickness and remaining dimensions. Due to this fact the dispersion curves, and all data, and considerations presented earlier are applied here.

Regarding the implementation of phased arrays, so that undesirable side lobes are not generated, the phased array pitch should be less than or equal to half of the wavelength to be excited (Giurgiutiu, 2008):

$$pitch \leq \frac{\lambda}{2} \tag{22}$$

Since the array pitch is equal to the dimension of one of its PZT transducers plus the spacing in between consecutive elements in the array, the tuning of such system (selection of excitation frequency and wavelength) should now be performed with relation to the array pitch and no longer with relation to the PZT element dimensions. However, it is still true that the amplitude of generated Lamb waves is increased when the dimension of the PZTs used approaches half wavelength of excited waves. The logical conclusion is that the array pitch should be equal to half of the wavelength (to avoid considerable side lobes) of waves to be activated and the PZT transducer dimension should be as close as possible to the phased array pitch, reducing spacing in between consecutive elements to a minimum. Nonetheless, neighbouring elements should not be in contact (and not even their bonding material). If contact existed, the actuation of one element would be passed to the adjacent element, creating actuation noise and interference on the later PZT. To enforce this condition while trying to minimize the spacing in between consecutive elements of the array, and due to practical aspects, a spacing of $3mm$ was considered to account for $1mm$ around each transducer to be potentially occupied by its bonding material and $1mm$ of effective spacing. This means that it was considered that the PZT elements have a diameter equal to the phased array pitch (equal to half of the wavelength to be excited) minus $3mm$.

4.2 Tuned Lamb Waves: Mode, Frequency and Transducer Selection

The tuning of the phased array system was performed in the same manner as presented earlier. However, the pitch of the phased array must now be considered, instead of the PZT transducers diameter. Tuning the phased array system for the S_0 wave can be achieved by selecting an activation frequency and correspondent array pitch (equal to half of S_0 wavelength to be excited, corresponding to the frequency selected, to enhance S_0 actuation and sensing and not to generate side lobes), and PZT transducers diameter, such that the PZT diameter is a multiple of A_0 wavelength to be excited (to decrease actuated A_0 amplitude and decrease its sensing), again corresponding to the frequency selected (or at least as close as possible to a multiple). To refer that the higher that multiple is, the more the activated A_0 amplitude is decreased.

A tuning curve is now obtained in the following manner: based on the wavelength vs. frequency curves, for all frequencies the S_0 wavelength was halved, with these values being subtracted by $3mm$ (the spacing in between consecutive elements in the array) to obtain the optimum PZT dimensions (to excite and sense those S_0 wavelengths at their corresponding frequencies) as before, the resulting values were then divided by the A_0 wavelengths excited at the same frequencies. Afterwards, the same process as described earlier was applied: the absolute values of the difference between such results and the nearest integers were analyzed and for an easier representation and assessment of those results, those differences were doubled and the results were subtracted to unity (values to be presented close to one represent optimum conditions and values close to zero the worse cases). The resulting tuning curve for the phased array is presented in Figure 18.

According to this curve, activation frequencies around $150kHz$ (and cor-

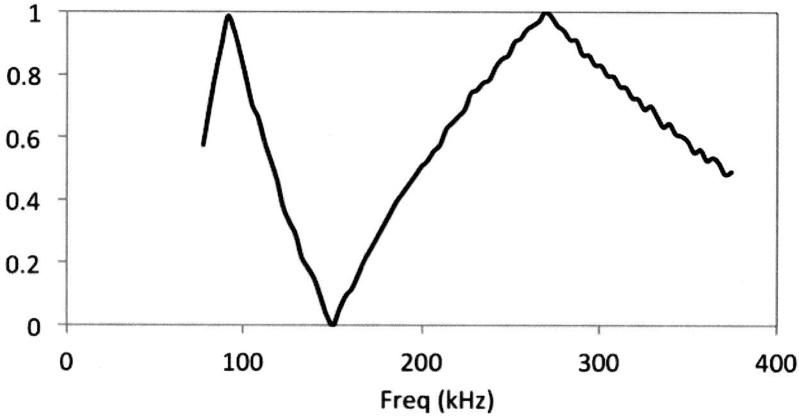

Figure 18. Lamb wave tuning for the phased array system

respondent wavelengths, array pitch and PZT transducers diameters) should be avoided, while frequencies around $100kHz$ and $275kHz$ give the best results. It must be remembered however, that the application of smaller wavelengths is desirable for the detection of smaller damage dimensions, what is achieved at higher frequencies. The frequency of excitation should then be selected to be $275kHz$. However, due to their availability the same PZT transducers applied before in all experiments were used with a diameter of $8mm$. Considering the spacing in between neighbouring elements in the array of $3mm$, this corresponds to an array pitch of $11mm$ (which should be equal to half of the wavelength of the S_0 wave to excite). The corresponding actuation frequency is then $245kHz$ with the S_0 waves of that frequency presenting a propagation velocity of $5450m/s$, in the considered aluminum plate.

To refer that due to the same reasoning as presented earlier, the same actuation signal waveform is used for the activation of each PZT element in the phased array. Considering the actuation frequency of $245kHz$ (inner sine frequency) the actuation signal will have a total duration in time of $20.41\mu s$. According to what was referred to earlier and considering that the S_0 waves of that frequency present a propagation velocity of $5450m/s$, in the considered aluminum plate, a distance of $56mm$ around the PZT transducers (phased array) will not be inspected.

4.3 Number of Elements in the Array

To be able to inspect an entire component, the scans performed by a phased array must be repeated, with the wave front being steered into different directions in the component in each scan. Consequently, different time delays must be applied between the actuation of consecutive array elements in the different scans. To define the directions (and related time delays) that must be considered to enable the inspection of the entire component the aperture of the generated wavefronts must be taken into account. The aperture angle is centred around the propagation direction of the wave front generated by the phased array. It defines the region where such propagating wave front will be effectively formed and then the useful inspection region for each scan direction considered.

Such aperture ($\Delta\alpha$) is dependent on the total length of the phased array (l), the phased array pitch and number of elements in the array (n) and on the excited Lamb wave wavelength (λ) (Giurgiutiu, 2008):

$$\Delta\alpha \simeq 0.886\frac{\lambda}{l} \simeq \frac{\lambda}{n \cdot pitch}(rad) \tag{23}$$

This relation is valid for all angles of the propagation direction of the array generated wavefront (in an isotropic plate), particularly for angles approaching the perpendicular to the linear array. It is predicted that due to the influence of the array elements into the excitation/propagation of the wavefronts these will present lower amplitudes (decreasing inspection capability) and will be more spread in space when their propagation direction approaches the direction of the array (increasing the aperture in those directions). Then the aperture calculated according to Equation 23, can be regarded as the minimum aperture. If such value is considered as the angle interval to be established in between consecutive inspection directions, it will be guaranteed that the entire component can be inspected. Simultaneously, the overlap of inspection regions will be promoted and inspection capabilities will be enhanced.

Theoretically, constructive interference in between generated waves from different transducers in the array is augmented (the amplitude of generated wavefronts is increased), when the number of generated waves and then the number of transducers is increased in the array. However, the wavefronts to be generated will be more spread in space in those conditions. As can be observed from the above equation, such results in a lower aperture of the array. As a compromise, it was decided to implement an array with seven elements.

4.4 Phased Array Actuation System

Regarding the phased array actuation system, a configuration based on a master circuit controlling the phased activation of different slave circuits was implemented. Each slave circuit, when activated by the master, generates the actuation signal to one PZT transducer in the array. The master circuit consists of a simple Micro Controller Unit (MCU). The MCU was selected considering its processor frequency, the output frequency it is capable of and number of output pins. The number of output pins limits the maximum number of slave channels that one MCU is able to control however, two MCUs can be connected to communicate with each other, in order to expand the number of channels to control. The processing and maximum MCU output frequencies determine the minimum time delays that it is capable to apply for phased activation of the slave channels. A Texas Instruments (TI) MSP430F2012 MCU with $16MHz$ of processing frequency, $4MHz$ of output frequency and two output ports, with eight and two pins respectively, was selected the same MCU used in the developed PCB before, for the PZT network SHM system. For the selection of the minimum time delay that the MCU would be able to apply, in this case $250ns$ (corresponding to an output frequency of $4MHz$), a simple exercise was followed. The time delay to be applied in between consecutive actuations (Δt) corresponding to the inverse of output frequency (f_{out}) is equal to the difference in distance travelled by the consecutive generated waves (Δd) divided by the activated waves propagation velocity (V):

$$\Delta t = \frac{1}{f_{out}} = \frac{\Delta d}{V} \tag{24}$$

Through simple geometrical relations depicted in Figure 19, the difference in travelled distances can be related to the phased array pitch and the direction of propagation of the activated wavefront (α) by:

$$cos(\alpha) = \frac{\Delta d}{pitch} \tag{25}$$

Combining the previous two equations:

$$\Delta t = \frac{pitch \cdot cos(\alpha)}{V} \tag{26}$$

Knowing that for a wave:

$$V = \lambda \cdot f_{wave} \tag{27}$$

Assuming that an optimized phased array is implemented such that its pitch is half of the wavelength of Lamb waves being generated:

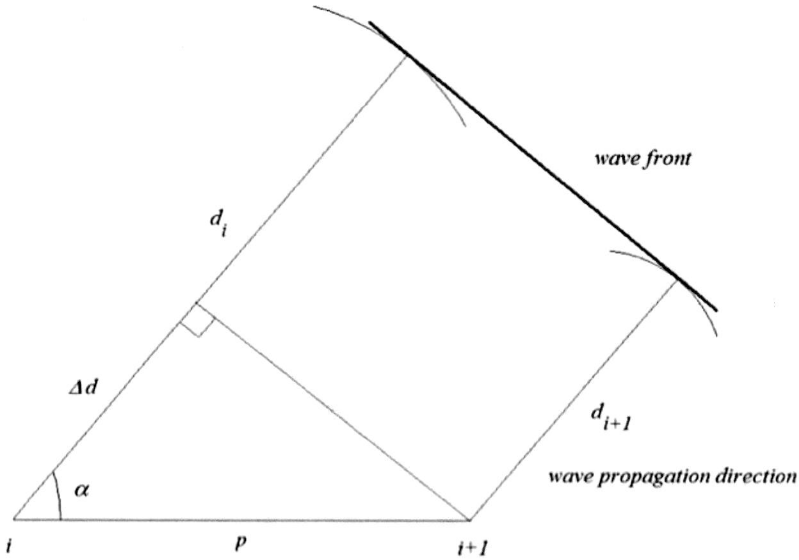

Figure 19. Phased array beamforming

$$\lambda = 2 \cdot pitch \tag{28}$$

Replacing this equality in Equation 27 and the result in Equation 26 , the following relations can be obtained:

$$\Delta t = \frac{cos(\alpha)}{2f_{wave}} \rightarrow f_{out} = \frac{2f_{wave}}{cos(\alpha)} \tag{29}$$

To generate a wave front propagating in a direction perpendicular to the linear array, the master MCU must output simultaneously the activation signals to all slave circuits without any delays applied in between actuations. This means that the MCU must output all pins simultaneously. This is not problematic. The smallest non zero time delays to be applied correspond to scanning directions (to wavefront propagation directions) closest to that normal direction. If it is desirable to inspect the entire component, for the different inspected regions in different scans to be contiguous, the different inspection directions must be spaced by no more than the array aperture. To achieve this the scanning directions closest to the normal of the linear array to be selected will be separated from that normal by the aperture angle. It is then necessary to calculate a minimum aperture that will correspond to the minimum time delays that the master will be able to apply. Equation 23 can be simplified by assuming an optimized phased array, i.e., Equation 28, resulting in:

$$\Delta\alpha = \frac{2x0.886}{n}(rad) \tag{30}$$

Considering the application of a phased array with a maximum number of seven elements, the array aperture is calculated to be 0.2476 rad, or 14.19°. Considering a little superposition of adjacent inspection regions (for safety), the closest angle to 90° will then be 77°. With Equation 23, the maximum wave frequency that can be applied in the inspection method for the output frequency of the master MCU of 4MHz will be $450kHz$. Such value is higher than the upper limit of $400kHz$ established for the actuation frequency to apply in this system.

It must be noted that the processing frequency of the MCU is also important for the correct implementation of the method since the error in time between the desired time for output and the time at which such output occurs in reality is in the worse situation 20% of its clock period (then $12.5ns$), usually below 10% of the clock period ($6.25ns$).

According to the aperture determined for the array of 14.19°, and as referred for safety reasons, the maximum angle interval in between consecutive directions of inspection was established to be 13°. The inspection direction angles, or the angles at which is intended that the wavefronts will propagate were selected to be 6°, 19°, 32°, 45°, 58°, 70°, 80°, 90° and the symmetric directions with relation to the normal of the array. Angles are measured anti-clockwise with relation to the direction of the linear array, from the right. With the selection of these directions some overlap in between the regions to inspect is promoted. Particularly, the 6° direction was selected to enable the inspection of the region laterally adjacent to the phased array while avoiding the destructive interference of the PZT elements in the array to the propagation of the wavefront in the array direction.

With the selected inspection angles, knowing the propagation velocity of the S_0 waves (and wavefronts) to generate and the linear phased array pitch, the corresponding time delays to be introduced in each scan (for each propagation direction angle) were determined. These delays were then implemented in the developed code to use in the master MCU.

For the design of the slave circuits for the actuation of each PZT in the array, it was also considered a MCU to generate the digital signal corresponding to the actuation waveform (the bit trail). One of the main advantages of the use of such MCU is that as in the case of setting the time delays for the different scans through the use of the master MCU, it can be programmed so that the actuation signal waveform can be modified, particularly regarding its frequency. This will enable the application of the inspection technique to different materials in the future. The MCUs used in the slave channels were similar to the one used for the master circuit. The number of pins was more than sufficient from the available 10 pins, five pins are needed at the most, one for input from the master, two to output the inner sine function and the modulating window digital signals and between one to three pins to control output switches (depending on the configuration selected by the user). With the designed and implemented technique to generate the actuation signal, the MCU is capable to generate signals with frequencies up to $2MHz$ (half of its maximum achievable output frequency of $4MHz$). The small errors in time introduced in the signal output are related to the MCU processor frequency as it was the case for the master MCU. Some of the tests being performed to the components of the phased actuation system, using a bread board are depicted in Figure 20.

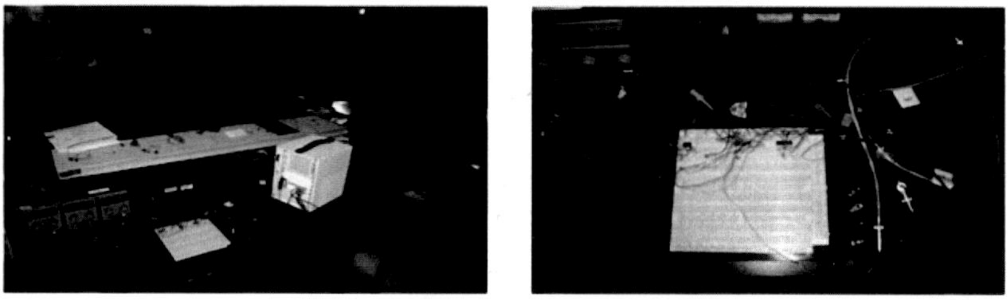

Figure 20. Components of the phased actuation circuit being tested

The slave circuit experimental (non miniaturized prototype) board (and system diagram), with DIP (through hole) larger electronic components is depicted in Figure 21.

4.5 Damage Detection Algorithms

Regarding the damage detection algorithms, two different approaches were followed concurrently. The first was based on the application of the phased array sensing principle referred in the literature review, which is applied by the majority of works in this field. In this approach, since the inspection direction is known, the phased array sensor signals are shifted in time for the corresponding relative time delays. Afterwards, the signals are added with the resulting signal being searched for a potential damage reflection. With this procedure, incoming reflection waves in the inspection

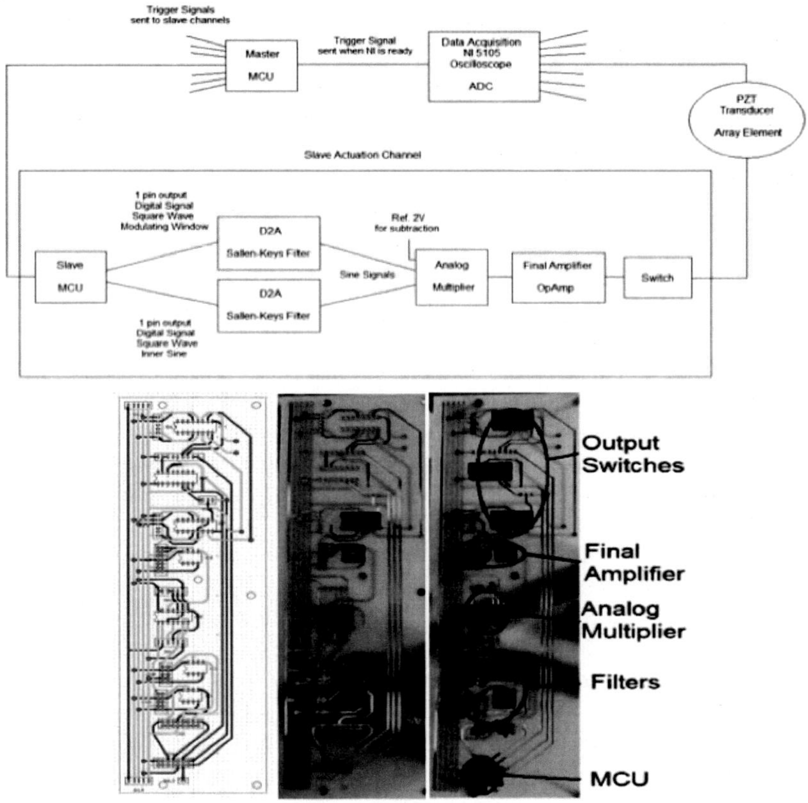

Figure 21. Slave circuit

direction will be enhanced, increasing SNR. After encountering a potential damage reflection damage detection, its Time of Flight is retrieved from the sensor signals (time between the wavefront generation and detection of such reflection). Knowing the wave propagation velocity (equal for generated wave/wavefront and reflections), the total distance travelled by the generated and reflected waves is calculated. The distance between the phased array and damage is then determined, as half of the total distance travelled by the generated and reflection waves. Knowing the inspection direction, damage location is then determined. The disadvantage of this approach is that it is based on the approximation that the incoming damage reflection to the phased array is a (linear) wavefront and not a curved wave, as it is in reality. This approximation gives acceptable results for the determination of damage location when the damage is sufficiently far from the array and the section of its reflected wave arriving to the phased array is approximately linear. However, when damages are close to the array, damage location is not accurate enough.

This approach can be used with either a single or a phased array actuation and can be applied in a supervised or unsupervised method. In the supervised approach, the (potential damage) reflections are searched for directly in the sensor signals obtained from scans for the current health condition of the component being inspected. There is no need in this mode of previously saved signals for a reference state (undamaged or with known damages, to monitor new damages and damage growth) and then no subtractions or comparisons are done. Since no subtractions are performed, the current signals to be analyzed will present all reflections from boundaries and other intrinsic geometric or material discontinuities in the component. To distinguish a potential damage reflection from these reflections a supervised approach is required.

In the unsupervised approach (as used for the networks experiments) with the execution of subtractions between current and reference signals potential damage reflections are enhanced (with relation to the disappearing boundary reflections) and easier to detect in the sensor signals. Besides this, the unsupervised method presents advantages towards a more global automation of the damage detection process.

To refer that the application of the supervised approach, depending on the direct detection of potential damage reflections in the current sensor signals, requires such reflections to present considerable amplitudes in order to be identified from the remaining noise in the signal (higher SNR). To achieve this, an increased actuation power and/or the enhancement of sensor signals (through the previously referred phased array sensing method) are required. To note that, due to this aspect, this approach will present better results when a phased array based scan/inspection is applied. In the previous development of transducer networks, due to the characteristic individual actuation, it was verified that the application of this method was not possible low SNR with the reflections being less easily directly detected in the sensor signal analysis.

Either in the previous method or on the second one, to present next, to decrease the random influence of noise in sensor signals, as performed for the PZT networks and for the same reasons, each scan was repeated consecutively fifty times (with the wavefront propagating in the same direction). This was performed for each and all scans executed (for the entire component, considering all inspection directions) both for the reference and actual health conditions, saving all corresponding sensor signals. After performing the multiple repeated scans, sensor signals corresponding to each transducer and for each scan direction are averaged, maximized and minimized for each and all times, as it was performed in the network SHM system. Based in a Gaussian distribution approach new maximums and minimums corresponding to a 90% probability that the signal will be contained in such range, are determined, for all times. Signal bands are then obtained for all sensor signals corresponding to the actual and reference conditions of the component, for all scans, i.e., inspection directions. Specifically for the unsupervised approach, the comparison of sensor signals for current and reference states, for each of the different scans (and for each sensor), is then based on the verification of when the current health state corresponding signal band differs from the reference band by more than 50% of its width.

In the damage detection procedure to be applied, as it was done before, all possible reflections are considered. To further contribute for the reduction of the influence of (random) noise and resulting false indications, the possible damage locations to be determined for each reflection, for each sensor signal and for each inspection scan (for every deployment of the array for the same and different directions of inspection, or directions of propagation of the wavefront) are superimposed being equivalent to perform the intersection of all possible damage locations and forming a damage index. In this manner, it is predicted that the indication of damage positions will be enhanced with relation to false positive since the probability of the intersection of the random false indications resulting from (random) noise to fall in the same position is reduced. In this process bandpass filters are also applied to sensor signals, around the elected frequency for the application of the method (actuation and excited waves and reflections frequency) as before.

The second procedure was derived from the Time of Flight determination of damage reflections considering the excited wavefront and the wave reflection generated by the damage (in this case considered as a curved circular wave propagating and not as a linear wavefront). Please refer to the procedure depicted in Figure 3.

For each transducer in the array and for each scan (inspection direction) the sensed signal (band) is searched for potential damage reflections. In this procedure the ToF determined concurrently by the first method is also used. As referred previously, the ToF of all reflections are considered, in this case until the ToF of the reflection of the farthest boundary (corner) with relation to the phased array is reached since the damage is contained in the component.

Since the signals from the individual sensors are considered in this method, although the phased actuation is applied, the reflections to be searched for will not be directly enhanced by the phased array sensing method. Since it will be prone to give better results with such expected smaller reflection amplitudes it was considered the application of an unsupervised approach with this method, i.e., considering the comparison/subtraction to a baseline or reference health state corresponding signals (obtained in previous scans and saved), and then determining the outliers.

In this damage detection technique, the known direction of propagation of the wavefront (predetermined by the phase delays) or the predetermined inspection direction is marked for each one of the sensors in the array for the specific scan. All the Time of Flights considered are then related with the corresponding total traveled distance of the activated wavefront and reflected wave, through the known wave propagation velocity. As referred before, a damage distance corresponding to half of the total traveled distance is determined and marked in the inspection direction lines for each and all PZTs Figure 3, corresponding to the intersection point of the line representing the wavefront propagation direction and the line marked as wavefront at t_{max} (in light blue). Departing from each of those initial points and through the Time of Flight, parabolas are determined for each PZT. Each parabola is symmetric with relation to the inspection direction line passing through the corresponding transducer. It has its focus point in the corresponding transducer considered and its directrix parallel to the generated wavefront at a distance from the transducer corresponding to the Time of Flight previously determined considering the known wavefront (wave) propagation velocity. The intersection of all the parabolas will determine (enhancing) the damage location, eliminating (or decreasing the relative importance of) the false indications.

For the same reasons, the same method presented earlier was applied, regarding the generation of an identical mesh and corresponding array. In this array, the point in the first line and first column corresponds to the left corner in the boundary closer to the phased array with the columns representing the points in a direction parallel to the phased array. The same process is then followed regarding now the determined parabolas (instead of circles and ellipses, as before). To decrease the processing time for the calculations performed corresponding to a certain inspection direction/region (to a certain direction of propagation of the generated wavefront) only the points in the mesh in a region delimited by the inspection region (in between, or delimited by the dashed yellow lines in Figure 3) and by the wavefront passing in the initially determined points (wavefront at t_{max} in Figure 3, marked in light blue) are considered. As referred before, the inspection region is determined by the aperture angle of the phased array, centered in the propagation direction of the corresponding scan. This approach revealed to be still time consuming. A second approach was developed based on the different delays, predetermined, and set to be applied by the actuation system and corresponding propagation directions of the excited wavefronts. For each of such set (inspection) directions, the distance of each point in the mesh to a considered transducer in the phased array is determined and summed to the distance of such mesh point to a line parallel to the excited wavefront, passing by the centre PZT in the array. This value is then introduced in a new array created similarly as the one referred before. This process is repeated with the generation of seven arrays, one for each transducer and each inspection direction. With the known wave propagation velocity each ToF determined (which includes the ToF of the excited wavelength plus the ToF of the potential damage reflection) for all detected potential damage reflections in a sensor signal for a determined inspection direction is translated to a total travelled distance (by the wavefront and potential damage reflection). A range of distances obtained from such total travelled distance, plus or minus $5mm$ (related with mesh spacing) is established. It is then verified if each distance in the array corresponding to such sensor and inspection direction falls in that range for all potential distances and for all sensors and scans (considering that the different directions of inspection were selected to promote a slight superposition of the different corresponding inspection regions). If this condition is verified, a unity value is introduced in the array in the corresponding position and the subsequent process referred for the first approach is then executed with the final determination of the probability of damage existence for each candidate point. This is the approach that was used in the end since it enabled a considerable reduction of processing time.

4.6 Damage Detection: Metallic Plates

The phased array system was then experimentally tested in the $1.5m$ x $1.5m$ x $2mm$ aluminum plate subject to different boundary conditions ranging from totally supported (to increase Lamb waves propagation damping) to simply supported and riveted in the boundaries. The experimental setup is presented in Figure 22. The experiments were conducted first in a laboratory setting and afterwards in an aircraft maintenance hangar with no surrounding noise control and limited temperature control. Experiments were performed with the introduction (cumulatively) of surface and through the thickness circular holes and cuts (with different orientations) with a maximum dimension of 1mm, to simulate damage, as before. Damages were not inflicted around the phased array, respecting the non inspected distance of

$56mm$.

Figure 22. Phased array experimental setup

A total of 46 damages were introduced cumulatively in different positions in the entire aluminum plate. From these, 17 damages were simulated by through the thickness holes, 7 damages by surface holes, 9 damages by surface cuts, 3 oriented in a radial direction considering the center as the phased array, 2 oriented perpendicular to the local radial direction, 1 at 45° from the local radial direction and 3 randomly oriented, at an angle (around 30° and 60°) with the local radial direction and 13 damages by through the thickness cuts, 5 oriented in a radial direction considering the center as the phased array, 2 oriented perpendicular to the local radial direction, 2 at 45° from the local radial direction and 4 randomly oriented, at an angle with the local radial direction).

The simulated damages were successfully detected, even considering damages imposed in the direction of application of the phased array. The only exceptions were damages created behind other damages previously introduced, i.e., in the same radial direction with relation to the phased array (originating in the phased array) and also 60% of the cuts aligned with wavefront propagation directions, since their thickness was inferior to $1mm$.

The software for automated inspection previously developed for networks (in LABVIEW ®with embedded MATLAB ®codes for signal processing and to implement the damage detection algorithms) was adapted for the phased array system. Subsequently the software for networks and the software for phased arrays were integrated as two modules of an inspection software that can be applied for both network and phased array configurations.

A typical results window from the code is shown in Figure 23. In this plot the superimposed solution (tomography based) is presented to the user as the plot of the resultant, final matrix described and obtained through the method explained in the end of the previous section. In this particular case an existing through the thickness circular hole of $0.5mm$ in diameter was enlarged to 1mm in diameter (in the position in the plate presented as dark red in the plot). In these plots the most probable region where damage exists appears in dark red and the lowest probable areas in dark blue, passing by light red, orange, yellow, green and light blue. This result is presented to prove the capability of the system to monitor damage growth (in steps inferior to $1mm$) since the baseline used, i.e., the reference health condition of the plate considered for comparison/subtraction was for an existing damage of $0.5mm$ then enlarged by another $0.5mm$.

In Figure 23, it is also presented the results given by the code obtained from the values of the final matrix representing the six areas with highest probability of damage existence (normalized to one, by the damage index/probability of the most probable area for damage existence). In these results the coordinate system's origin is positioned in the center of the phased array (in the middle of the lower boundary in the figure), with the X axis being parallel to the array direction (pointing to the right side in the figure) and the Y axis being perpendicular to the array (pointing upwards).

In this plot, the remaining colors represent the consideration by the system of false positives due to noise (interfering more, usually, with boundary and other damage reflections) and also due to the intrinsic and inevitable generation of other waves (with smaller amplitude) than the wavefront and corresponding generated reflections in the damage (where mode conversion could also happen). Those additional waves/wavefronts include S_0 wavefronts of smaller amplitude with relation to the main wavefront (being generated and propagating before and after such wavefront, due to the actuation signal morphology) with the same propagation direction and velocity and include also A_0 waves and wavefronts (different mode first antisymmetric Lamb wave mode) propagating with different velocities and then in different directions than the S_0 wavefronts. The existence of other S_0 waves/wavefronts (propagating just before and after the main wavefront) and corresponding damage reflections will be responsible for the determination of different ToFs and then for shifting the determined damage position along the inspection direction. The existing A_0 waves/wavefronts (propagating in different directions and with different propagation velocities) will be responsible for the determination of damage positions in directions apart from the inspection direction.

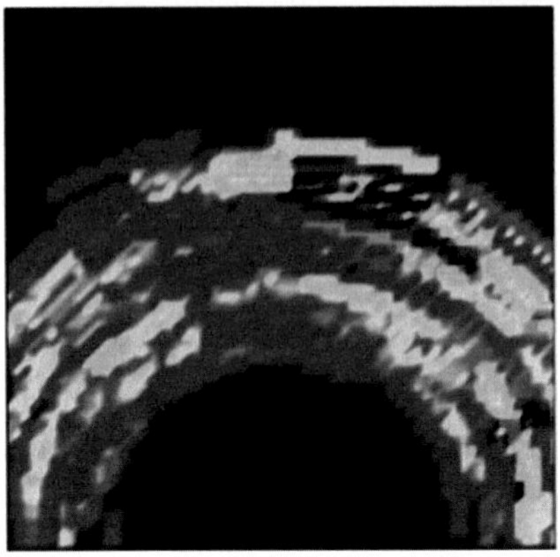

X (m)	Y (m)	Probability
0.26	1.05	1
0.22	1.07	0.82
0.625	0.3	0.6
0.72	0.72	0.55
0.55	0.9	0.5
0.5	0.8	0.45

Figure 23. Damage location contour output

5 SHM using Fiber Optic Sensors

Fiber Bragg Gratings (FBGs) are becoming increasingly more popular for such applications due to their advantages such as their size, immunity to electro magnetic interference, multiplexing potential and absolute reading. FBGs can be used to measure strain and temperature among other properties. Another useful sensor is an etched fiber sensor (EFS). This type of sensor is quite simple; a small section of optical fiber is etched to expose the core. When this etched region is surrounded by resin (fluid) flowing through the mold, the light power that is transmitted through the optical fiber changes hence enabling the sensor to detect fluid (Ahn et al., 1995). Since both EFS and FBG sensors allow light to pass there is potential for them to be multiplexed on one single fiber.

Many researchers have studied the use of a single type of embedded fiber optic sensor however there is less exploration into the use of a combination of sensor types on a single optical fiber. Most of the work that has been done involves the combination of FBG and extrinsic Fabry-Perot interferometer (EFPI) sensors (Kang et al., 2000). EFPI sensors cannot be multiplexed on a single fiber; this severely limits their use when multiple sensors are required. Since ingress/egress issues of fiber optics are not trivial it is highly desirable to minimize the number of instances by placing the largest practical number of sensors on one single fiber. If optical fiber is embedded for the use of sensors in one particular stage of a composite material's life (such as in service) then it is only logical to optimize the use of the fiber and combine sensors that could be used during other stages such as manufacturing.

Here, two FBG sensors and three EFS sensors multiplexed on a single fiber are embedded in a fiberglass panel with a purpose built RTM mold that has a glass viewing window that allows for visual monitoring of the resin injection process. The sensors are monitored during injection to detect the presence of resin and indicate possible dry spots. The viewing window in the mold is used to confirm the sensor readings. Two EFS sensors are also embedded in a semicircular tube to demonstrate their applicability to 3D surfaces such as those found in aerospace applications. The second part of this study involves the use of embedded FBG sensors to detect strain in the composite. Specimens are prepared and tested in a material test machine to determine the strain sensitivity of the FBGs.

To produce composite components with embedded fiber optics a resin transfer moulding laboratory-scale apparatus has been designed and built. The apparatus has the flexibility of accommodating different mold designs and thicknesses, with the feature of a glass viewport to allow for visual monitoring of resin flow during the injection process. It has been tested by producing composite parts with different geometries such as flat panels, hollow and foam cored square and semicircular tubes made from various types of reinforcements. The general layout of the experimental apparatus is shown in Figure 24. The apparatus can be separated into seven separate components: the injection system, injection valve, mold, manipulating/clamping fixture, catchpot, vacuum pump and temperature controller. It can be described as a clamshell system with the mold mounted on it. For a mold to be used with this apparatus it must be $533mm$ X $850mm$. Any thickness is possible with minor modifications to the clamping system.

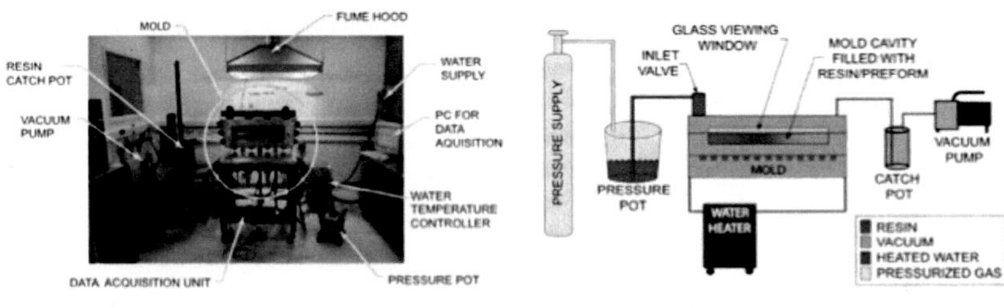

Figure 24. Layout (left) and schematic of RTM apparatus (right)

Fiber optic ingress/egress is one of the most important issues for the application of embedded fiber optic sensors in real composite components (Green and Darvish, 1996) and has been addressed by various researchers (Udd, 1995),(Kang et al., 2000),(Green and Darvish, 1996) however little information on ingress/egress with RTM is available in the literature. The closed nature of the RTM process as well as the extremely fragile nature of optical fiber makes the ingress/egress of FBG sensors into the mold challenging. Sealing issues also arise due to the extremely small diameter of the optical fiber. When optical fiber is embedded with the in-plane method, the ingress/egress point of the optical fiber is located at the edge of the composite. This eliminates the possibility of trimming the outer edges of the composite to size, a very common practice in the industry. To remove a composite part from a mold it must be removed normal to the mold, therefore the embedded fibers must enter and exit through the mold so that upon removal, the fibers are not severed. A novel through thickness ingress/egress method has been developed, which can overcome the limitations of the in-plane method and be applicable to closed mold processes such as RTM.

A novel method to achieve a through thickness ingress/egress that is applicable to pressurized injection molding such as RTM has been developed. Two major obstacles were overcome when developing this technique, sealing the optical fiber and protecting the fiber as it entered the mold. Optical fiber is quite delicate and must maintain a minimum bend radius before it fractures. When a through thickness fiber ingress technique is used the fiber sees an abrupt 90° bend as it travels through the mold and into the thin composite part as shown in Figure 25 (left). This is inherent to any through thickness ingress/egress technique.

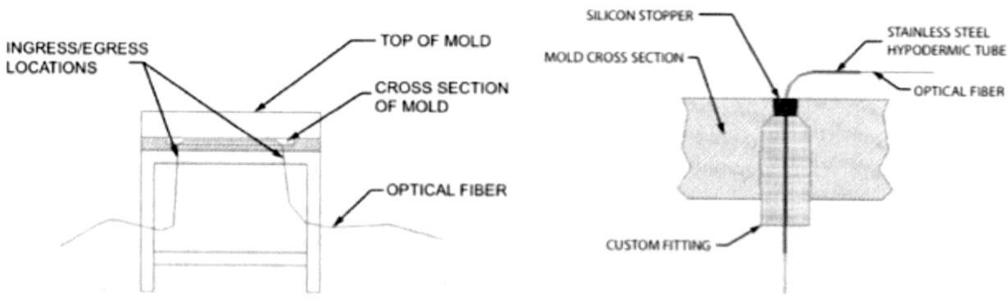

Figure 25. Path of fiber for through thickness ingress/egress (left), and the schematic of fiber sealing (right)

To protect the fiber with minimal disturbance to the composite material a thin hypodermic tube is placed around the fiber. This protects the fiber through the radius of the bend as well as reinforces the fiber at the ingress/egress point once the part is removed from the mold. As one would imagine it is difficult to seal around something as small as an optical fiber that has an outer diameter of $250\mu m$ without permanently caulking or bonding the fiber into the mold. A tapered silicone stopper was used to seal around the hypodermic tube as shown in Figure 25 (right). A custom fitting is used to keep the stopper and fiber in place. Figure 26 shows the panel and semicircular tube with embedded optical fiber.

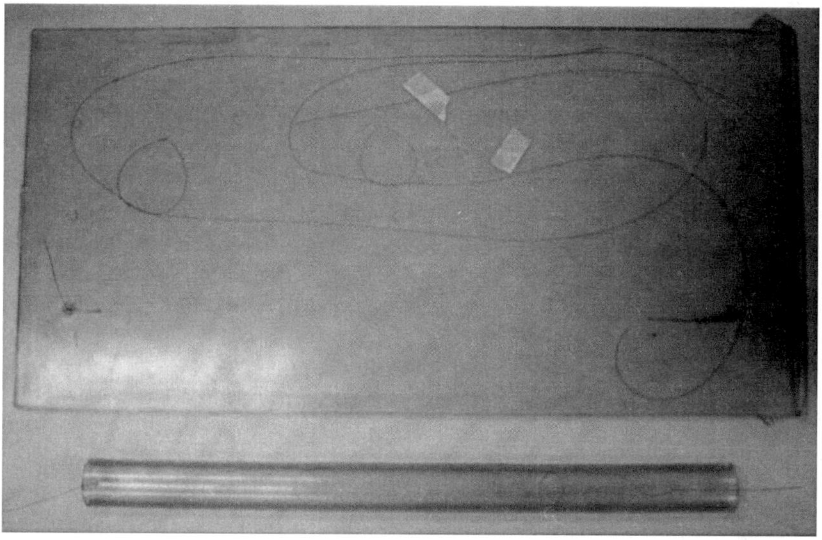

Figure 26. A panel and semicircular tube with embedded optical fiber

The novelty of this technique lies in the use of the tapered silicon stopper and custom fitting. One technique briefly reported by (Kosaka et al., 2003) involves the use of a plastic plug that seals the fiber into the mold and remains bonded to the surface of the composite once it is demolded. The detraction of this technique is that the plastic can become debonded from the composite while in service thereby severing the fiber and rendering the system useless. Also, the fiber must be sealed to the plastic, likely with a caulking that requires time to cure and cannot removed or adjusted if required. The technique developed and described here overcomes these detractions by using a tapered silicon stopper that applies pressure to the hypodermic tube thereby sealing it instantly without any caulking. This allows the fiber to be adjusted at any point prior to injection. Since the stopper is silicon it is easily removed after molding. This technique can be applied to a mold of any thickness over $10mm$ by simply adjusting the length of the fitting. This modularity comes in useful due to the wide variation of RTM molds. The fitting also allows the injection pressure to be quite high since it is threaded into the mold making this technique applicable to higher-pressure injection techniques such as SRIM and thermoplastic injections.

5.1 Fiber Bragg Grating Sensors

Fiber Bragg Gratings (FBGs) are becoming increasingly more popular for many applications due to their previously stated advantages. They have been used to measure properties such as displacement, strain, temperature, pressure, humidity and radiation dose among others (Dennison and Wild, 2008). FBGs were first demonstrated by (Hill et al., 1978). Embedded FBGs can be used for three distinct purposes during the life of a resin transfer molded part. An array of FBGs in a mold can monitor the mold filling process and can ensure that the mold is completely filled with resin as demonstrated by (Novo et al., 2000). Once injected the resin must go through a specific time-temperature cure cycle. In complicated 3-D parts with varying thicknesses and surface areas, different regions of resin cure at different rates and to varying degrees across the part. FBGs have been used to monitor the cure throughout the part (Kang et al., 2000),(Dewynter-Marty et al., 1998),(Murukeshan et al., 2000),(Jung and Kang, 2007). Once in service the embedded FBGs can be used in a variety of ways to monitor the health of the part thereby reducing maintenance cost and service time while increasing safety (Leng and Asundi, 2003),(Sundaram et al., 2005),(Guo, 2007).

A fiber Bragg grating is a segment of a single mode optical fiber core with a periodically varying refractive index in the axial (longitudinal) direction and commonly created using a high intensity UV laser (Othonos et al., 1999). It allows a broad band of light to pass through while reflecting back a narrow band based on a wavelength known as the Bragg wavelength. The reflected wavelength depends on the grating pitch (spacing between the refractive index variations) and the variation in refractive index. The periodic modulation of the refractive index at the grating location will scatter the light traveling inside the fiber core. Out of phase scattered waves will form destructive interference thereby canceling each other and in phase light waves will add up constructively forming a reflected spectrum with a center wavelength known as the Bragg wavelength.

5.2 Etched Fiber Sensors

Etched fiber sensors are an excellent low cost method of detecting the presence of resin. Operation of etched fiber sensors is quite simple requiring only standard equipment found in most optics lab. Essentially, a light source launches light into one end of the fiber optic and an optical power meter measures the light intensity at the other end. When resin makes contact with the sensor there is a sudden, sustained drop in the transmitted light intensity (Lim, 2000). The amount of light lost through the etching is a function of the refractive index of the medium it is in (Hect, 2006). This type of sensor has been used to measure the permeability of fiber perform materials by Ahn, S. H. et al (Green and Darvish, 1996) and Lim, S. T. et al (Novo et al., 2000) among others.

The theory behind the etched fiber sensor is very basic; it pertains to the fundamental principles of light transmission through optical fiber. The sensor consists of a small section of optical fiber roughly 3-5mm long with the cladding removed leaving the fiber core exposed. Light is contained in the core of an optical fiber by the cladding. As light travels through the core of a fiber it is continuously bouncing off the cladding. When the light reaches the section of fiber where the cladding is removed a small portion of it escapes while most is transmitted (Eum et al., 2007). When resin contacts the core its refractive index being higher than that of the glass causes more light to escape. Therefore the amount of light transmitted through the fiber is less.

A novel variation of the basic sensor is used in this study. This variation involves looping the etched sensor to create a bend in the etched portion of the fiber. This is done to allow more light to escape from the sensor while leaving some of the cladding on the core to physically protect it. The looped variation is a more robust version of the sensor that is easier to handle and implement. Another benefit of looping the fiber is that the sensitivity can be tuned by adjusting the radius of the loop. As the radius of the loop increases the amount of transmitted light increases. This option is desirable when multiple sensors are used on a single strand of fiber and minimum light loss is desired so that all sensors can make readings. Figure 27 shows the sensor with a Canadian quarter for reference (23.81mm diameter); note the etched section in the upper right portion of the loop.

5.3 Sensor Characterization

To demonstrate the capabilities and multiplexing potential of FBGs and etched fiber sensors, experiments were performed in the aforementioned RTM apparatus using the described ingress/egress technique. The experiments validated the ability of the EFS sensors to detect the presence of resin during the RTM process and the FBG sensors to be used to detect strain all while multiplexed on a single fiber. To explore the versatility and potential for different applications of these sensors three different specimens were pro-

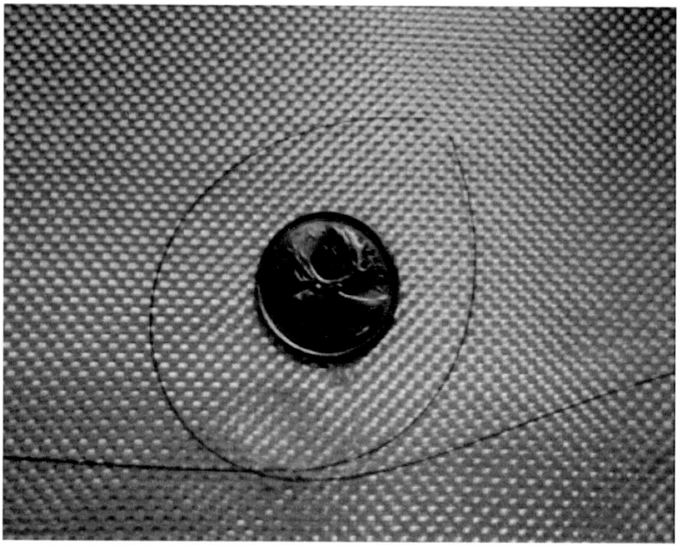

Figure 27. Looped etched fiber sensor

duced; a quasi 2D flat panel and two 3D hollow tubes with a semicircular cross section.

The flat panel was composed of 18 plies of 200gsm plain weave glass fiber cloth and Huntsman Renfusion 8601 epoxy resin, with dimensions of 610mm X 305mm X 3mm. An optical fiber containing three EFS sensors and two FBG sensors was embedded on the upper ply of the panel.

Two semicircular tube specimens were produced; one composed of five layers and the other composed of six layers of 200gsm plain weave glass fiber cloth and Huntsman Renfusion 8601 epoxy resin fully enclosing a steel mandrel that is removed after molding. The tube has an outer radius of 38mm, length of 500mm and wall thickness of 1.5mm. In the specimen composed of five layers, an optical fiber containing two EFS sensors is embedded between the fourth and fifth ply on the curved surface of the specimen. The specimen composed of six layers contains two FBGs embedded between the fourth and fifth ply on the curved surface of the specimen. The inner layer is considered the first layer in these specimens increasing towards the outside of the tube.

5.4 Flow Monitoring

Flow monitoring experiments were performed during the resin transfer molding process. The general procedure was similar for both specimens and involved embedding the optical fiber in the mold, closing it and injecting the resin. The optical fiber was connected to a photodetector and measurements were logged throughout the injection process.

5.5 Flow Monitoring of Panel Specimen

An optical fiber containing three EFS sensors (EFS#1,2,3) and two FBG sensors (FBG#1,2,3) was embedded into the mold on the upper ply of the laminate, the mold was closed and the resin was injected. Etched fiber sensor readings were recorded with a photodetector and data logger while the FBGs were manually observed with an OSA. A simple circuit was used to interrogate the sensors as shown in Figure 28. A BBS light source is connected to one end of the optical fiber, the fiber runs through the mold, a 50/50 coupler is connected to the other end of the fiber with one branch of the coupler going to the photodiode and the other to an OSA.

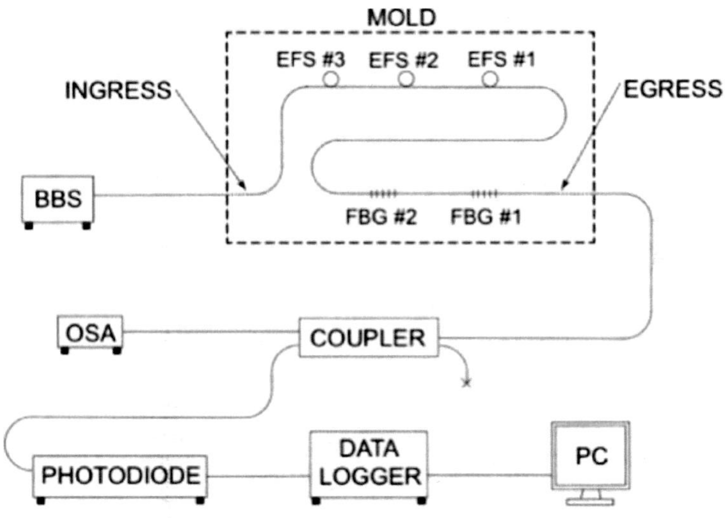

Figure 28. Interrogation system

Figure 29 shows the sensors through the glass viewing window prior to injection. Figure 30a shows the resin approaching EFS #1 while Figure 30b

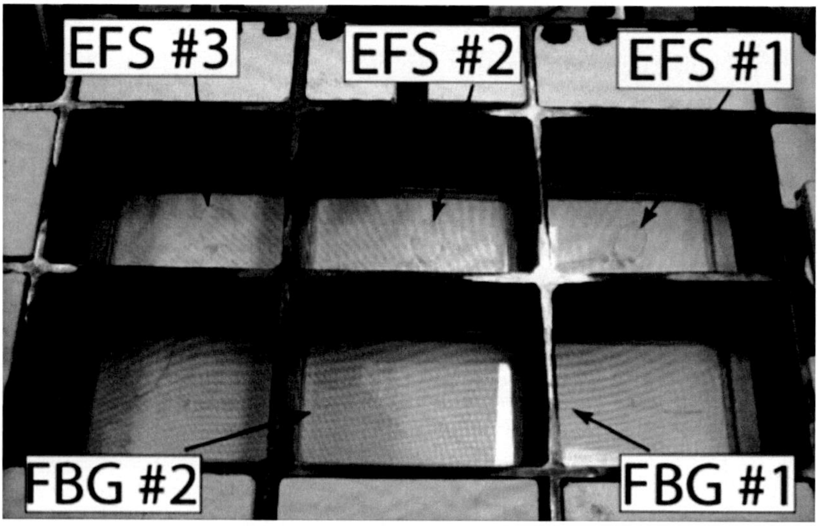

Figure 29. Sensors positioned in RTM prior to injection (mold closed)

shows the resin just after the sensor is saturated and the transmitted light intensity is reduced. This occurs roughly 2.4 minutes into the injection. Figure 30c shows the mold midway through the injection.

Figure 31 shows a plot of the photodiode voltage output vs. injection time. It can clearly be seen that as the resin reaches the first sensor at roughly 2.4 minutes there is a sharp and sustained drop in the transmitted light. Another drop occurs at roughly 7.7 minutes when the resin reaches the second sensor and again at 12.2 minutes when the resin reaches the third sensor. Once the mold was saturated and the injection was complete, the light source was turned off to ensure that the readings could be differentiated from a change in minimal transmitted light and no transmitted light. When the light source was off the photodiode output was zero. The light source was turned back on and the transmitted light intensity was the same as before it was turned off therefore indicating that light was indeed being transmitted. During this time the FBG sensors were manually observed with the OSA to ensure that they were still functional. Due to the arrangement of the sensors on the fiber (FBG before EFS) no change in power was noticed when resin contacted the EFS sensors therefore demonstrating that the EFS do not effect the FBG sensors as long as they are situated after the FBGs on the fiber. The FBG output is not included here however this technique has been thoroughly researched by others such as (Novo et al.,

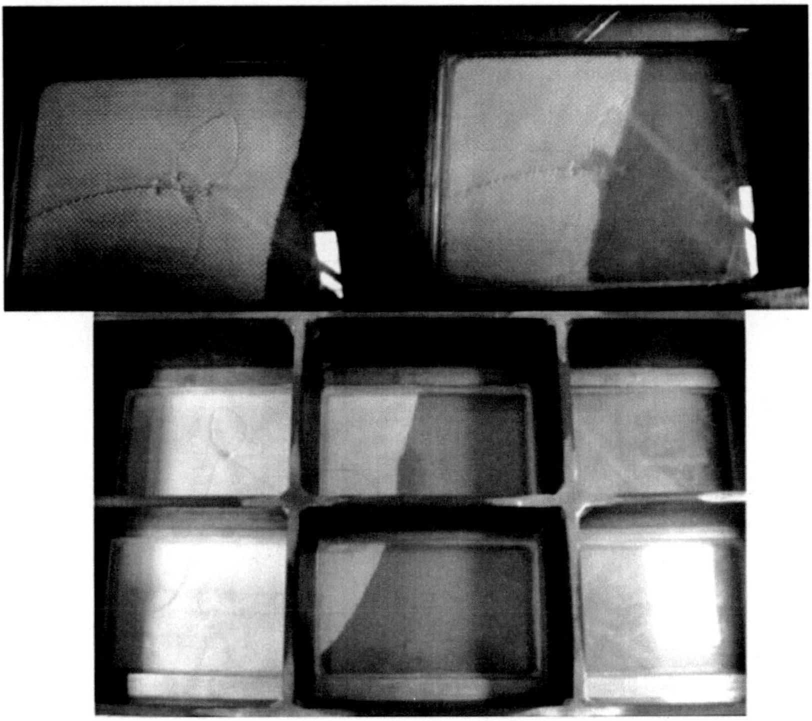

Figure 30. CW from top left: a) resin approaching sensor, b) resin just after contacting sensor, c) mold midway through injection

2000) and (Eum et al., 2007),(Eum et al., 2008).

5.6 Flow Monitoring of 3D Semicircle Specimen

To demonstrate the versatility and applicability of these sensors to more realistic structures, an optical fiber containing two EFS sensors was embedded in a hollow tube with a 3D semicircle cross section using the aforementioned RTM apparatus. The sensors were embedded between the fourth and fifth layers on the curved surface of the semicircle. The flat side of the semicircle is visible through the viewing window however the curved surface is not. Due to the relatively small radius of the tube and uniform resin flow it can be estimated with a high level of confidence that the resin is in contact with the sensors and their output can be verified. This is a more realistic experiment for these sensors since the majority of industrial molds

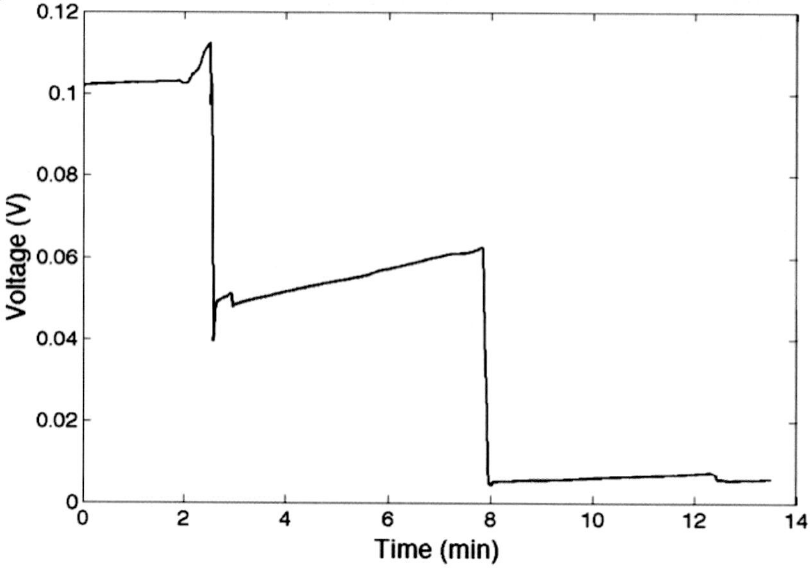

Figure 31. Plot of photodiode output vs. injection time for panel

are not translucent.

Figure 32. Semicircular tube with embedded fiber optic sensors

The sensors were interrogated in a manner similar to that shown in Figure 28. Figure 32 shows the specimen with the sensors visible. Data was recorded during the injection and is shown in Figure 33 showing resin arrival at the first and second EFS at roughly 2.9 and 4.0 minutes respectively.

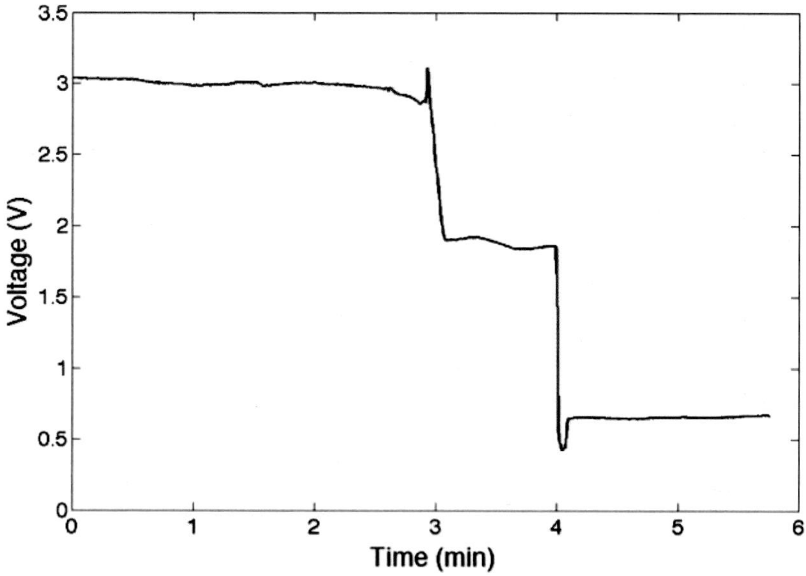

Figure 33. Plot of photodiode output vs. injection time for tube

5.7 Strain Sensitivity Characterization

For FBGs to be used as strain sensors after they are embedded in a composite material the relationship between the shift in the Bragg wavelength and the axial strain in the sensor must be known. This relationship is known as the sensitivity or the strain gage factor, S. Embedded FBG strain gages experience additional radial strain field in application, therefore once integrated into a structure, every gage must be recalibrated (Fan and Kahrizi, 2005).

The strain gage factor for an electrical foil strain gage can be described as the relationship between the resistance of the wire, R, axial strain εa and the change in resistivity of the wire (Wheeler and Ganji, 2004). The expression for the strain gage factor is given in (31).

$$S = \frac{dR/R}{\varepsilon_a} \tag{31}$$

For an FBG, the gage factor would be the relationship between the change in the Bragg wavelength and axial strain. The expression for this is shown in (32).

$$S = \frac{\delta \lambda_B}{\varepsilon_a} \quad (32)$$

To determine the gage factor for both 2- and 3-D specimens they were tested in an MTS 810 material test machine. A 6mm grid, 120Ω strain gage from Omega Engineering Inc. was bonded to the surface directly beside the FBG. The strain gage was connected to a Measurements Group Instrument Division P-3500 strain indicator and the FBG was hooked up to an Ando AQ6331 OSA, JDS Uniphase broadband light source and a 50:50 optical coupler. Figure 34 shows the 2D specimen in the MTS machine with the instrumentation equipment (left) and a close up of the 3D specimen (right).

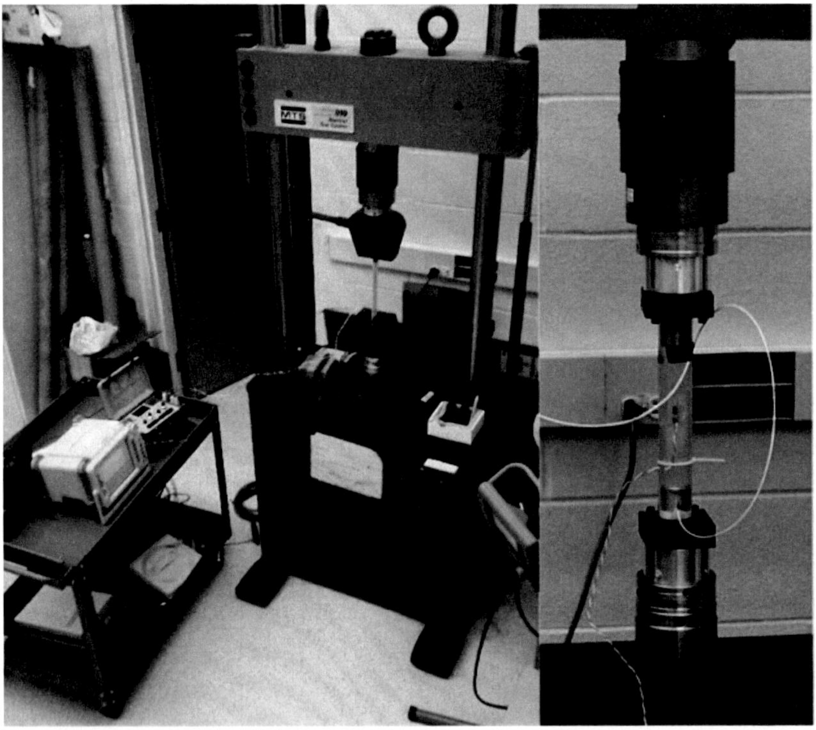

Figure 34. 2D specimen in MTS machine with instrumentation equipment (left) and 3D specimen loaded in MTS machine (right)

5.8 Strain Sensitivity of Panel Specimen

To determine the gage factor for the panel a tensile specimen was cut out of the RTM'd panel. The specimen has dimensions of $38.1mm$ X $300mm$ X $3mm$. The embedded FBG is located directly in the middle of the specimen along the length and width. Small pieces of sand paper were bonded to the ends of the specimen to help the test machine grip without slipping.

The specimen was loaded into the test machine and a tensile load was applied at a rate of $0.051mm/min$ up to a maximum of roughly $2.88mm$. Readings were taken at roughly $0.254mm$ intervals starting at zero strain. The test was repeated four times to ensure the repeatability of the results. Figure 35 shows a plot of the collected data, namely, Bragg wavelength vs. measured strain.

A linear line was fit to the data with an equation of $y = 0.001141x + 1541.04$ with an R^2 value of 0.999 obtained from four test runs. ¿From this equation we can ascertain that $S = 1.141 \times 10^{-3}$ or every $1.141pm$ shift in the Bragg wavelength is equal to one micro strain.

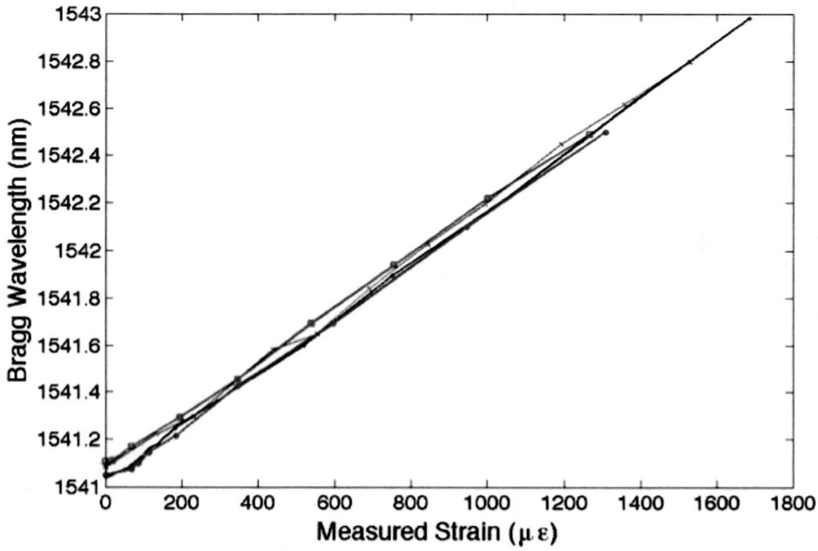

Figure 35. Bragg wavelength vs. strain gage measured strain for 2D panel

5.9 Strain Sensitivity of Semicircular Specimen

To determine the gage factor for the semicircular tube a special fixture was designed to restrain the specimen in the aforementioned MTS machine for compression testing. Semicircular blocks, $38mm$ long with the same cross sectional dimensions as the mandrel were fastened to adapter plates that interface with the test machine. The tube was cut at it's mid-length to produce two specimens and simplify the testing process. Each specimen has one FBG at the midspan.

The specimen was slid onto the fixture and a compression load was applied then released. This was repeated twice to ensure repeatability. The strain gage output (in microstrain) and Bragg wavelength (in nm) was recorded as the load was applied and released. The recorded data is plotted and shown in Figure 36.

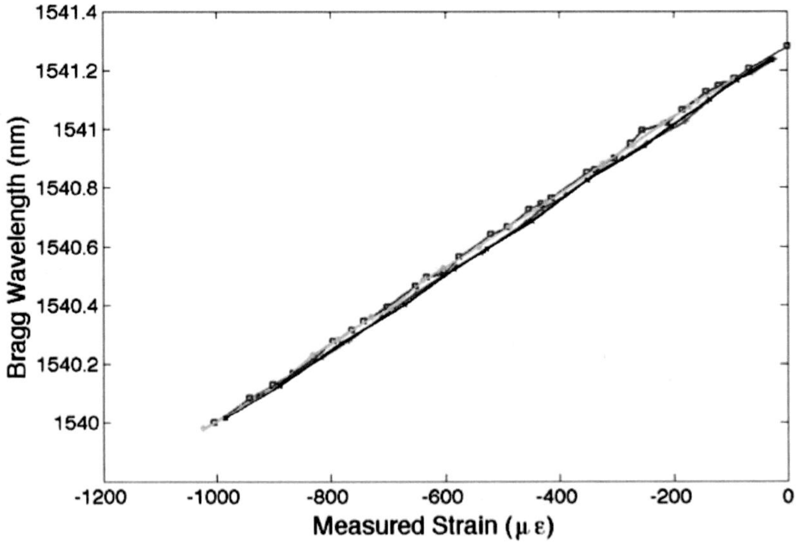

Figure 36. Bragg wavelength vs. strain gage measured strain for 3D structure

A linear line was fit to the data and an equation of $y = 0.001288x + 1541.28$ with an R^2 value of 0.999 was obtained from four test runs (two loading, two unloading). From this equation we can ascertain that $S = 1.288e-3$ or every $1.288pm$ shift in the Bragg wavelength is equal to one micro strain.

As mentioned earlier the strain sensitivity of the embedded FBG is

$1.141pm/\mu\varepsilon$ for the flat panel specimen and $1.288pm/\mu\varepsilon$ for the semicircular specimen. These sensitivities deviate from the commonly referenced value of $1.2pm/\mu\varepsilon$ (Othonos et al., 1999) for a bare FBG with typical material properties by -4.9% and 7.3% for flat panel and semicircular specimens respectively. These values fall within the range of deviation reported by Fan, Y., et al (Fan and Kahrizi, 2005). Factors such as variations in the material and manufacture of each individual FBG and experimental error in the processing and testing phase such as misalignment of FBG or strain gage with the axis of the specimen may contribute to this deviation. Another factor which may contribute to the variation in sensitivity between the two geometries is the mismatch of mechanical properties between the FBG fiber and host structure, particularly their Poisson ratio in the lateral direction as embedded sensors experience an additional radial strain field when load is applied along the sensor direction (Fan and Kahrizi, 2005).

6 Concluding Remarks

Piezo transducer network and linear phased array SHM systems have been developed for damage detection and location in metallic and composite plate structures. The SHM systems were based on the first symmetric mode (S_0) of Lamb waves. The most important advantages and simultaneously the main difficulties in the application of the S_0 mode that were overcome here are related to its high propagation velocity. The advantages are that, with the application of this wave mode and its high propagation velocity, this wave and its reflection suffer a small interference from lower propagating waves (modes). Furthermore, due to their deformation pattern they are more prone to detect any types of damage in the components, being easier to deal with, while presenting small amplitude propagation damping. Regarding obstacles that were overcome, due to such high propagation velocity, difficulties emerge in the generation of wavefronts resulting from the constructive interference of waves generated by the single actuators in the phased array and particularly in the precise establishment of the small time delays required for the phased actuation. One other difficulty is related with the small amplitude of damage reflected waves. This negative aspect is diminished by the use of the phased array and beamforming, increasing the amplitude of actuation and then the amplitude and SNR of reflections. Furthermore, with the application of a phased array, an entire component can be inspected by performing several scans in which the inspection direction is defined and changed from scan to scan. In this manner the inspection region is delimited for each scan, resulting in an intensification of inspection effort in such a region, decreasing simultaneously the complexity and increasing the precision of the inspection process. Also in the actuation of Lamb waves, their dispersive behaviour, i.e., the dependency of their propagation velocity to their frequency, must be accounted for. This means that the actuation signal waveform must excite one single frequency, to obtain a wave with a particular frequency, propagating with a unique velocity.

A dedicated actuation system, required for the phased generation of the waves and wavefront was developed. This system is based on a master MCU circuit, controlling the phased activation of slave channels. The master MCU has programmed the time delays to be applied for the phased activation for each and all scan (directions) to be implemented. Furthermore, in that code the number of scans and scan repetitions are also determined. The slave circuits/channels, connected to each PZT in the array, generate the correct actuation signal waveform. In these circuits, an output switch was applied to match the impedance of actuation and acquisition systems, what is a usual problem in the current existing systems. With the use of such switch, measured wave amplitudes were increased significantly.

Bibliography

S. H. Ahn, W. I. Lee, and G. S. Springer. Measurement of the Three-Dimensional Permeability of Fiber Preforms Using Embedded Fiber Optic Sensors. *Journal of Composite Materials*, 29(6):714–733, April 1995.

C. R. Dennison and P. M. Wild. Enhanced sensitivity of an in-fibre Bragg grating pressure sensor achieved through fibre diameter reduction. *Measurement Science and Technology*, 19(12):125301, October 2008.

V. Dewynter-Marty, P. Ferdinand, E. Bocherens, R. Carbone, H. Beranger, S. Bourasseau, M. Dupont, and D. Balageas. Embedded fiber Bragg grating sensors for industrial composite cure monitoring. *Journal of Intelligent Material Systems and Structures*, 9(10):785–787, 1998.

USA Department of Defense DOD. Airplane damage tolerance requirements. *MIL-A-83444, Washington DC, USA*, 1987.

S. Eum, K. Kageyama, H. Murayama, I. Ohsawa, K. Uzawa, M. Kanai, and H. Igawa. Process/health monitoring of turbine blades with FBG sensors with multiplexing techniques. *Proceedings of SPIE*, 7004, June 2008.

S. K. Eum, K. Kageyama, H. Murayama, I. Ohsawa, K. Uzawa, M. Kanai, and H. Igawa. Resin flow monitoring in vacuum-assisted resin transfer molding using optical fiber distributed sensor. In *Behavior and Mechanics of Multifunctional and Composite Materials 2007*, pages 65262T–65262T–8. SPIE, April 2007.

Y. Fan and M. Kahrizi. Characterization of a FBG strain gage array embedded in composite structure. *Sensors and Actuators a-Physical*, 121 (2):297–305, June 2005.

V. Giurgiutiu. *Structural health monitoring with piezoelectric wafer active sensors.* Elsevier Academic Press, Boston, USA, 2008.

A. K. Green and S. Darvish. Practical terminations for optical fibres embedded in composite materials. In *Proceedings of SPIE*, 1996.

Z. S. Guo. Strain and temperature monitoring of asymmetric composite laminate using FBG hybrid sensors. *Structural Health Monitoring*, 6(3): 191–197, 2007.

J. Hect. Understanding fiber optics. *Pearson Education Inc., Upper Saddle River, New Jersey*, 2006.

K. O. Hill, Y. FujiiI, D. C. Johnson, and B. S. Kawasaki. Photosensitivity in Optical Fiber Waveguides - Application to Reflection Filter Fabrication. *Applied Physics Letters*, 32(10):647–649, 1978.

K Jung and T. J. Kang. Cure monitoring and internal strain measurement of 3-D hybrid braided composites using fiber Bragg grating sensor. *Journal of Composite Materials*, 41(12):1499–1519, June 2007.

H. K. Kang, J. W. Park, and C. Y. Ryu. Development of fibre optic ingress/egress methods for smart composite structures. *Smart Materials and Structures*, 9:149–156, 2000.

T. Kosaka, K. Osaka, S. Nakakita, and T. Fukuda. Fiber optic strain monitoring of textile gfrp during rtm molding and fatigue test by using embedded fbg sensors. *Proceedings of SPIE*, 5056, 2003.

J. Leng and A. Asundi. Structural health monitoring of smart composite materials by using EFPI and FBG sensors. *Sensors and Actuators a-Physical*, 103(3):330–340, 2003.

S. T. Lim. An analysis of the three-dimensional resin-transfer mold filling process. *Composites Science and Technology*, 2000.

V. M. Murukeshan, P. Y. Chan, and L.S. Ong. Cure monitoring of smart composites using fiber Bragg grating based embedded sensors. *Sensors and Actuators a-Physical*, 79:153–161, 2000.

C. Novo, O. Frazão, A. N. Costa, A. Vieira, N. Correia, I. Dias, F. M. Araujo, and A T Marques. Progression monitoring of the flow front in RTM process using fibre Bragg grating sensors. *Society of Photo-Optical Instrumentation Engineers (SPIE) Conference Series*, 4185:808–811, 2000.

A Othonos, K Kalli, and G. E. Kohnke. *Fiber Bragg Gratings: Fundamentals and Applications in Telecommunications and Sensing.* Artech House Inc., Boston, Massachusetts, 1999.

R. Sundaram, G. M. Kamath, and N. Gupta. Structural health monitoring of co-cured composite structures using FBG sensors. In *Proceedings of SPIE*, 2005.

E Udd. *Fiber optic smart structures.* John Wiley & Sons, Inc., New York, 1995.

A. J. Wheeler and A. R. Ganji. Introduction to engineering experimentation. Pearson/Prentice Hall Inc., Upper Saddle River, New Jersey, 2004.

Vibration-Based Damage Diagnosis and Monitoring of External Loads

Claus-Peter Fritzen[‡], Maksim Klinkov[‡] and Peter Kraemer[‡]

[‡] Institute of Mechanics and Control Eng. - Mechatronics, University of Siegen, Siegen, Germany

Abstract This book chapter deals with vibration-based SHM methods which enable a structure to deliver information about its current health status and help to predict its future remaining lifetime. Therefore, the structure has to be equipped with diagnostic sensor networks and sensors for monitoring of the loads. Computational algorithms are used to analyze these response data and perform a self-diagnosis with the goal that the structure can release early warnings about any critical health state, locate and classify type and size of damage. This paper intends to give an overview and point out recent developments of vibration-based methods for SHM. All these methods have in common that a structural change due to a damage results in a more or less significant change of the dynamic behavior. Unfortunately, also changing environmental and operational conditions have a non-negligible influence on the system behavior. Therefore, a diagnostic system must be able to distinguish between damage and other influences, learning the system behavior under varying conditions in the healthy state by means of classification techniques. Moreover, for a prognosis of the remaining useful life, the knowledge of the time-history of external deterministic or random loads is essential. An overview is given on currently used methods and a robust observer-based algorithm is presented in detail for on-line reconstruction of the unknown external loads from the vibration response data. Different practical examples show the applicability of the methods.

1 Introduction

One possibility to classify SHM methods is to divide them into local and global methods. This classification is based on the goals of the monitoring strategy. If we wish to monitor a small area where it is known that the structure is highly loaded (the "hot spots") and where it is likely that e.g. fatigue damage or cracks might occur, the sensor elements are concentrated densely in this section. Very often local SHM methods use ultrasonic waves for the diagnosis. High excitation frequencies with their corresponding small wavelengths allow to detect small damage (cracks, delaminations, debondings, etc.) with a size of the order of the wavelength. On the opposite side, if the purpose is to monitor a structure or components as a whole (e.g. like bridges or wind turbines towers), without having a predefined area of potential damage, global methods can be applied which use informations from a globally distributed, rather coarse sensor network. Global methods usually work with vibrations in the lower to the middle frequency range. These vibrations are often generated from the ambient excitation from (wind loads, traffic, etc.) so that no extra shaking systems are required. Consequently, due to the lower frequency range, we obtain less sensitivity with respect to small damage. For many applications, the sensitivity is still high enough to find damage of a tolerable size. Another useful systematic was introduced by Rytter (Rytter, 1993). He defined four levels on the damage assessment scale: Level I: Damage detection; Level II: Damage localization; Level III: Damage quantification and Level IV: Prognosis of remaining service life. Level I only provides information that damage is present. For many practical applications this is sufficient. Challenges on this level are: to quantify the probability of detection (POD) for SHM methods, how to select sensitive features, to detect damage in a rather early stage without getting false alarms and how to separate effects resulting from damage from those coming from changes in environmental conditions. At level II the location(s) of single or multiple damage sites are determined. On level III the extent of damage is evaluated. For this purpose a model must be available to describe the effect of damage (by means of parameters like crack length, size of a delamination or stiffness decrease etc.) on the dynamic behavior. Some authors, (Sohn et al., 2004; Worden and Dulieu-Barton, 2004), include the determination of the type of damage as an additional step between levels II and III. The most sophisticated level is the step from diagnosis to damage prognosis with the predictions of the remaining lifetime. This requires the combination of the global structural model with local damage models to predict the evolution of damage, e.g. fatigue crack growth, or probabilistic failure models based on previously failed or damaged components (Inman et al., 2005; Farrar et al., 2003; Dettmann and Soeffker, 2011). The knowledge of the past history of external loads on a structure, resulting e.g. from extreme events, allows us to update earlier life-time predictions. Literature overviews can be found in (Doebling et al., 1996; Natke and Cempel, 1997; Sohn et al., 2004; Staszewski et al., 2004; Inman et al., 2005; Balageas et al., 2006; Farrar and Worden, 2007).

2 Modeling and Diagnosis of Damaged Systems

2.1 General Non-Linear Dynamical Model

We describe the dynamics of a general non-linear, time-varying, damaged structure by the spatially discrete and coupled system of the non-linear equation of motion (1) and the non-linear evolution of damage (2) in the following way:

$$\boldsymbol{M}(\boldsymbol{q}, \boldsymbol{\theta}_d, \boldsymbol{\theta}_e)\, \ddot{\boldsymbol{q}} + \boldsymbol{g}(\boldsymbol{q}, \dot{\boldsymbol{q}}, \boldsymbol{\theta}_d, \boldsymbol{\theta}_e) = \boldsymbol{f}_{op}(t, \boldsymbol{\theta}_d, \boldsymbol{\theta}_e) + \boldsymbol{f}_{test}(\boldsymbol{\theta}_d, \boldsymbol{\theta}_e, \boldsymbol{u}(t)) \tag{1}$$

$$\dot{\boldsymbol{\theta}}_{\boldsymbol{d}} = \boldsymbol{\Gamma}(\boldsymbol{\theta}_d, \boldsymbol{\theta}_e, \boldsymbol{q}, \dot{\boldsymbol{q}}, t) \tag{2}$$

$$\boldsymbol{y}(t) = \boldsymbol{h}_{\boldsymbol{out}}(\boldsymbol{\theta}_d, \boldsymbol{\theta}_e, \boldsymbol{q}, \dot{\boldsymbol{q}}, t) \tag{3}$$

$$\boldsymbol{f}_{test}(t) = \boldsymbol{h}_{\boldsymbol{in}}(\boldsymbol{\theta}_d, \boldsymbol{\theta}_e, \boldsymbol{u}(t)) \tag{4}$$

where $\boldsymbol{M}$ is the mass matrix, $\boldsymbol{g}$ the force vector of elastic forces, damping forces, etc. depending on the displacements $\boldsymbol{q}$, the respective velocities and time t. The external load vector $\boldsymbol{f}$ is split into operational loads $\boldsymbol{f}_{op}$ and (if applicable) into test loads $\boldsymbol{f}_{test}$. The number of degrees of freedom (dof) is m. Eq.(1) also allows to express non-linear effects of damage like stiffness variation due to an opening or closing crack depending on the instantaneous deformation (Seibold et al., 1996). The non-linear function $\boldsymbol{\Gamma}$ describes the evolution of the damage parameters $\boldsymbol{\theta}_d$ (e.g. crack length, play, loss of stiffness, loss of mass etc.). The parameter vector $\boldsymbol{\theta}_e$ in Eqs.(1)-(4) indicates the influence of environmental and operational conditions (e.g. temperature, humidity, rotational speed) on the equation of motion. The two differential Eqs.(1),(2) interact due to their coupling in the mechanical displacements, velocities and parameters. For example, larger amplitudes of vibration $\boldsymbol{q}$ will cause larger stresses in the structure and hence accelerate the growth of damage. By means of Eq.(2) it is possible to perform a damage prognosis and estimate the residual service life (Farrar and Lieven, 2007; Vachtsevanos et al., 2006). A compilation of different methods for modeling the change of stiffness of structural elements due to damage was presented e.g. by (Ostachowicz and Krawczuk, 2001). The evolution of the damage on one hand and the dynamics of the structure on the other hand usually take place on two different time scales. Compared to the vibrations of the structure the evolution of damage is usually considered to be a rather slow process so that we can assume that $\boldsymbol{\theta}_d$ keeps constant during the short time span of data acquisition. Model-based methods require accurate computational models. Thus, model-updating (Natke, 1992; Friswell and Mottershead, 1995) is an important step to improve the quality of the initial model before it is used for damage identification. Eq.(3) delivers a relation between the internal model state variables ($\boldsymbol{q}$, $\dot{\boldsymbol{q}}$, $\boldsymbol{\theta}_d$) and other parameters (like $\boldsymbol{\theta}_e$) with the output variables $\boldsymbol{y}(\mathrm{t})$ like strains, accelerations, voltages, etc. which we wish to compare with a corresponding output of a measurement device. The fourth equation describes the transformation of an input signal $\boldsymbol{u}(t)$ (e.g. an input voltage) into forces or moments, e.g. in the case of piezo actuators. In Eq.(3),(4) the damage parameters $\boldsymbol{\theta}_d$ can also include sensor or actuator faults. In the case of a fault of a piezo actuator, e.g. there might be a degradation of the adhesive described by $\boldsymbol{\theta}_d$ so that only a reduced force or moment is transfered to the structure.

As we can see in Eqs.(1)-(4), we can only describe the dynamical behavior correctly, if we consider the dependency of the system properties on the environmental/operational influences so that we are able to separate them properly from damage effects for a reliable diagnosis. From this point of view, it makes sense to measure the relevant parameters $\boldsymbol{\theta}_e$, too.

2.2 Feature Extraction

Usually, we do not use the time domain system responses $\boldsymbol{y}(t)$ directly to compare two system states. In fact, we try to extract so-called features $\boldsymbol{f}_y$ like statistical quantities or an intermediate condensed model from the raw data, which are arranged in a vector and which should be representative for the system under consideration within a range of interest. The feature vector is

$$\boldsymbol{f}_y(\boldsymbol{\theta}_d, \boldsymbol{\theta}_e) = FE(\boldsymbol{y}(t, \boldsymbol{\theta}_d, \boldsymbol{\theta}_e)) \tag{5}$$

FE is a feature extraction function applied to the raw data $y(t)$. Such features can be mean values, variances, maximum values, crest factors or other statistical characteristics of signals, correlation functions, spectral densities, frequency response functions, impedance spectra, eigenfrequencies and mode shapes or any other feature that makes sense in this context. An important point is the sensitivity of the features with respect to a damage parameter. The sensitivity of the i-th feature with respect to the j-th damage parameter is expressed by the first partial derivative

$$s_{ij} = \frac{\partial \boldsymbol{f}_{yi}(\boldsymbol{\theta}_d, \boldsymbol{\theta}_e)}{\partial \boldsymbol{\theta}_{dj}} \tag{6}$$

If the features are very sensitive then we get significant changes of the damage parameters even for small changes of the features. The sensitivity matrix $\boldsymbol{S} = (s_{ij})$ contains all single sensitivity elements, where the rows corresponds to the features, the columns to the parameters. On the opposite side, we wish to get low sensitivities of the features with respect to $\boldsymbol{\theta}_e$.

2.3 Residual Vector and Damage Indicator

To classify the health state of a structure at different time instants, we compare the feature vectors $\boldsymbol{f}_y$ at instant "1" with an earlier state "0" e.g. representing the undamaged state (usually called baseline) or an earlier damaged state. One possibility is simply to subtract the feature vectors like

$$\boldsymbol{r}(\boldsymbol{\theta}_d, \boldsymbol{\theta}_e) = \boldsymbol{f}_{y,1}(\boldsymbol{\theta}_{d1}, \boldsymbol{\theta}_e) - \boldsymbol{f}_{y,0}(\boldsymbol{\theta}_{d0}, \boldsymbol{\theta}_e) \tag{7}$$

Other possibilities to define residuals are correlations of feature vectors or projections of the features into another space so that the fault free case results in a zero residual vector. The latter case will be treated in the chapter about the Null-Space Based Fault Detection Method (NSFD). To express the deviations between different states in one scalar quantity, we can introduce a damage indicator

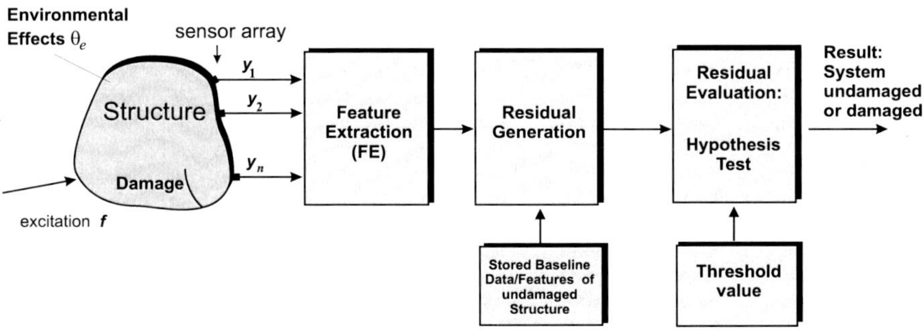

Figure 1: Schematics of chain from measurement to data-based decision

$$DI(\boldsymbol{\theta}_d, \boldsymbol{\theta}_e) = \boldsymbol{r}^T \, \boldsymbol{W}_r \, \boldsymbol{r} \tag{8}$$

where $\boldsymbol{W}_r$ is a symmetric, positive definite weighting matrix which is introduced to take a proper scaling of the different elements of the feature vector into account. The inverse of the covariance matrix $\boldsymbol{\Sigma}$ of the features is a reasonable choice: $\boldsymbol{W}_r = \boldsymbol{\Sigma}^{-1}$. Finally, the damage indicator should mainly depend on the damage parameters $\boldsymbol{\theta}_d$ and not on the environmental parameters. In a damage detection context (Level I), the Damage Indicator DI should deliver information for decision-making: is damage present or not? DI measures an overall distance between two states considering the chosen features. In a framework of hypothesis testing, we compare DI to a certain threshold value $DI_{th}(\alpha)$ which we can set deterministically or preferably derive it from the statistical distribution of the DI values, see

Decision from test	Reality	
	Ho true	Ho false
Ho Retained	Decision correct with probability P=1- α	Error of 2. kind, ß-error, P(H0|H1)=ß, Missed hit
Ho Rejected	Error of 1. kind, α-error, P(H1|Ho)=α, False alarm	Decision correct with power P=1- ß

Figure 2: Comparison of test decision and truth with different kinds of errors

Figure 1. Choosing a confidence level, e.g. α of 0.01 or 0.05 which corresponds to the error probability or the probability of false alarm (PFA), we are able to determine $DI_{th}(\alpha)$ and then decide whether we retain or reject the so-called null-hypothesis H_0 of an undamaged structure with certain statistical confidence. The alternate hypothesis H_1 corresponds to the damaged system. DI_{th} is set so that the probability $P(DI \leq DI_{th}) = 1 - \alpha$. If $DI > DI_{th}$ we reject the null-hypothesis and consider the system not to be healthy, otherwise we retain it. Figure 2 shows the different combinations between our decision and the real health state of the system. If the system is damaged (H_1), the probability $P = 1 - \beta$ of a correct decision, in the statistical literature also called power (discriminatory power), corresponds to the probability of detection (POD). The POD-curve $POD(\theta_d)$ is important to quantify the capability of a method or device to detect a damage of a certain extent, e.g. cracks with a certain length, with a desired reliability. The relation between POD and PFA (and therefore between α and β) is determined by the so-called Receiver-Operating-Characteristic (ROC), see e.g. (Vachtsevanos et al., 2006) for details.

2.4 Model-Based Diagnostics vs. Data-Based Damage Detection

In the previous sections we have dealt with a rather general form of the dynamic model including the modeling of the interaction of the dynamical behavior with a possible damage. If we are able to describe the reality with sufficient accuracy using such models, we can use them for diagnostic purposes. Especially, these models allow us to describe explicitly the dependence of measurable features on the damage parameters $\boldsymbol{\theta}_d$. The problem then is an inverse problem: determine the damage parameters $\boldsymbol{\theta}_d$ under consideration of varying environmental and operational conditions $\boldsymbol{\theta}_e$ given measured features vectors respectively their changes or the measured input-output data. Whether this problem can be solved depends strongly on the available information. These tasks tend to be ill-posed resulting in problems with the existence and uniqueness of the solution and its robustness against measurement noise.

The inverse problem can be formulated as estimating the parameters by minimizing the function J:

$$\hat{\boldsymbol{\theta}}_{\boldsymbol{d}} = \underset{\theta_d}{\operatorname{argmin}}[J \mid J = \boldsymbol{\epsilon}^T \, \boldsymbol{W}_\epsilon \, \boldsymbol{\epsilon} + \Delta\boldsymbol{\theta}_d^T \, \boldsymbol{W}_{\boldsymbol{\theta}} \, \Delta\boldsymbol{\theta}_d \; ; \; \Delta\boldsymbol{\theta}_d = \boldsymbol{\theta}_d - \boldsymbol{\theta}_{\boldsymbol{d}0}] \qquad (9)$$

which is achieved by reducing the error term $\boldsymbol{\epsilon}$ on one hand and by keeping

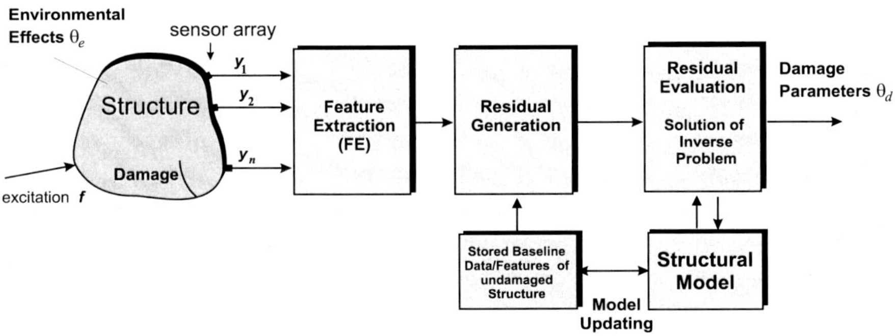

Figure 3: Schematics of chain from measurement to model-based decision

the changes of the parameter vector small on the other hand. Figure 3 shows the basic principle. The weighting matrices $\boldsymbol{W}$ determine the influence of the two antagonistic terms. The term on the left-hand side of Eq.(9) functions as regularization term. If we start from the undamaged state, then $\boldsymbol{\theta}_{\boldsymbol{d}0} = 0$. For a linearized approach, this error term can be defined e.g. as

$$\boldsymbol{\epsilon} = \boldsymbol{r}^{(meas)} - \boldsymbol{r}^{(model)} = \boldsymbol{r}^{(meas)} - \boldsymbol{S}(\boldsymbol{\theta}_{d0}, \boldsymbol{\theta}_e) \, \Delta\boldsymbol{\theta}_d \qquad (10)$$

with $\boldsymbol{r}$ according to Eqn.(7) and $\boldsymbol{S}$ is the sensitivity matrix. More general

$$\begin{aligned}
\boldsymbol{\epsilon} &= \boldsymbol{r}^{(meas)} - \boldsymbol{r}^{(model)} \\
&= \left(\boldsymbol{f}_{y,1}^{meas}(\boldsymbol{\theta}_e) - \boldsymbol{f}_{y,0}^{meas}(\boldsymbol{\theta}_e)\right) - \left(\boldsymbol{f}_{y,1}^{model}(\boldsymbol{\theta}_e, \boldsymbol{\theta}_d) - \boldsymbol{f}_{y,0}^{model}(\boldsymbol{\theta}_e, \boldsymbol{\theta}_{d0})\right) \qquad (11) \\
&= \left(\boldsymbol{f}_{y,1}^{meas}(\boldsymbol{\theta}_e) - \boldsymbol{f}_{y,1}^{model}(\boldsymbol{\theta}_e, \boldsymbol{\theta}_d)\right) - \left(\boldsymbol{f}_{y,0}^{meas}(\boldsymbol{\theta}_e) - \boldsymbol{f}_{y,0}^{model}(\boldsymbol{\theta}_e, \boldsymbol{\theta}_{d0})\right)
\end{aligned}$$

This means that the changes of the features observed in the measurement must be equal to the feature changes predicted by the model. In the case that we have a perfect match of measurement and model in the undamaged state: $\boldsymbol{f}_{y,0}^{meas}(\boldsymbol{\theta}_e) = \boldsymbol{f}_{y,0}^{model}(\boldsymbol{\theta}_e, \boldsymbol{\theta}_{d0})$ the second difference term becomes zero and Eq.(11), reduces to

$$\boldsymbol{\epsilon} = \boldsymbol{f}_{y,1}^{meas}(\boldsymbol{\theta}_e) - \boldsymbol{f}_{y,1}^{model}(\boldsymbol{\theta}_e, \boldsymbol{\theta}_d) \tag{12}$$

In the case of the linear approach, we get the well-known Extended Weighted Least Squares (EWLS) method, see (Natke and Cempel, 1997):

$$(\boldsymbol{S}^T \boldsymbol{W}_\epsilon \boldsymbol{S} + \boldsymbol{W}_\theta)\, \Delta\boldsymbol{\theta_d} = \boldsymbol{S}^T\, \boldsymbol{W}_r \boldsymbol{\epsilon} + \boldsymbol{W}_\theta\, \boldsymbol{\theta}_{d0} \tag{13}$$

which can be solved iteratively. (Link et al., 2007) has presented an interesting multi-model approach adapting the model of the undamaged and damaged system simultaneously. Special attention has to be paid to the problem of the high dimensionality of the parameter space, which has to be reduced as far as possible (Balageas et al., 2006; Fritzen et al., 1998; Fritzen and Bohle, 1999c). The reduced subset of the dominant parameters must be able to describe the damage scenario and should finally concentrate on those parameters corresponding to the damaged sub-region(s) of the structure. The parameter reduction strategy is performed in two steps as described in (Fritzen and Bohle, 1999c). As result only those (few) parameters are considered which yield a significant contribution to reduce the error $\boldsymbol{\epsilon}$.

If such reliable physics-based models are not available with sufficient precision, because there are unknown physical effects or modeling is too complicated or too expensive, we still can work with data-based modeling for fault detection. These data-based methods extract a model solely from the measured data or features, such as AR or ARMA models. The lack of physical background means that we cannot predict the changes of the features due to changes of $\boldsymbol{\theta}_d$ or $\boldsymbol{\theta}_e$. Therefore the influence of $\boldsymbol{\theta}_e$ must be compensated by measuring the changes of the features of the undamaged system due to their changes and learn how the system behaves under certain conditions. This can be done e.g. by neural networks. In this way, we gather the necessary statistical information and determine when the system starts to deviate from "normal" operating condition due to evolving damage. As another alternative, we have the possibility to judge the system's behavior solely on the basis of measured data, e.g. define threshold values for signals.

3 Diagnosis of Linear Systems

3.1 Equation of Motion and State Space Representation

Let us assume that the dynamics of the structure can be described by the linear equation of motion with m degrees of freedom

$$\boldsymbol{M}(\boldsymbol{\theta}_d,\boldsymbol{\theta}_e)\ddot{\boldsymbol{q}}+\boldsymbol{C}(\boldsymbol{\theta}_d,\boldsymbol{\theta}_e)\dot{\boldsymbol{q}}+\boldsymbol{K}(\boldsymbol{\theta}_d,\boldsymbol{\theta}_e)\boldsymbol{q}=\boldsymbol{f}_{op}(t,\boldsymbol{\theta}_d,\boldsymbol{\theta}_e)+\boldsymbol{f}_{test}(\boldsymbol{\theta}_d,\boldsymbol{\theta}_e,\boldsymbol{u},(t)) \tag{14}$$

where $\boldsymbol{M}$, $\boldsymbol{C}$ and $\boldsymbol{K}$ are the m x m mass, damping and stiffness matrix, respectively. $\boldsymbol{q}$ is the displacement vector and $\boldsymbol{f}$ again represents the external loads (operational and test loads). We can split the system matrices into a contribution "o" representing the undamaged state at a certain reference condition and changes "Δ" resulting from the damage parameters $\boldsymbol{\theta}_d$ and from environmental parameters $\boldsymbol{\theta}_e$:

$$\begin{aligned}&\left(\boldsymbol{M}_{\mathbf{0}}+\boldsymbol{\Delta M}(\boldsymbol{\theta}_d,\boldsymbol{\theta}_e)\right)\ddot{\boldsymbol{q}}+\left(\boldsymbol{C}_{\mathbf{0}}+\boldsymbol{\Delta C}(\boldsymbol{\theta}_d,\boldsymbol{\theta}_e)\right)\dot{\boldsymbol{q}}+\ldots\\&\ldots+\left(\boldsymbol{K}_{\mathbf{0}}+\boldsymbol{\Delta K}(\boldsymbol{\theta}_d,\boldsymbol{\theta}_e)\right)\boldsymbol{q}=\boldsymbol{f}_{op}(t,\boldsymbol{\theta}_d,\boldsymbol{\theta}_e)+\boldsymbol{f}_{test}(\boldsymbol{\theta}_d,\boldsymbol{\theta}_e,\boldsymbol{u},(t))\end{aligned} \tag{15}$$

In order to distinguish between the effects of $\boldsymbol{\theta}_d$ and $\boldsymbol{\theta}_e$, we must know the dependency from each quantity, otherwise we are not able to discriminate the damage $\boldsymbol{\theta}_d$ finally. A large number of methods is based on the state space representation:

$$\dot{\boldsymbol{x}}(t)=\boldsymbol{A}_c\,\boldsymbol{x}(t)+\boldsymbol{B}_c\,\boldsymbol{u}(t) \tag{16}$$

$$\boldsymbol{y}(\boldsymbol{t})=\boldsymbol{C}_c\,\boldsymbol{x}(t)+\boldsymbol{D}_c\,\boldsymbol{u}(t) \tag{17}$$

where the matrices $\boldsymbol{A}_c$, $\boldsymbol{B}_c$, $\boldsymbol{C}_c$ and $\boldsymbol{D}_c$ are the system, input, measurement and feed-through matrix, respectively of the continuous-time state space model having appropriate dimensions. The arguments $\boldsymbol{\theta}_e$, $\boldsymbol{\theta}_d$ have been omitted here for simplicity. For the linear equation of motion, we obtain the relations

$$\boldsymbol{x}=\begin{bmatrix}\boldsymbol{q}\\\dot{\boldsymbol{q}}\end{bmatrix}\;;\;\boldsymbol{A}_c=\begin{bmatrix}\mathbf{0}&\boldsymbol{I}\\-\boldsymbol{M}^{-1}\,\boldsymbol{K}&-\boldsymbol{M}^{-1}\,\boldsymbol{C}\end{bmatrix}\;;\;\boldsymbol{B}_c=\begin{bmatrix}\mathbf{0}\\\boldsymbol{M}^{-1}\end{bmatrix} \tag{18}$$

The corresponding time discrete state space equation is

$$\boldsymbol{x}_{k+1}=\boldsymbol{A}_D\,\boldsymbol{x}_k+\boldsymbol{B}_D\,\boldsymbol{u}_k \tag{19}$$

$$\boldsymbol{y}_{k+1}=\boldsymbol{C}_D\,\boldsymbol{x}_k+\boldsymbol{D}_D\,\boldsymbol{u}_k \tag{20}$$

with the transformed system and input matrix $\boldsymbol{A}_D$ and $\boldsymbol{B}_D$:

$$\boldsymbol{A}_D=\exp(\boldsymbol{A}_c\cdot\Delta t)\;;\;\boldsymbol{B}_D=\int_0^{\Delta t}\exp(\boldsymbol{A}_c\cdot\tau)d\tau\cdot\boldsymbol{B}_c \tag{21}$$

and Δt being the sampling time. The matrices of the measurement equation remain unchanged: $\boldsymbol{C}_D=\boldsymbol{C}_c$; $\boldsymbol{D}_D=\boldsymbol{D}_c$.

3.2 Damage Identification in the Modal Domain

Modal quantities (natural frequencies, damping and mode shapes) can be extracted by means of classical modal analysis methods using input-output measurements resulting from special modal tests or by output-only methods which utilize the ambient excitation from wind, traffic loads, etc. (Peeters and DeRoeck, 1999, 2001; Kraemer and Fritzen, 2010b). In the sense of Eq.(5) the determination of the modal data can be considered as feature extraction from the raw data $\boldsymbol{y}(t)$. The use of modal data for system identification is discussed extensively in (Natke, 1992; Friswell and Mottershead, 1995). The modal quantities can be either real or complex.

Under the assumption that the system is undamped or only lightly damped, characteristic features of the system are the natural frequencies (eigenvalues $\lambda_i = \omega_i^2$) and the (real) normal mode shapes $\boldsymbol{\phi}_i$ determined by the solution of the eigenvalue problem:

$$[\boldsymbol{K}(\boldsymbol{\theta}_d, \boldsymbol{\theta}_e) - \omega_i^2 \boldsymbol{M}(\boldsymbol{\theta}_d, \boldsymbol{\theta}_e)]\, \boldsymbol{\phi}_i = \boldsymbol{0}. \tag{22}$$

In this case, the feature vector is:

$$\boldsymbol{f}_y^T = \left[\omega_1^2, \omega_2^2, \omega_3^2, \cdots, \boldsymbol{\phi}_1^T, \boldsymbol{\phi}_2^T, \boldsymbol{\phi}_3^T, \cdots\right] \tag{23}$$

The example of a 3 dof oscillator in Figure 4 is used to demonstrate the

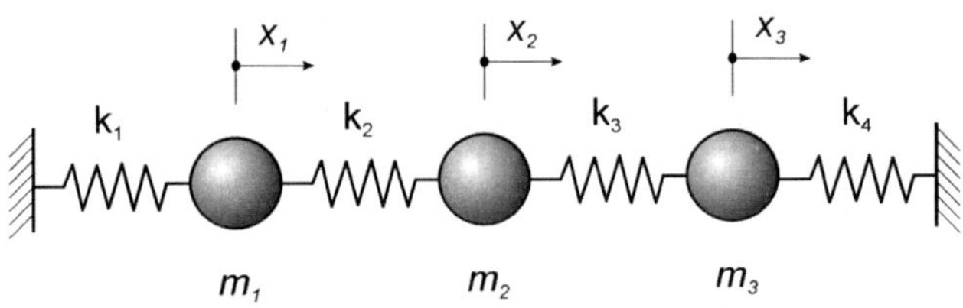

Figure 4: 3 dof oscillator to demonstrate the effect of superimposed temperature variation and damage

superimposed change of the first two natural frequencies due to temperature and damage. Assuming that the Young's modulus E is decreasing quadratically with temperature as adapted to material data $E(T)$, Figure 5(a), the natural frequencies will globally drop linearly with temperature. The temperature is changed randomly between 0 and $50^o C$. Also measurement uncertainty is simulated by adding some random noise. If damage occurs locally (only a reduction in stiffness of one of the springs), this will also result in a certain pattern of the first two frequencies. It can be seen that the three cases (a) random temperature variation only, (b) random

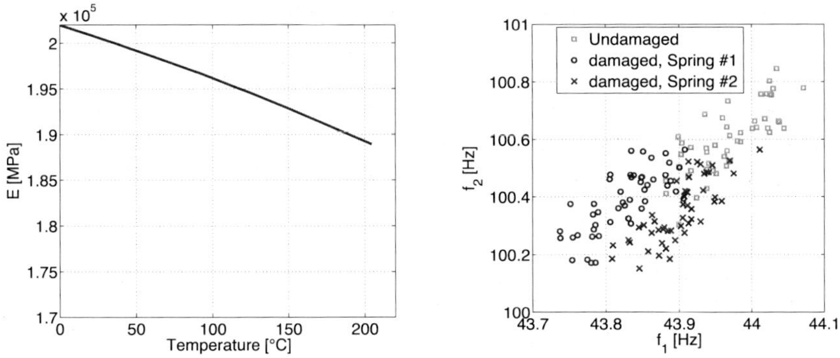

(a) Temperature dependent modulus of elasticity for a steel

(b) First and second natural frequency as a result of temperature variation and damage

Figure 5: Influence of temperature and damage on vibration behavior

temperature variation and damage at spring no. 1, (c) random temperature variation and damage at spring no. 2 deliver frequencies ranges for f_1 and f_2 which are partially overlapping and therefore a clear separation between the cases of undamaged and damaged systems is not possible without further information, see Figure 5(b).

Eigenfrequency and mode shape residuals. First approaches used the resonant frequencies of a structure only, e.g. (Cawley and Adams, 1979). A structural (e.g. stiffness) change produces a characteristic shift of the eigenfrequency spectrum, which is used to deduce the parameters causing this change, expressed as residual: $r_{\omega,i} = \Delta\omega_i^2/\omega_i^2$. Problems using eigenfrequencies alone arise in the case of symmetric structures or if the measured and model spectra cannot be assigned properly. The mode shapes introduce spatial information about the damage. Local changes of stiffness result in a local change of the mode shapes curvature and hence in the mode shapes themselves. For a mode no. i the change of the eigenvector is $\boldsymbol{r}_{\phi,i} = \boldsymbol{\phi}_{1,i} - \boldsymbol{\phi}_{0,i}$ ("1" denotes the current state, "0" the reference state). Correct pairing of the "1" and "0" eigenvalues/eigenvectors and correct scaling of the mode shapes is obtained by applying the Modal Assurance Criterion (MAC) and the Modal Scale Factor (MSF), see e.g. (Friswell and Mottershead, 1995; Balageas et al., 2006; Heylen et al., 1998). From the author's point of view using frequencies and mode shapes simultaneously should be preferred. The model allows also to calculate the first order partial derivatives (modal sensitivities, expressed by the sensitivity matrix $\boldsymbol{S}$) and can be determined by the methods of Fox and Kapoor or by Nelson, see (Friswell and Mottershead, 1995; Natke, 1992). Higher mode shapes are more sensitive to local parameter changes, however problems may arise from the sensor distribution if the sensor network is not dense enough to properly describe the mode shape. Furthermore, it is more difficult to accurately predict the higher frequencies and mode shapes. In the case that damping models are available, damping parameters can also be used for the purpose of damage identification. The methods based on model-updating using modal data or FRFs in combination with the very efficient parameter selection strategies, see e.g. (Fritzen and Bohle, 1999c), as described in the previous sections, have been applied to different large-scale structures, such as the I40-bridge (Fritzen and Bohle, 1999a,b), the Z24-bridge (Bohle, 2005; Fritzen and Kraemer, 2009; Balageas et al., 2006) or the "Steelquake"-structure, see (Fritzen and Bohle, 2001, 2003).

Energy considerations and curvature mode shapes. Stubbs, Kim and Farrar (Stubbs et al., 1995) developed a formula from strain energy expressions to determine a change of the flexural stiffness EI due to damage in beams or beam-like structures. Each element/substructure is tested to discover a local change $\Delta(EI)$ from a local change of the curvatures (second spatial derivatives) of the mode shape functions for the reference state "0" and the damaged state "1". (Maeck, 2003) uses a regularization technique to reduce the errors from the numerical curvature calculation. A more general approach was presented by Ladeveze and Reynier (Ladeveze and Reynier, 1989) with the MECE concept. As a result they also get indicators from strain energy expression pointing out the most erroneous locations of the reference model indicating structural changes due to damage. A sensitivity-based approach using modal kinetic energies (MKE) was proposed in (Fritzen and Bohle, 2004; Bohle, 2005).

3.3 Electro-Mechanical Impedance Method (EMIM)

In contrast to the modal methods, the impedance method works in a medium to high frequency range (typically $\Omega > 20kHz$). At high frequencies, the displacement field shows shorter wave lengths leading to a higher sensitivity to small local damage compared to the lower frequency range. Structural changes due to damage are observed by changes in the impedance spectrum of the electro-mechanical system. A detailed overview on the EMIM is given by (Park et al., 2003; Giurgiutiu and Lyshewski, 2004; Giurgiutiu, 2008). The impedance of the EM-system is determined by the input voltage V and the current I through the piezo-electric actuator: $Z_{EM}(\Omega) = \frac{V(\Omega)}{I(\Omega)}$, Figure 6. Instead of using expensive impedance analyzers, (Peairs et al., 2004, 2005) have proposed a simplified circuit where an output voltage is measured over an auxiliary resistor. The EMIM makes use of the self-sensing properties of the piezo-element which is sensor and actuator at the same time. The idea is that the damage in the structure changes the structural impedance and hence the resulting spectrum of the coupled EM-system:

$$Z_{EM}(\Omega, \boldsymbol{\theta}_d, \boldsymbol{\theta}_e) = \frac{1}{j\Omega\, C_{PZT}} \left(1 - \kappa^2 \frac{Z_{struct}(\Omega, \boldsymbol{\theta}_d, \boldsymbol{\theta}_e)}{Z_{struct}(\Omega, \boldsymbol{\theta}_d, \boldsymbol{\theta}_e) + Z_{PZT}(\Omega, \boldsymbol{\theta}_d, \boldsymbol{\theta}_e)}\right)^{-1} \tag{24}$$

where C_{PZT} is the zero load capacitance and κ is an electro-mechanical cross coupling coefficient of the piezo-electric transducer, Z_{struct}, Z_{PZT} are the mechanical impedances of the structure and the piezo-element, respectively. The current spectrum $Z_{EM,1}$ (in most cases its real part) is compared to the stored reference spectrum $Z_{EM,0}$ of the undamaged system. Alternatively, also the admittance spectrum can be used: $Y_{EM}(\Omega) = \frac{1}{Z_{EM}(\Omega)}$. The impedance method is a qualitative method, therefore, Level I-detection can be performed. The PZT-actuator disperses its energy into the structure so that a localization (Level II) is possible if the wave field is mainly concentrated around the actuator and if the damage lies within the area of influence of the actuator. This effect is highly dependent on the material damping. The lower the damping is, the worse are the localization properties due to the reflections from the boundaries. To get quantitative results about the extent of damage (Level III), a coupled EM-model must be incorporated. The EMIM can also be used for self-diagnostics of the piezo-elements, see e.g. (Park et al., 2006). Own experiments, Figure 7, show the change of the imaginary part of the admittance spectrum (called susceptance). As can be seen, the admittance changes with temperature ($\boldsymbol{\theta}_e$) and also with damage ($\boldsymbol{\theta}_d$) which is in this case the degradation of the adhesive layer between PZT-element and structure. (Sepehry et al., 2011) have treated temperature compensation using neural networks.

3.4 Time Domain Methods

The formulation of the diagnosis problem in the time domain especially in state space notation is frequently used in control and automation theory to identify faults in general technical systems (Gertler, 1998). These ideas have also been used in the context of structural diagnosis and vibration problems. (Fassois and Sakellariou, 2007) give an overview on time-domain methods. Some representatives are discussed in the following sections.

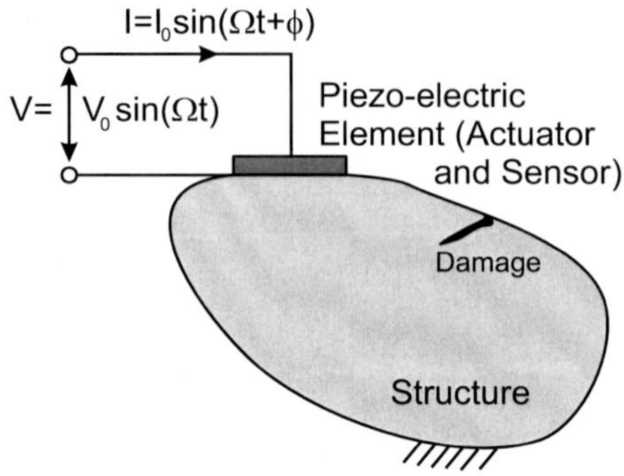

Figure 6: Measurement of the impedance spectrum of the coupled electromechanical system

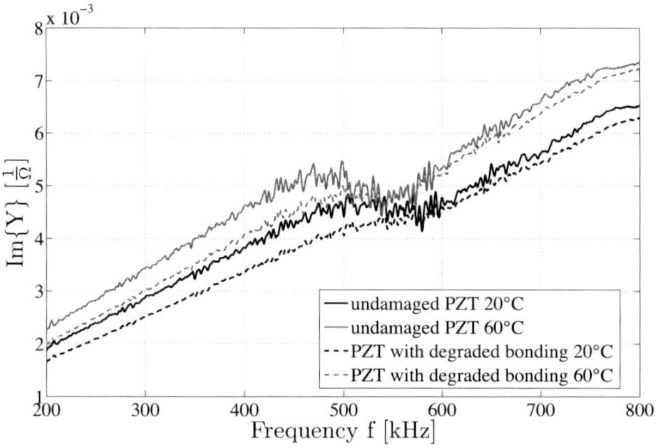

Figure 7: Measured admittance spectrum for PZT sensor/actuator on aluminium plate under different temperature and health conditions

Nullspace-based fault detection method (NSFD). Basically this method has been developed by (Basseville et al., 2000, 2001). It is a Level-I method and works under output-only conditions assuming that the system excitation is random Gaussian white noise. However, practical experience shows that the method yields good results also under less restrictive conditions. Using the premise that the excitation of the structure is unknown but normally distributed, the origin of the damage detection approach in this paper is derived from the equations of motion (in discrete time state space notation) for a stochastically excited linear system

$$\begin{aligned} \boldsymbol{x}_{k+1} &= \boldsymbol{A}_D \boldsymbol{x}_k + \boldsymbol{w}_k \\ \boldsymbol{y}_k &= \boldsymbol{C}_D \boldsymbol{x}_k + \boldsymbol{v}_k \end{aligned} \tag{25}$$

where $\boldsymbol{x}$ is the state vector and $\boldsymbol{y}$ contains the multivariate time data of the measured sensor signals. $\boldsymbol{A}_D$ is the discrete state space matrix, $\boldsymbol{C}_D$ is the measurement matrix, $\boldsymbol{w}$ and $\boldsymbol{v}$ are the process and measurement noise, respectively. $\boldsymbol{w}$ represents the unknown random system excitation. k is one time instant of the time series of length n_t. The unbiased auto- and cross-correlations between the different sensor signals

$$\hat{\boldsymbol{R}}_r = \frac{1}{n_t - r - 1} \sum_{k=1}^{n_t - r} \boldsymbol{y}_{k+r} \boldsymbol{y}_k^T \tag{26}$$

can be arranged in one matrix with the following Block-Hankel form

$$(\boldsymbol{H}_{\alpha,\beta})_{r-1} = \begin{bmatrix} \hat{\boldsymbol{R}}_r & \hat{\boldsymbol{R}}_{r+1} & \cdots & \hat{\boldsymbol{R}}_{r+\beta-1} \\ \hat{\boldsymbol{R}}_{r+1} & \hat{\boldsymbol{R}}_{r+2} & \cdots & \hat{\boldsymbol{R}}_{r+\beta} \\ \vdots & \vdots & \ddots & \vdots \\ \hat{\boldsymbol{R}}_{r+\alpha-1} & \cdots & \cdots & \hat{\boldsymbol{R}}_{r+\alpha+\beta-2} \end{bmatrix}. \tag{27}$$

For $r = 1$ the Hankel matrix can be expressed as the product of the observability and controllability matrix with the considered number of time shifts α, β respectively. The identification of the reduced system of order p requires a singular value decomposition of the Hankel matrix $\boldsymbol{H}_0 \approx \boldsymbol{U}_p \boldsymbol{S}_p \boldsymbol{V}_p^T$:

$$\hat{\boldsymbol{A}}_D = \boldsymbol{S}_p^{-1/2} \boldsymbol{U}_p^T \boldsymbol{H}_1 \boldsymbol{V}_p \boldsymbol{S}_p^{-1/2}. \tag{28}$$

The system eigenfrequencies and the modal damping ratios are identified by means of the eigenvalues of the state space matrix $\hat{\boldsymbol{A}}_D$. Multiplying the eigenvectors of $\hat{\boldsymbol{A}}_D$ with the system output matrix $\hat{\boldsymbol{C}}_D = \boldsymbol{E}_p^T \boldsymbol{U}_p \boldsymbol{S}_p^{1/2}$ (with $\boldsymbol{E}^T = [\boldsymbol{I} \;\; \boldsymbol{0} \;\; \cdots \;\; \boldsymbol{0}]$) provides the complex mode shapes of the system. Instead of using these features, namely the eigenfrequencies, damping ratios and mode shapes directly, an approach based on projected system residuals was proposed in (Basseville et al., 2006) and (Fritzen et al., 2003a). This residual turned out to be very sensitive to small structural changes. The column vectors of the matrix of left singular vectors obtained from SVD: $\boldsymbol{U}_p = \boldsymbol{K}^T$ span the null space (or left kernel space) of the reduced Hankel matrix of the undamaged structure with $\boldsymbol{H}_{ref} = (\boldsymbol{H}_{\alpha,\beta})_0$ so that

$$\boldsymbol{K}_{ref} \boldsymbol{H}_{ref} = \boldsymbol{0} \tag{29}$$

$\boldsymbol{K}_{ref}$ is stored as reference information about the undamaged system and will be used to generate the residuals from the incoming new data sets (index n) which can be compressed to one damage indicator $\boldsymbol{\epsilon}_n = \boldsymbol{K}_{ref}\boldsymbol{H}_n$ as feature:

$$DI_n = \boldsymbol{\varsigma}_n^T \hat{\boldsymbol{\Sigma}}^{-1} \boldsymbol{\varsigma}_n \text{ with } \boldsymbol{\varsigma}_n = vec\left(\boldsymbol{\epsilon}_n\right) \text{ and } \hat{\boldsymbol{\Sigma}} = \frac{1}{N_\Sigma - 2} \sum_{q=2}^{N_\Sigma} \boldsymbol{\varsigma}_q \boldsymbol{\varsigma}_q^T. \tag{30}$$

vec in Eq.(30) is the stack operator, arranging the matrix column-wise in one vector. N_Σ is a number of data sets belonging to the undamaged structure used to define the covariance matrix $\hat{\boldsymbol{\Sigma}}$. If new measurement data are taken from the undamaged structure and the reference Hankel matrix is composed of the data of the undamaged system, the residual $\boldsymbol{\varsigma}$ should be close to zero and vary only within certain statistical bounds due on the measurement errors. If damage occurs, the residuals $\boldsymbol{\varsigma}$ and therefore DI should differ significantly from zero. For a more detailed discussion of the method see (Basseville et al., 2000, 2001), an applications of the theory within the smart structures concept is described in (Fritzen et al., 2002, 2003b). In the latter paper, it was shown by controlled experiments in an oven how changes of the system response due to temperature variations can be compensated. The topic of variations of the environmental parameters $\boldsymbol{\theta}_e$ will be treated in more detail in a later subsection. In (Aimin et al., 2005) different formulations of the NSFD method are compared to principal component analysis (PCA).

Residual generation using Kalman Filters and ARMAX models. The idea of using Kalman Filters (KF) for damage detection is to first identify a multi-input multi-output state space model from measured reference data sets and then to determine the corresponding Kalman gain matrix. When applying new data sets to the KF, it produces residuals (or also called innovations in the KF context) by testing the nominal model against these new measurement data sets. If the actual data set stems from a damaged system, however the KF was designed for the reference model, the misfit shows up in a change of the statistical properties of the residuals. This change can be used to perform a detection that a change of the mechanical structure has occurred (Level I). First, this method was presented in the automation literature, e.g. (Mehra and Peschon, 1971). In (Fritzen and Mengelkamp, 2003) the KF was applied to detect a delamination in a CFRP plate. In (Seibold et al., 1996), an Extended Kalman Filter (EKF) was used to determine the crack depth of an opening and closing crack in a rotating shaft exhibiting nonlinear dynamic behavior. Sohn et al. (Sohn et al., 2001) have demonstrated the use of the Auto-Regressive (AR) and AR with Exogenous input (ARX) time-series models in a two-stage procedure which is quite similar to the KF approach. Here, the basic idea is to identify an ARX reference model from data sets of discrete time series representing the undamaged structure in a first stage. This model represents the dynamic behavior of the undamaged system (reference state). In the second stage this model is tested with new data sets. The resulting residual error is statistically evaluated. As long as the feature extracted from the residuals lies within a defined range of the statistical variation there is no evidence that the structure has changed its physical properties. The ratio $h = (\sigma/\sigma_{ref})$ is defined here as damage sensitive feature where σ and σ_{ref} are the standard deviations of the current and the reference residual time series, respectively. This method provides a Level I test. An appropriate threshold value h_{th} has to be chosen for h. (Kraemer, 2011) used vector autoregressive (ARV) models to extract modal data and to perform damage detection from output-only data. Further aspects of outlier analysis were discussed by (Sohn et al., 2001).

Example of damage detection of an airplane panel using the NSFD method. This example treats a stiffened shell structure from an Airbus A320. The purpose was to find out whether damage in the stringers can be detected and localized. There was no change of the environmental parameters during the test. The method used here is a combination of the self-sensing capabilities of the piezo-elements (as usually used for the impedance method) and the evaluation of the output voltages using the NSFD method, see (Fritzen et al., 2007). This means that only one channel is used to build up each Hankel matrix for the nine sensors. The structure was excited and measured consecutively with the nine PZTs (P1 to P9, see Figure 8), each PZT acting as actuator and sensor according to the self-sensing principle. The positions of the sensors are displayed in Figure 8. The PZT-elements were excited by a broadband noise signal over a period of 10 s and the response (current) was used to build up a Hankel matrix. The total number of measurements for each PZT was 15, 10 for the undamaged and 5 for the damaged state of the structure. Damage, see Figure 8, was introduced by cutting into one of the stringers of the shell structure between the PZT nos. 9 and 6, but closer to PZT 9. The NSFD-Method (see Eqs. 25 to 30) was applied to the output signals of each sensor separately. So, the nine positions can be tested one by one. The results are displayed in Figure 9. The NSFD localized the damage correctly close to the sensor 9 in measurements no. 11 to 15. The application of ARX models yielded quite similar results with a significant change for sensor 9 only.

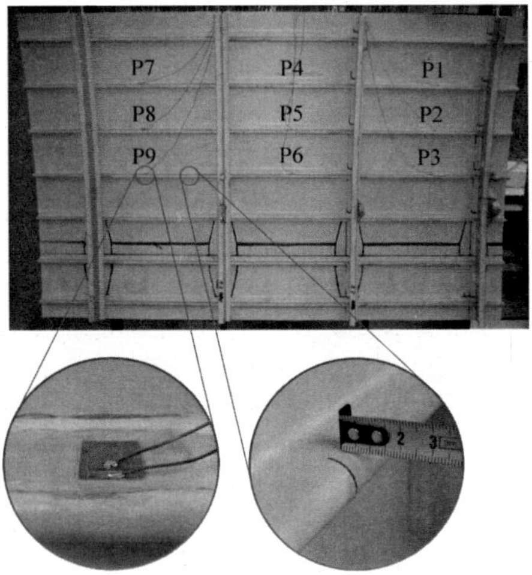

Figure 8: Damage Detection in stringers of A320 fuselage panel

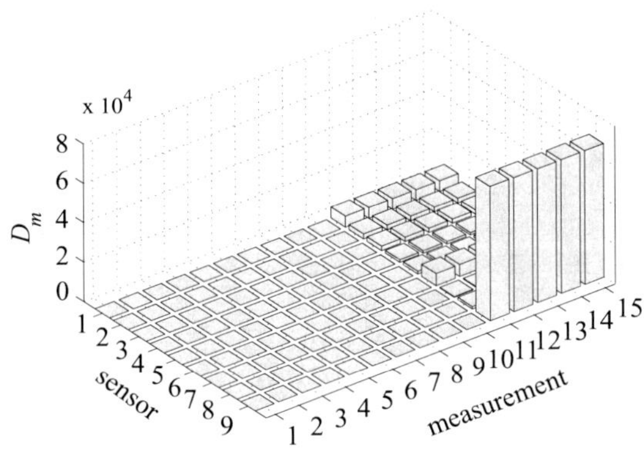

Figure 9: Results for damage indicators with NSFD-method

4 Data-based Damage Detection under Changing Environmental and Operational Conditions (EOC)

4.1 Overview of Methods

As shown before, Environmental and Operational Conditions (EOC) play an important role when dealing with long term monitoring, because they can complicate damage detection. The raw measurement signals can stem from vibration-based or wave propagation-based tests and have to be reduced to the significant features. This is necessary in order to be able to distinguish between influences on the signal driven by environmental changes and influences driven by damage. Different approaches with their advantages and disadvantages were already published before. Depending on the chosen strategy, the methods can be classified in such using the measurements of the EOC itself or not, see (Sohn, 2007) and (Kraemer, 2011). Methods for damage detection under changing operational conditions were developed by (Sohn et al., 2002) using a combination of AR-ARX (AR models with exogenous inputs) models with Non Linear Principal Component Analysis (NLPCA). (Peeters et al., 2001) used ARX models to compensate the temperature effects on bridge eigenfrequencies. (Kullaa, 2005) applied missing data analysis or factor analysis in (Kullaa, 2002) to eliminate the environmental effects from damage sensitive features. In (Deramaeker et al., 2008) the EOC-effects on damage indicators provided by modal filters are diminished by means of factor analysis. (Yan et al., 2005) proposed a local PCA and in (Moll et al., 2008) classification techniques were used for structural damage diagnosis under changing environmental conditions. In (Mujica et al., 2006) a Self-Organizing Map (SOM) is used to train features from different damage scenarios at different environmental conditions. (Kraemer et al., 2011) used SOMs to compensate the EOC effects on damage features like modal data and coefficients of autoregressive modells. A systematic review of the methods for compensation of environmental conditions can be found in (Sohn, 2007). In this chapter fuzzy classification techniques, see also (Kraemer, 2011), are proposed to reduce the effects of EOC on damage indicators provided by the Null-Space based Fault Detection (NSFD) method.

4.2 Environmental and operational conditions for wind energy plants

The large spectrum of EOC effects on the structural dynamics can be well exemplified by means of Offshore Wind Energy Plants (OWEP). Here, the changing EOCs (wind velocity, wind direction, temperature, orientation of the nacelle, rotational speed of the blades and atmospheric conditions) strongly influence the dynamic behavior of the plant. Sometimes, also the boundary conditions are changing (e.g. by ground erosion). For example, the wind velocity, the wind direction and the height of waves affect the excitation level of the structure and blades (e.g. at low wind speed only a small number of lower modes is excited). It is also known from bridges that a high wind velocity can cause changes in the eigenfrequencies and a varying temperature affects the dynamic behavior of structures (Sohn, 2007). These effects appear also in case of OWEPs (Kraemer, 2011). Since the structure of the OWEP is not perfectly symmetric along the tower axis (caused by transformers, pumps, etc. which are placed along the tower), the distribution of mass/mass moment of inertia changes due to the orientation of the nacelle and with them also the dynamic properties of the system (Kraemer and Fritzen, 2010a). Also the eigenfrequencies of the blades depend nonlinearly on the rotational speed of the rotor, see (Bir, 2008), and can change with the blades' pitch angle. Regarding excitation mechanisms, OWEPs are mainly excited stochastically by wind and waves. But also the angular speed of the rotor plays a major role in the structural excitation. In that case the structure is periodically excited by the stall-effect appearing in the instant when a blade passes the tower. The mentioned periodic excitation can be amplified by additional flap moments generated by mass differences between the blades. The structural excitation induced by the mentioned stall-effect is stronger than the one induced by a high wind speed level, see also (Kraemer and Fritzen, 2010a). Further periodical excitation occurs as an OWEP park-effect, when one plant stands in the slipstream of the other. Transient excitations of the structure and blades can occur at small wind speeds. In this case the rotation of the nacelle and the pitching of the blades in order to adapt to the wind conditions are the dominating sources of excitation. Further transient excitation may arise due to sudden crosswinds and brake manoeuvres.

4.3 Compensation of environmental effects on damage detection approach using fuzzy classification

In the following subsection, we use the Null-Space based Fault Detection (NSFD) method and follow the idea presented in (Kraemer, 2011; Fritzen and Kraemer, 2011). The approach requires the measured EOCs. The used set of features in the training or learning phase (when damage is not present) for a certain measurement no. i is formulated as follows:

$$\mathbf{\Theta}_i = \left[\boldsymbol{\theta}_{e,i},\ DI_i^{NSFD}\right]. \tag{31}$$

$\boldsymbol{\theta}_{e,i}$ is a set of measured environmental and operational parameters (temperature, wind speed, nacelle position, etc.) which influence the dynamics of the structure and thus the damage indicator described in Eq.(30). In order to find a limited number of representative references for the undamaged state of the structure and the influence of different EOC on the damage indicator, at first the indicators are calculated by means of one reference data set as described in Eq.(30). The representative EOC can be established by the minimization of the following objective function (Balasko et al., 2005):

$$J = \sum_{j=1}^{c} \sum_{i=1}^{n_c} (w_{ji})^m \, d_{ji}^2 \text{ with } d_{ji}^2 = \left\| \boldsymbol{\Theta}_i - \boldsymbol{\mu}_{c_j} \right\|^2 . \tag{32}$$

c in Eq.(32) is an a priori given number of references, n_c is the total number of data used in the training phase, d is the Euclidian distance between the features and the references given by the centers of the data $\boldsymbol{\mu}_{c_j}$. w_{ji} are weights which define the membership of $\boldsymbol{\Theta}_i$ to $\boldsymbol{\mu}_{c_j}$. m is a fuzziness factor defined in most of cases as $m = 2$. This kind of feature clustering is known as soft partition or fuzzy classification. The constraints of the optimization procedure are:

$$w_{ji} \in [0 \; 1], \; \sum_{j=1}^{c} w_{ji} = 1 \text{ and } 0 < \sum_{i=1}^{n_c} w_{ji} < n_c \text{ for } 1 \leq i \leq n_c, \; 1 \leq j \leq c. \tag{33}$$

The meaning of the constraints are: the weights are scaled between 0 and 1; the sum of the weights is normalized to one; there are no empty classes and it is not possible that all features belong only to one class. The constraints are implemented in the optimization procedure by means of Lagrange multipliers λ:

$$\overline{J} = \sum_{j=1}^{c} \sum_{i=1}^{n_c} (w_{ji})^m \, d_{ji}^2 + \sum_{i=1}^{n_c} \left[\lambda_i \left(\sum_{j=1}^{c} w_{ji} - 1 \right) \right] \tag{34}$$

and the derivatives of the Eq.(34) with respect to the unknowns (the weights and the class centers) yields the optimum for the weights

$$w_{ji} = \frac{1}{\sum_{k=1}^{c} \left(d_{ji}^2 / d_{ki}^2 \right)^{\frac{1}{m-1}}} \tag{35}$$

and also for the centers (references)

$$\mu_{ji} = \frac{\sum_{i=1}^{n_c} (w_{ji})^m \, \boldsymbol{\Theta}_i}{\sum_{i=1}^{n_c} (w_{ji})^m} . \tag{36}$$

The weights and the centers are iteratively updated, as long as $J_I - J_{I-1} < \epsilon$, where I is an iteration index and ϵ is a predefined allowed convergence error. After the classification, Hankel matrices $(\boldsymbol{H}_{ref})_i^{EOC}$ for each data set i will be calculated from the Hankel matrices of the data at the centers $(\boldsymbol{H}_{ref}^{(\boldsymbol{\mu}_{c_j})})$

with the corresponding weights

$$(\boldsymbol{H}_{ref})_i^{EOC} = \sum_{j=1}^{c} w_{ji} \boldsymbol{H}_{ref}^{(\boldsymbol{\mu}_{c_j})}. \tag{37}$$

The weights and consequently the membership of one calculated DI in the presence of two simultaneous EOC-effects are illustrated in Figure 10. The damage indicator can now be defined by the corresponding left kernel

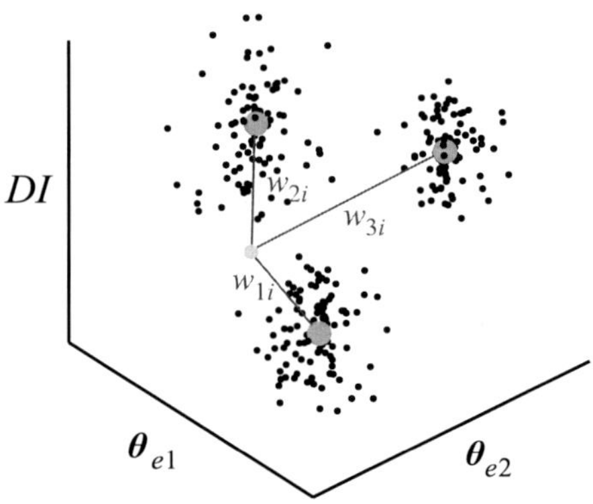

Figure 10: Example of $\boldsymbol{\Theta}$-membership for three classes

space $(\boldsymbol{K}_0)_i^{EOC}$ of the aforementioned Hankel matrices $(\boldsymbol{H}_{ref})_i^{EOC}$ of the undamaged system:

$$DI_i = \boldsymbol{\varsigma}_i^T \left(\hat{\boldsymbol{\Sigma}}^{-1}\right)^{EOC} \boldsymbol{\varsigma}_i \text{ with } \boldsymbol{\varsigma}_i = vec\left[(\boldsymbol{K}_{ref})_i^{EOC} \boldsymbol{H}_i\right]. \tag{38}$$

where $\boldsymbol{H}_i$ is built up with data from the current measurement. The covariance matrix $\boldsymbol{\Sigma}^{EOC}$ is calculated from the residuals of the reference data at the c centers analogously to $\boldsymbol{\Sigma}$ in Eq.(30).

The training procedure is closed by setting a statistical threshold for the damage indicators (based on the generalized extreme value distribution). The mathematical procedure is not described here due to the lack of space, see (Kraemer, 2011) for details.

In the test phase of the algorithm the features consists only of measured EOCs ($\boldsymbol{\Theta}_n = \boldsymbol{\theta}_{e,n}$). The features from the undamaged or damaged structure are classified w.r.t. the centers obtained in the learning phase. The following weights are obtained:

$$w_{jn} = \frac{1/d_{jn}^2}{\sum_{j=1}^{c} 1/d_{jn}^2}. \tag{39}$$

The damage indicator of data set n will be now calculated using Eq.(38) for $i = n$ and compared to the threshold established in the training phase.

4.4 Example: Wind turbine

Figure 11, left, shows the eigenfrequencies of the 5MW prototype offshore wind energy plant M5000-2, see Figure 11, right, built by AREVA-Wind in Bremerhaven, Germany. The modal data are calculated by means of Operational Modal Analysis (OMA), (Kraemer and Fritzen, 2010a). The sensor signals are provided by eight accelerometers positioned along the tower. The signals were measured simultaneously with a sample rate of 50Hz and a time length of 10 minutes per data set. The upper part of Figure 11(left) shows the OMA results for one data set at a high wind excitation of 15.5 m/s. In the lower part some results of the long time monitoring between the years 2007 and 2009 are considered, resulting in 3824 data sets.
The gray intervals in Figure 11 represent monitoring bounds for the vibration modes. The most stable poles, displayed by black circles, result e.g. from the edge- or flapwise blade-vibration (B) or from vibration interaction between tower and blades measured on the tower. In the case of a spinning rotor, additional "stable poles", coming from the rotational frequency and its higher harmonics, will complicate the accurate interpretation of the real modes. The black points in the lower part of Figure 11 show the mean value of the stable poles for the 3824 measurements in the prescribed intervals. The maximum variation of tower eigenfrequencies in the intervals is 8%, mainly resulting from EOC variations. Since the vibration direction depends strongly on the orientation of the nacelle, the tower bending modes (TB) appear always along (ANO) or lateral (LNO) to the nacelle orientation. The results of the OMA shows that the modes of the plant were treated as coupled tower, blade and drive train modes influenced by the EOCs. Their appearance is strongly dependent on the excitation level. Therefore it will be impossible to identify small damages without compensation of the EOC effects on the features. Since no damage occurred at the M5000-2, a laboratory structure and a realistic simulation model of the real plant were used to validate the damage detection approach.
The structure, shown in Figure 12, left, is excited by a shaker using random signals. The rotation of the frame on the structure simulates the position of the nacelle on the tower of an OWEP. Eight accelerometers measure the system responses (sample frequency: 4800Hz). The damage is induced by loosening of 1, 2, 3 or all 4 nuts respectively (corresponding to grades 1, 2, 3 and 4 of damage) at one foot of the structure. The total number of measurements is 130. The first 90 measurements are acquired in the undamaged

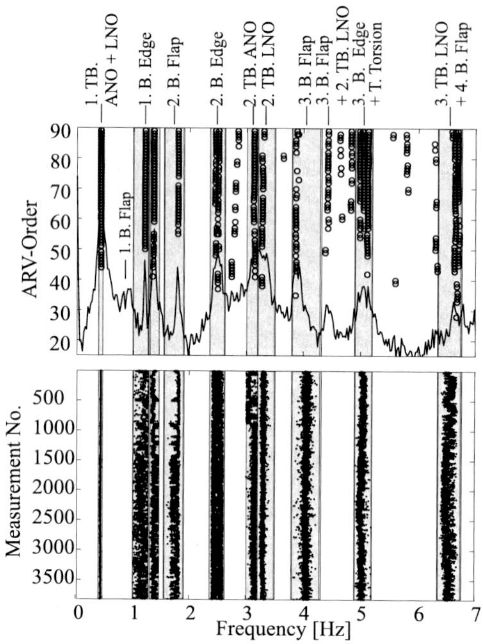

Figure 11: Results of OMA (left); 5MW wind energy plant (right)

state of the structure. The power spectral densities (PSD) of the signals for different damage levels are displayed in Figure 12, right. These show that a damage occurring at the stiff tripod, "under the water level" where no sensors are placed, is not well revealed by spectral analysis. The results of the NSFD method with only one reference shows that the damage cannot be detected, see Figure 13, left. After classification, the damage detection method shows correct results (Figure 13, right). Also small damages by loosening of only one nut can be well identified. For simulation purposes the model in Figure 14 was used. Eight sensor signals (accelerations) were simulated by means of a beam model of M5000-2 with 396 dofs. The structure was excited with loads calculated from measured wind velocities. The loads act at the top of the nacelle. During the excitation, the position of the nacelle and the direction of wind velocity relative to the nacelle are permanently changing. The position and the measurement directions of the sensors are also displayed in Figure 14. The used sample rate was again 50Hz and the measurement duration for one data set was 10 minutes. 1412 data sets for the whole period were taken. In the first 1008 data sets, no

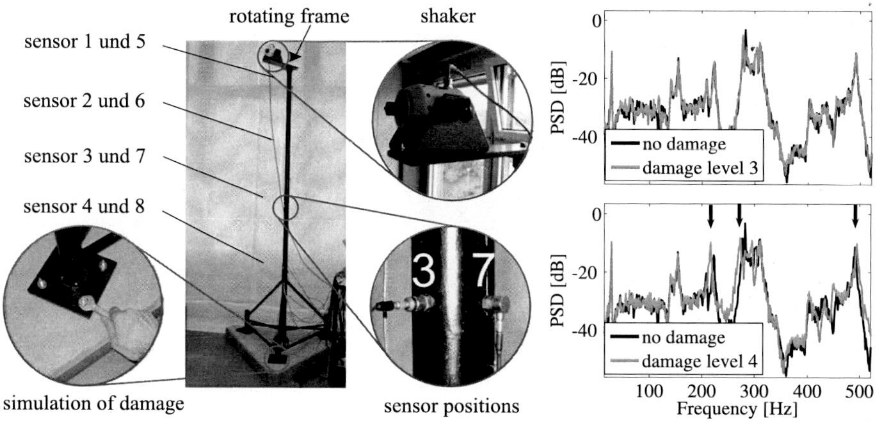

Figure 12: Laboratory structure with simulated damage (left) and PSD of signals from sensor 3 (right)

damage occurred and for the last 404 data sets a progressive damage under the "water level" at the indicated position in Figure 14 was simulated. The simulated damage extent was a reduction of the stiffness in the mentioned element of 1, 5, 10 and 25% (101 data sets for each damage level). The position of the nacelle and the wind velocity for the system without and with damage are displayed in Figure 14, right.
The results of the damage detection approach with only one reference are shown in Figure 15, left. This reveals only the highest damage of 25% stiffness reduction. After the application of the classification based approach, 15 representative references for the EOC effects on the indicators, obtained from 500 training sets, are used. The system was tested with the remaining 912 data sets. The new calculation of the indicators by means of the new references reveals very clearly all damages in data no. 1009 to 1412.

5 Reconstruction of External Loads

5.1 Problem overview and previous work

Buildings, wind turbines, airplanes or stadiums (during concerts or football matches) are some of the examples where the dynamics of a structure should be taken into consideration for reliable construction. In particular, both, material properties and load characteristics, are essential parameters when assessing the fatigue life of a structure or its components. Therefore, the time history of external forces is an important quantity in the forecast of

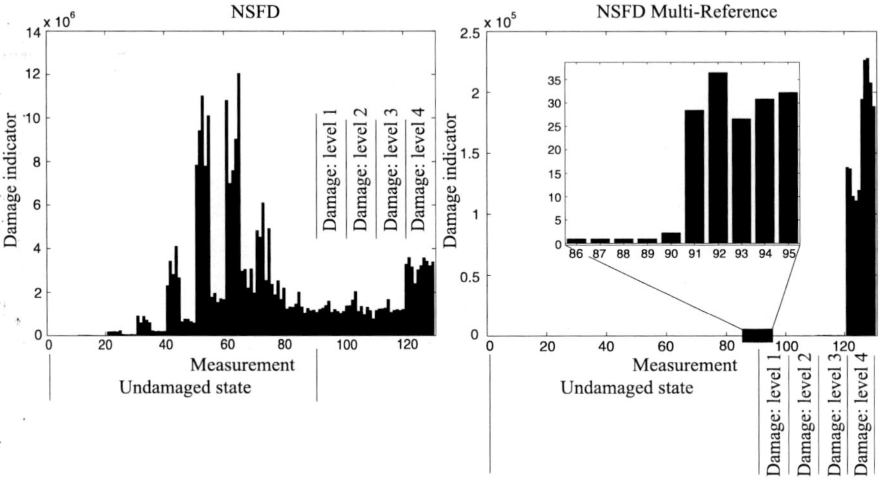

Figure 13: Results of damage detection without (left) and with (right) EOC compensation

the remaining lifetime, (Johannesson, 1999). In many practical applications, however, the measurement of external loads is limited or not possible due to sensor limitations or the unknown nature of the external load, see Figure 16 (a)) . Even when load measurement transducers are available, they may still alter the properties of the system (electronic circuits) or intrude in the load path (contacts and joint). Furthermore, the load localization problem becomes an important issue, especially in the case of impacts, since the aircraft industry started to use new composite materials more extensively. Impacts cause delaminations in composite materials which might not be found by visual inspection, thereby leading to structural failure. Accordingly, the process of estimating loads can be partitioned into the following subparts: 1) reconstructing load time history assuming that the location of a load is known a priori; and 2) determining the locations of applied forces. Load history identification has been the subject of extensive studies in the course of the last two decades. Generally, all force reconstruction methods that have been put forth by engineers can be summarized in three main groups (Uhl, 2007): 1. Deterministic methods (based on the structural model and measured signals), 2. Stochastic methods (statistical models), 3. Artificial intelligence-based methods (neural networks). Deterministic approaches are based on the system model, and therefore yield results which are strongly dependent upon the accuracy of identified or analytically devel-

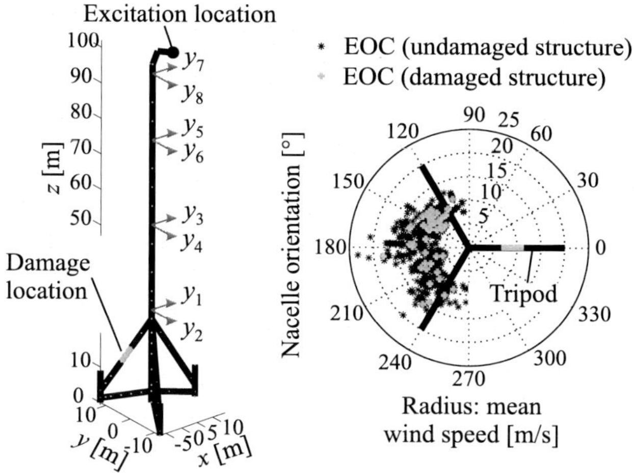

Figure 14: Model of OWEP (left); measured EOCs (right)

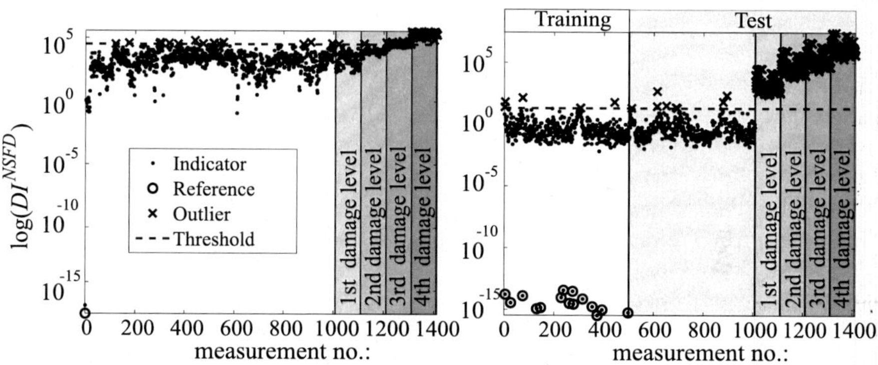

Figure 15: Damage detection without (left) and with EOC compensation (right)

oped models. A further classification within this group can be carried out on the basis of the signal analysis procedure applied for load estimation; such procedures include frequency-domain methods (Adams and Doyle, 2002) and (Hillary and Ewins, 1984) or time-domain methods (Genaro and Rade, 1998) and (Jankowski, 2009). Time-domain approaches are mainly based

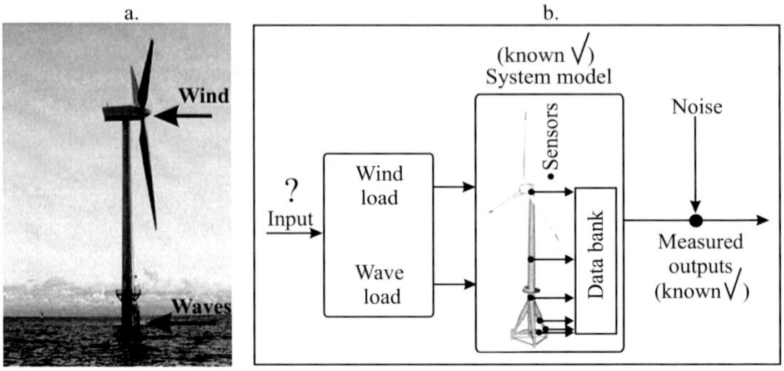

Figure 16: Offshore wind energy plant (a) and block diagram of the load reconstruction problem (b)

on the deconvolution of the (Duhamel-Integral)

$$\boldsymbol{y}(t) = \int_0^t \boldsymbol{H}(t-\tau)\boldsymbol{F}(\tau)d\tau\,, \tag{40}$$

where the vector $\boldsymbol{F}(t)$ represents the external excitation forces, $\boldsymbol{y}(t)$ the system responses and $\boldsymbol{H}(t)$ the structural impulse response function, respectively. Accordingly, the force reconstruction problem is posed as follows: given some measurements $\boldsymbol{y}(t)$ and known system impulse response functions $\boldsymbol{H}(t)$, determine the loads $\boldsymbol{F}(t)$. It should be noted that either $\boldsymbol{y}(t)$, $\boldsymbol{H}(t)$ or both may not be known with perfect precision.

In the frequency domain, information regarding the Frequency Response Function (FRF) $\boldsymbol{H}(\omega)$ of the investigated structure and the spectrum of measured responses $\boldsymbol{Y}(\omega)$ are both required. On the basis of this information, the deconvolution of Eq. (40) is carried out by means of a simpler algebraic equation

$$\hat{\boldsymbol{F}}(\omega) = \boldsymbol{H}^{-1}(\omega)\boldsymbol{Y}(\omega)\,, \tag{41}$$

where $\hat{\boldsymbol{F}}(\omega)$ is estimated force spectrum. Finally, force histories in the time domain are obtained by the inverse Fourier Transformation. Stochastic methods are attempts to establish statistical correlations between output and input. In order to get such correlations, it is necessary to conduct measurements of the inputs and outputs while the structure is being operated. The most useful statistical model for load identification is the regression model. Remarks regarding the applicability of the regression model to load identification can be found in (Trujillo and Busby, 1997). A dedicated procedure for regression model inversion is presented in (Uhl, 2007). Techniques based on artificial intelligence can be realized using Neural Networks (NN) algorithms (Uhl and Petko, 2006) and (Uhl and Chudzikiewicz, 2002), fuzzy algorithms (Goral et al., 2002) or evolutionary algorithms (Uhl and Pieczara, 2003). These methods require a learning process in order to find the relation between the inputs and outputs. It is necessary for the learning process to conduct direct measurements of loads and system responses. In some cases this might be impossible, but numerical simulations of the system behavior can be applied for the learning process of intelligent system, (Uhl, 2002).

The force reconstruction problem is mainly solved by using indirect measurement techniques, which include the transformations of related measured quantities such as acceleration, velocity, position or strain, see Figure 16 (b). These transformations generally lead to a so-called inverse problem or deconvolution problem, where the system properties $\boldsymbol{H}(t)$ and responses $\boldsymbol{y}(t)$ are known while the excitations $\boldsymbol{F}(t)$ are unknown, see Eq. 40. It is well known that inverse problems are often 'ill-posed' in the mathematical sense, that is one of 1) the existence, 2) the uniqueness, or 3) the stability of solution is violated (Hansen et al., 1999; Jacquelin et al., 2003). If this inversion can be done, the system itself becomes its own force sensor, i.e. a "virtual force sensor" which would solve the problem of the load history estimation. A variety of methods were elaborated to overcome the above mentioned difficulties. Stevens (Stevens, 1987) gave an excellent overview of this topic, which includes some earlier studies on inverse analysis of external forces, and (Inoue et al., 2001) wrote a comprehensive summary about impact force estimation techniques. Some of these methods are based on the frequency or impulse response functions or use regularization and dynamic programming in the time domain, as proposed by Trujillo and Busby (Trujillo and Busby, 1997) and applied by Doyle (Doyle, 1989; Adams and Doyle, 2002) and Nordstroem (Nordstroem, 2005). Others use an integration of measured acceleration signals (Elliott et al., 2005) and (Carne et al., 1992) or can only determine the sum of all forces and moments applied to the center of mass (Sum of Weighted Acceleration Technique SWAT) (Genaro and Rade, 1998). Most of the authors assume the prior knowledge of the force location and require first to record the system responses and then do the force history reconstruction. The regularization techniques or a time shift of the collected measurement data in non-collocated case (Steltzner and Kammer, 1999) and (Kammer and Steltzner, 2001) (sensors and loads positions are not collocated) are used to overcome the ill-posedness of the inverse problem. Other researchers have treated the same problem from the control engineering point of view (Soeffker et al., 2003; Corless and Tu, 1998) by using observer properties that are able to estimate the states and unknown inputs or external disturbances. Related to SHM, algorithms which are able to estimate the external forces and their locations online or quasi-online, and which are also capable to do this for nonlinear or time variant mechanical systems, are of special interest. Among them are approaches by (Doyle and Kannal, 1997; Doyle, 1999), (Gopalakrishnan and Doyle, 1994; Doyle and Martin, 1996) who have treated the two following problems: 1) the impact force history reconstruction and 2) the impact force location estimation primarily from the point of wave propagation theory. For the former, several model-based approaches which perform deconvolution of Eq. (40) or an inversion as in Eq. (41), were applied. In order to achieve this, the knowledge of the wave propagation and the model (spectral element model or finite element model) was always required. Doyle has shown that for the finite beam type of structures, such as that in Figure 17 , the measured signals (accelerations, strains or displacements) must undergo a preprocessing procedure where all effects of wave reflections (which always occur in finite structures and, in the case of impact load types, represent the redundant information) and signal trend (especially for accelerometers) must be removed from the acquired data and only then transformed into a frequency domain. Additional difficulties which Doyle had to deal with were 1) loss

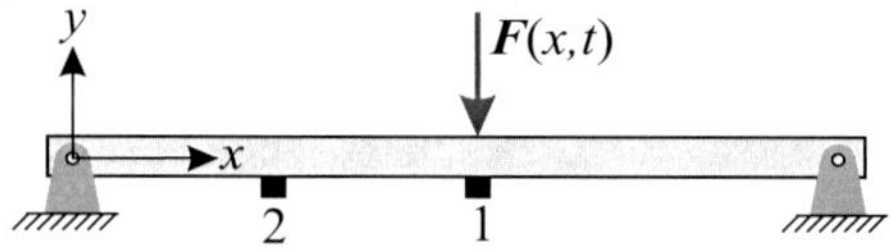

Figure 17: Schematic beam excited by unit force with sensors placed on positions 1 and 2

of information (zeros or no information available at some frequencies) and 2) inherited oversensitivity (sharp spectral peaks) in the function $\boldsymbol{H}(\omega)$. These either make the inversion in Eq. (41) problematic or make $\hat{\boldsymbol{F}}(\omega)$ (at these frequencies) diverge from the real force $\boldsymbol{F}(\omega)$ due to mismatch between $\boldsymbol{Y}(\omega)$ and $\boldsymbol{H}(\omega)$. To overcome the above problems, Doyle used more than one sensor simultaneously together with the addition of small noise which has a regularization property, so that Eq. (41) was modified to the following form:

$$\hat{\boldsymbol{F}}(\omega) = \hat{\boldsymbol{F}}_1(\omega) = \left[\frac{\boldsymbol{H}_1^*(\omega)\boldsymbol{Y}_1(\omega) + \boldsymbol{H}_2^*(\omega)\boldsymbol{Y}_2(\omega)}{\boldsymbol{H}_1^2(\omega) + \boldsymbol{H}_2^2(\omega) + \boldsymbol{R}(\omega)} \right], \tag{42}$$

where $(.)^*$ indicates a complex conjugate and $\boldsymbol{R}(\omega)$ is a small amount of random noise. Addition of noise to the denominator does not replace the loss of information but rather removes potential singularity at those frequencies where $\boldsymbol{H}(\omega)$ tend to be zero. The calculated force spectra were then mapped into the time domain by inverse Fourier transform to get $\boldsymbol{f}(t)$.

Determination of force location. For the problem of force localization, (Doyle and Kannal, 1997) used the advantages of the techniques proposed by (Whiston, 1984) and extended it by applying a Genetic Algorithm (GA) for global search in a complex frame structure. Here again the model (spectral element model or finite element model) and at least two sensors have been used to estimate the force location. The essence of this approach is the usage of estimated forces from recorded measurements and modeled frequency response function $\boldsymbol{H}(\omega)$ for different positions of applied impact. When the guessed position in $\boldsymbol{H}(\omega)$ is nearly correct, then the force reconstructions from both sensors is very similar. As a consequence Doyle used a correlation or difference of these forces as an objective function for GA. (Seydel and Chang, 2001a,b) considered the identification of the impact location and impact load history for a stiffened composite plate by using a smart sensor layer (a distributed array of sensors glued to the plate) as illustrated in Figure 18. For the force history reconstruction, first the position on the plate where the force is applied was determined. This information is very important because the authors used a mathematical model (flexural equations for an orthotropic plate) for impact history reconstruction which is built on the assumptions that the force location is known in advance. In contrast to (Doyle and Kannal, 1997), Seydel and Chang worked in the time domain and used the information acquired from the area of search (only 4 sensors in the vicinity of impact were used, as shown by the rectangle of the four nearest sensors in Figure 18 for impact location identification. As the impact time was not exactly known, the authors utilized only differences between calculated Times Of Arrival (TOA) for these sensors and then tried to minimize the cost function of the following form:

$$\min_{(x,y)}(\boldsymbol{J}) = \sum_{i=2}^{4}\left(\underbrace{(t_i^{\mathrm{toa}} - t_1^{\mathrm{toa}})}_{\text{measured differences}} - \underbrace{\left[\frac{E_{(x,y)} - E_{si}}{\boldsymbol{c}_p(\theta_i)} - \frac{E_{(x,y)} - E_{sl}}{\boldsymbol{c}_p(\theta_l)}\right]}_{\text{estimated differences}}\right)^2 \tag{43}$$

where t_i^{toa} is the calculated time of arrival for the ith sensor. Estimation of t_i^{toa} from the recorded time signals was done with the help of the double peak technique which is a selection of the minimum before a global maximum. $E_{(x,y)}$ is the estimated impact location on the plate, $E_{si(x,y)}$ is the actual

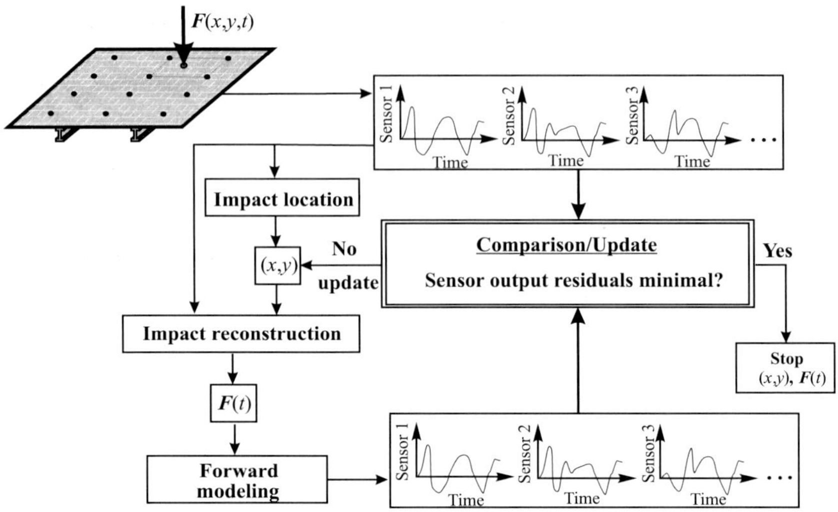

Figure 18: Overview of the model-based impact identification scheme (Seydel and Chang, 2001a)

location of the ith sensor. Sensor 1 is always defined as the one with the smallest TOA. $c_p(\theta_i)$ is the phase velocity of the flexural wave travelling in the orthotropic plate with a given direction θ_i. The phase velocity and the angle are determined analytically as described in (Doyle, 1997). After the impact location was found iteratively by minimization of the cost function in Eq. (43), the model based optimization technique was applied for final time history estimation. Figure 18 shows the principle block diagram of force location estimation and force history reconstruction. The optimization problem was solved by a smoother filter algorithm developed by (Tracy and Chang, 1998a,b).

Inverse structural filters. (Steltzner and Kammer, 1999; Kammer and Steltzner, 2001) have proposed a non-causal Inverse Structural Filter (ISF) which takes the structural response data $\boldsymbol{y}_k$ as an input and returns the estimate of the input forces $\boldsymbol{u}_k$ as an output. They considered a discrete-time invariant system in state space domain

$$\boldsymbol{x}_{k+1} = \boldsymbol{A}_D \boldsymbol{x}_k + \boldsymbol{B}_D \boldsymbol{u}_k \tag{44}$$

$$\boldsymbol{y}_k = \boldsymbol{C}_D \boldsymbol{x}_k + \boldsymbol{D}_D \boldsymbol{u}_k. \tag{45}$$

Thus, based on homogeneous initial conditions, Eq. (45) was stepped forward and the output/input relationship was established as:

$$\boldsymbol{y}_k = \sum_{i=0}^{k} \boldsymbol{H}_i \boldsymbol{u}_{k-i}\,, \tag{46}$$

where the Markov parameters $\boldsymbol{H}_i$ (i.e. $\boldsymbol{H}_0 = \boldsymbol{D}_D$, $\boldsymbol{H}_i = \boldsymbol{C}_D \boldsymbol{A}_D^{i-1} \boldsymbol{B}_D$) were obtained either analytically or during standard vibration tests. Eq. (46) was manipulated to interchange the input and output, leading to an inverse structural system

$$\boldsymbol{x}_{k+1} = \hat{\boldsymbol{A}}_D \boldsymbol{x}_k + \hat{\boldsymbol{B}}_D \boldsymbol{y}_k \tag{47}$$

$$\boldsymbol{u}_k = \hat{\boldsymbol{C}}_D \boldsymbol{x}_k + \hat{\boldsymbol{D}}_D \boldsymbol{y}_k \tag{48}$$

with

$$\hat{\boldsymbol{A}}_D = \boldsymbol{A}_D - \boldsymbol{B}_D \boldsymbol{D}_D^{+} \boldsymbol{C}_D \qquad \hat{\boldsymbol{B}}_D = \boldsymbol{B}_D \boldsymbol{D}_D^{+} \tag{49}$$

$$\hat{\boldsymbol{D}}_D = \boldsymbol{D}_D^{+} = (\boldsymbol{D}_D^T \boldsymbol{D}_D)^{-1} \boldsymbol{D}_D^T \qquad \hat{\boldsymbol{C}}_D = -\boldsymbol{D}_D^{+} \boldsymbol{C}_D \tag{50}$$

so that the impulse response sequence was written for Eq. (48) in analogy to the forward system as in Eq. (46)

$$\boldsymbol{u}_k = \sum_{i=0}^{k} \boldsymbol{R}_i \boldsymbol{y}_{k-i}\,, \tag{51}$$

where, $\boldsymbol{R}_0 = \hat{\boldsymbol{D}}_D$, $\boldsymbol{R}_i = \hat{\boldsymbol{C}}_D \hat{\boldsymbol{A}}_D^{i-1} \hat{\boldsymbol{B}}_D$ are the inverse Markov parameters which are used for construction of causal ISF. Steltzner and Kammer have faced several difficulties including: (a) the Moore-Penrose pseudo-inverse requires that the number of sensors be greater than the number of reconstructed forces; moreover the matrix $\boldsymbol{D}_D$ (which requires acceleration measurements) should have a full column rank to ensure the existence of $\boldsymbol{D}_D^{+}$; (b) the inverse matrix $\hat{\boldsymbol{A}}_D$ can be unstable for structural systems with sensors at different locations than applied forces, that is a "non-collocated" sensors scenario, which can lead to divergence of Eq. (51). These problems were addressed by considering the complete absence of matrix $\boldsymbol{D}$ (all sensors considered as non-collocated), in such a way that the system in Eq. (45) was stepped forward before the inversion could take place

$$\boldsymbol{x}_{k+1} = \boldsymbol{A}_D \boldsymbol{x}_k + \boldsymbol{B}_D \boldsymbol{u}_k \tag{52}$$

$$\boldsymbol{y}_{k+1} = \boldsymbol{C}_D \boldsymbol{A}_D \boldsymbol{x}_k + \boldsymbol{C}_D \boldsymbol{B}_D \boldsymbol{u}_k \tag{53}$$

so that the non-causal expression (53) was used as a basis for the construction of ISF with "l-lead":

$$\boldsymbol{u}_k = \sum_{i=0}^{N_{R-1}} \boldsymbol{R}_i \boldsymbol{y}_{k+l-i}. \tag{54}$$

The "l step forward" is a non-causal lead integer accounting for the finite wave propagation time, from source to sensor, and the number N_{R-1} specifies the order of the ISF. Calculation of the inverse filter coefficients which are combined in matrix $\boldsymbol{R}$ is done with the help of Markov parameters that were cast into a matrix $\boldsymbol{H}$ and accounting the l-lead

$$\boldsymbol{R} = \begin{bmatrix} \mathbf{0} & \mathbf{0} & \boldsymbol{I} & \mathbf{0} \end{bmatrix} \boldsymbol{H}^{+}. \tag{55}$$

Here, the deconvolution of Eq. (40) is done in discrete time domain and normally requires a pseudo-inverse of the matrix built out of the Markov parameters. The beauty of this method is that the pseudo-inverse needs to be done only once. This approach enables an almost online load reconstruction for time invariant systems as well as load estimation in the case when the sensors and excitation positions are not the same (non-collocated). An individual sensor lead was proposed (Nordstroem, 2005) as an extension of this method. Disadvantages associated with the method include difficulties in choosing a proper non-causal lead l and high sensitivity to the measurement noise.

Dynamic programming approach. (Nordstroem, 2005) extended the Dynamic Programming (DP) method proposed by (Trujillo and Busby, 1997) for input estimation on linear time-variant systems (generally, a moving load and environmental changes are assumed) with measurements contaminated by noise. The starting point for the input reconstruction was a discrete time variant state space system as in Eq. (45) with a slight modification in the state vector which combined the states and inputs, as follows:

$$\begin{bmatrix} \boldsymbol{x}_{k+1} \\ \boldsymbol{u}_{k+1} \end{bmatrix} = \begin{bmatrix} \boldsymbol{A}_D & \boldsymbol{B}_D \\ \mathbf{0} & \boldsymbol{I} \end{bmatrix} \begin{bmatrix} \boldsymbol{x}_k \\ \boldsymbol{u}_k \end{bmatrix} + \begin{bmatrix} 0 \\ \Delta \boldsymbol{u}_k \end{bmatrix}. \tag{56}$$

Next, the search of a sequence $\Delta \boldsymbol{u}_k$ that minimizes the weighted least square error between measured $\boldsymbol{y}_k$ and estimated $\hat{\boldsymbol{y}}_k$ responses with Tikhonov regularization is carried out with the help of DP, see (Nordstroem, 2005):

$$\min_{\Delta u}(\boldsymbol{J}) = \sum_{k=1}^{N} (\boldsymbol{y}_k - \hat{\boldsymbol{y}}_k)^T \boldsymbol{W}_1 (\boldsymbol{y}_k - \hat{\boldsymbol{y}}_k) + \Delta \boldsymbol{u}_k^T \boldsymbol{W}_2 \Delta \boldsymbol{u}_k \tag{57}$$

here, the matrices $\boldsymbol{W}_1$ and $\boldsymbol{W}_2$ are considered to be symmetric and positive definite, containing weighting and regularization coefficients respectively. The DP is the recurrence algorithm that may be used to solve large least square problems and is appropriate for minimizing quadratic functions, and comprises two main steps: a backward and a forward sweep. The backward sweep heuristically establishes the input/output relationships at each discrete time step, whereas the forward sweep constitutes the calculation of the optimal inputs and states for the given initial conditions and relations that where established during the backward sweep. Recently an extended DP algorithm has been applied for identification of dynamic loads, see (Lourens et al., 2008, 2009).

Proportional integral observers (PIO). (Wojciechowski, 1978) introduced for the first time the Proportional Integral Observer which was an extension of the classical proportional Luenberger observer (Luenberger, 1971). In contrast to the Luenberger approach, a second loop with two gain matrices and an integrator is also used. (Soeffker et al., 1995) have studied the PIO in more detail and applied it to estimate nonlinearities, unmodeled dynamics or unknown inputs/disturbances for control purposes Figure 19(a). Initially, the PIO technique suffered from the disadvantage of amplifying measurement noise which was due to the usage of high gains in a manner analogous to a Luenberger observer. This problem was solved by

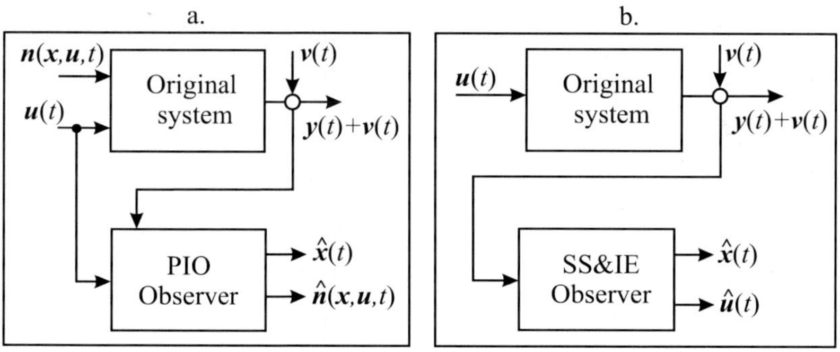

Figure 19: Input reconstruction principles with observer techniques (a) proportional integral observer (PIO) and (b) simultaneously state and input estimator (SS&IE)

(Krishna and Pousga, 2001) and the method was then adopted by (Soeffker and Krajcin, 2003) for contact force estimation as well as for crack detection in rotating machines. The modified PIO (MPIO) is equipped with an additional integrator which enables the attenuation of measurement noise. The gain matrices can be found by any conventional pole placement procedure. Nevertheless, the locations of nonlinearities/disturbances should be known prior to the observer design.

Simultaneous state and input estimator (SS&IE). (Ha et al., 2004; Ha and Trinh, 2004) established a new technique for Simultaneous State and Input Estimation (SS&IE), Figure 19(b), for a class of nonlinear systems for control purposes. In their work, the system nonlinearity is assumed to consist of known and unknown parts. To calculate the system observer, a general state space representation of the system is transformed in such way that the states are combined together with inputs. Then, the observer structure which includes only known nonlinearity is designed with the help of a Linear Matrix Inequality (LMI) algorithm. There are still two prerequisite conditions to be fulfilled in order to assure that such an observer exists. Among them is the continuity of the known nonlinear field (the known nonlinear function considered to be Lipschitz in its first argument) and that the number of sensors should be at least equal to or greater than the number of unknown inputs plus the number of nonlinear terms that do not satisfy the Lipschitz conditions.

Numerous attempts have been made by the researchers to employ the Kalman filter theory for simultaneous input and state estimation. Among them (Liu et al., 2000; Ma and Lin, 2000; Ma et al., 2003; Ma and Ho, 2004) have adopted the earlier work of (Tuan et al., 1996), who used the Kalman filter together with a recursive least squares algorithm, for the load estimation in both linear and nonlinear systems. The Kalman filter has been used initially to generate the residual innovation sequence by observation vector. Then a recursive least-squares algorithm computes the onset time histories of the excitation forces by utilizing the residual innovation sequence. (Deng and Heh, 2006) used the same approach for reconstruction of the loads distributed along the beam structure. In the recent work (Pan et al., 2010) has proposed to use the Kalman Filter with Unknown Inputs (KF-UI) for the estimation of the joint inputs and states. For this purpose Pan first combined the states and unknown inputs into an extended state vector. Then the weighted least squares problem has been solved for the extended state vector recursively. In comparison to other authors, who used always either displacement or velocity as measurement quantities, Pan's approach utilizes the more general measurement equation, which allows the usage of accelerometer sensors in estimation procedure.

5.2 Design of Simultaneous state and input estimator

The Simultaneous State and Input Estimation (SS&IE) method is based on the observer principle shown in Figure 20, proposed by (Ha and Trinh, 2004), and has been originally used in the field of control engineering. This approach is robust to noisy measurements and applicable for both linear and nonlinear systems. Generally, it can be categorized among deterministic algorithms, which use a mathematical model as a basis for observer calculation.

Initial Model Definition. The problem of observer construction can be stated as follows. Consider a dynamic system with n_q degrees of freedom, described by the second order differential equation of motion as:

$$\boldsymbol{M}_n\ddot{\boldsymbol{q}}(t) + \boldsymbol{D}_n\dot{\boldsymbol{q}}(t) + \boldsymbol{K}_n\boldsymbol{q}(t) + \boldsymbol{g}\left(\boldsymbol{q}(t),\, \dot{\boldsymbol{q}}(t),\, \boldsymbol{u}(t)\right) = \boldsymbol{B}_0\boldsymbol{u}(t) \tag{58}$$

here, $\boldsymbol{q}(t) \in R^{n_q}$ and $\boldsymbol{u}(t) \in R^m$ are generalized coordinates (in this case nodal) and inputs accordingly. $\boldsymbol{M}_n$, $\boldsymbol{D}_n$, $\boldsymbol{K}_n$ and $\boldsymbol{B}_o$ are the mass, damping, stiffness and force allocation matrices of appropriate dimensions. $\boldsymbol{g}\left(\boldsymbol{q}(t),\, \dot{\boldsymbol{q}}(t),\, \boldsymbol{u}(t)\right)$ is added to take into account nonlinear dependencies of the displacement, velocities and inputs. Eq. (58) can be transformed into state space notation with the following structure:

$$\begin{aligned} \dot{\boldsymbol{x}}(t) &= \boldsymbol{A}\boldsymbol{x}(t) + \boldsymbol{B}\boldsymbol{u}(t) + \boldsymbol{f}\left(\left(\boldsymbol{x}(t),\, \boldsymbol{u}(t)\right),\, \boldsymbol{y}(t)\right) \\ \boldsymbol{y}(t) &= \boldsymbol{C}\boldsymbol{x}(t) + \boldsymbol{D}\boldsymbol{u}(t) \end{aligned} \tag{59}$$

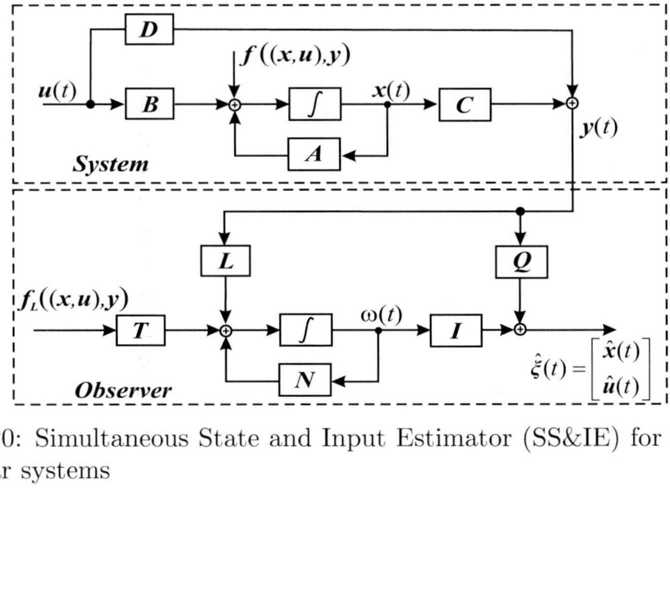

Figure 20: Simultaneous State and Input Estimator (SS&IE) for a class of nonlinear systems

where, $\boldsymbol{x}(t) \in R^n$ (with $n = 2n_q$), $\boldsymbol{u}(t) \in R^m$ and $\boldsymbol{y}(t) \in R^r$ are the states, unknown inputs and measured outputs, respectively.$\boldsymbol{A}_{n\times n}$, $\boldsymbol{B}_{n\times m}$, $\boldsymbol{C}_{r\times n}$ and $\boldsymbol{D}_{r\times m}$ are real constants of appropriate dimensions. $\boldsymbol{f}(\circ)$is a real non-linear vector function on R^n (in the case of a nonlinear system) which combines two portions as follows

$$\boldsymbol{f}\left((\boldsymbol{x}(t), \boldsymbol{u}(t)), \boldsymbol{y}(t)\right) = \boldsymbol{f}_L\left((\boldsymbol{x}(t), \boldsymbol{u}(t)), \boldsymbol{y}(t)\right) + \boldsymbol{W}\boldsymbol{f}_U\left((\boldsymbol{x}(t), \boldsymbol{u}(t)), \boldsymbol{y}(t)\right) \tag{60}$$

where $\boldsymbol{f}_L(\circ) \in R^n$ and $\boldsymbol{f}_U(\circ) \in R^d$ are known Lipschitz and unknown nonlinear vectors, respectively. The matrix $\boldsymbol{W}_{n\times d}$ is real and assumed to have full column rank d. The objective is to design an asymptotic observer to estimate the unknown states $\boldsymbol{x}(t)$ and unknown inputs $\boldsymbol{u}(t)$ from the measured output signals $\boldsymbol{y}(t)$. Therefore, a new state variable is introduced:

$$\boldsymbol{\xi}(t) = \begin{bmatrix} \boldsymbol{x}(t) \\ \boldsymbol{u}(t) \end{bmatrix}, \qquad \boldsymbol{\xi}(t) \in R^{n+m} \tag{61}$$

and matrices $\boldsymbol{E}$, $\boldsymbol{M}$ and $\boldsymbol{H}$ to combine the previous state space matrices:

$$\boldsymbol{E}_{n\times(n+m)} = \begin{bmatrix} \boldsymbol{I}_n & \boldsymbol{0}_{n\times m} \end{bmatrix} \tag{62}$$

$$\boldsymbol{M}_{n\times(n+m)} = \begin{bmatrix} \boldsymbol{A} & \boldsymbol{B} \end{bmatrix} \tag{63}$$

$$\boldsymbol{H}_{r\times(n+m)} = \begin{bmatrix} \boldsymbol{C} & \boldsymbol{D} \end{bmatrix} \tag{64}$$

Using the new state variable, the system is transformed into the algebraic differential equation:

$$\boldsymbol{E}\dot{\boldsymbol{\xi}}(t) = \boldsymbol{M}\boldsymbol{\xi}(t) + \boldsymbol{f}_L\left((\boldsymbol{x}(t), \boldsymbol{u}(t)), \boldsymbol{y}(t)\right) + \boldsymbol{W}\boldsymbol{f}_U\left((\boldsymbol{x}(t), \boldsymbol{u}(t)), \boldsymbol{y}(t)\right) \tag{65}$$

$$\boldsymbol{y}(t) = \boldsymbol{H}\boldsymbol{\xi}(t)\,. \tag{66}$$

The problem of the state and input estimator design for system Eq. (59) is now transformed into designing the observer for the generalized system in Eqs. (65), (66), such that the estimates converge asymptotically to the true states. For this purpose, a general minimal order observer structure is used with an additional nonlinear term:

$$\dot{\boldsymbol{\omega}}(t) = \boldsymbol{N}\boldsymbol{\omega}(t) + \boldsymbol{L}\boldsymbol{y}(t) + \boldsymbol{T}\boldsymbol{f}_L\left(\hat{\boldsymbol{\xi}}(t), \boldsymbol{y}(t)\right) \tag{67}$$

$$\hat{\boldsymbol{\xi}}(t) = \boldsymbol{\omega}(t) + \boldsymbol{Q}\boldsymbol{y}(t)\,, \tag{68}$$

where $\hat{\boldsymbol{\xi}}(t)$ represents the state estimation of $\boldsymbol{\xi}(t)$, $\boldsymbol{\omega}(t) \in R^{n+m}$ is the observer states vector, and matrices $\boldsymbol{N}$, $\boldsymbol{L}$, $\boldsymbol{T}$ and $\boldsymbol{Q}$ are of appropriate dimension. As input the observer uses measured signals from the original system, Eq. (65),(66), and the known nonlinear function in its dynamic equation. In addition, measurements $\boldsymbol{y}(t)$ are incorporated into the output observer Eq. (68) as a correction term.

Observer design procedure. The observer matrices $\boldsymbol{N}$, $\boldsymbol{L}$, $\boldsymbol{T}$ and $\boldsymbol{Q}$ need to be determined. To this end, a state estimation error is constructed first:

$$\boldsymbol{e}(t) = \hat{\boldsymbol{\xi}}(t) - \boldsymbol{\xi}(t)\,. \tag{69}$$

Using Eqs. (68) and (66), the error becomes

$$\boldsymbol{e}(t) = \boldsymbol{\omega}(t) + (\boldsymbol{QH} - \boldsymbol{I}_{n+m})\,\xi(t) \tag{70}$$

introducing the following relation

$$\boldsymbol{TE} + \boldsymbol{QH} = \boldsymbol{I}_{n+m} \tag{71}$$

the state error is transformed into

$$\boldsymbol{e}(t) = \boldsymbol{\omega}(t) - \boldsymbol{TE}\xi(t). \tag{72}$$

Taking the first derivative with respect to time and substituting expressions from Eqs. (65), (67) for $\boldsymbol{E}\dot{\boldsymbol{\xi}}(t)$ and $\dot{\boldsymbol{\omega}}(t)$ respectively, the error dynamic is obtained as:

$$\begin{aligned}\dot{\boldsymbol{e}}(t) = \boldsymbol{Ne}(t) + [\boldsymbol{NTE} + \boldsymbol{LH} - \boldsymbol{TM}]\,\boldsymbol{\xi}(t) - \boldsymbol{TW}\boldsymbol{f}_U\left(\boldsymbol{\xi}(t), \boldsymbol{y}(t)\right) \\ + \boldsymbol{T}\left[\boldsymbol{f}_L\left(\underbrace{(\boldsymbol{e}(t) + \boldsymbol{\xi}(t))}_{\hat{\boldsymbol{\xi}}(t)}, \boldsymbol{y}(t)\right) - \boldsymbol{f}_L\left(\boldsymbol{\xi}(t), \boldsymbol{y}(t)\right)\right].\end{aligned} \tag{73}$$

According to Eq. (73) the time dependant error converges asymptotically to zero if the following conditions hold:

$$\boldsymbol{TE} + \boldsymbol{QH} = \boldsymbol{I}_{n+m} \tag{74}$$

$$\boldsymbol{TW} = 0 \tag{75}$$

$$\boldsymbol{NTE} + \boldsymbol{LH} - \boldsymbol{TM} = 0\,. \tag{76}$$

Thus upon the satisfaction of conditions (74 - 76) the error dynamic expression (73) is reduced to:

$$\dot{\boldsymbol{e}}(t) = \boldsymbol{Ne}(t) + \boldsymbol{T}\left[\boldsymbol{f}_L\left(\hat{\boldsymbol{\xi}}(t), \boldsymbol{y}(t)\right) - \boldsymbol{f}_L\left(\boldsymbol{\xi}(t), \boldsymbol{y}(t)\right)\right] \tag{77}$$

which would be asymptotically stable if the matrix $\boldsymbol{N}$ is Hurwitz stable (the real part of all eigenvalues must be negative) and the following assumption holds:

Assumption 1. $\boldsymbol{f}_L(\circ)$ is assumed to satisfy the Lipschitz condition with a Lipschitz constant γ so that:

$$\left\| \boldsymbol{f}_L\left(\boldsymbol{\xi}(t), \boldsymbol{y}(t)\right) - \boldsymbol{f}_L\left(\hat{\boldsymbol{\xi}}(t), \boldsymbol{y}(t)\right) \right\| \leq \gamma \left\| \boldsymbol{\xi}(t) - \hat{\boldsymbol{\xi}}(t) \right\|, \quad \forall \boldsymbol{y}(t), \tag{78}$$

where, $\|\circ\|$ represents the norm and γ is a positive scalar referred later as Lipschitz constant. Combining Eqs. (74), (75) into a general one leads to:

$$\text{Condition 1} \quad \begin{cases} \boldsymbol{TE} + \boldsymbol{QH} = \boldsymbol{I}_{n+m} \\ \boldsymbol{TW} = \boldsymbol{0} \end{cases}. \tag{79}$$

In addition Eq. (76) can be partitioned into two parts by $\boldsymbol{TE} = \boldsymbol{I}_{n+m} - \boldsymbol{QH}$:

$$\boldsymbol{N} - [\boldsymbol{L} - \boldsymbol{NQ}]\,\boldsymbol{H} - \boldsymbol{TM} = \boldsymbol{0}. \tag{80}$$

Introducing a new matrix $\boldsymbol{F} = \boldsymbol{L} - \boldsymbol{NQ}$, expression (80) can be written in terms of the general condition:

$$\text{Condition 2} \quad \begin{cases} \boldsymbol{N} = \boldsymbol{TM} - \boldsymbol{FH} \\ \boldsymbol{F} = \boldsymbol{L} - \boldsymbol{NQ} \end{cases}. \tag{81}$$

Computation of the observer matrices $\boldsymbol{T}$ and $\boldsymbol{Q}$. Condition 1 also can be written in a more compact form:

$$\begin{bmatrix} \boldsymbol{T} & \boldsymbol{Q} \end{bmatrix} \boldsymbol{S} = \begin{bmatrix} \boldsymbol{I}_{n+m} & \boldsymbol{0} \end{bmatrix} \quad \text{where} \quad \boldsymbol{S} = \begin{bmatrix} \boldsymbol{E}_{n\times(n+m)} & \boldsymbol{W}_{n\times d} \\ \boldsymbol{H}_{r\times(n+m)} & \boldsymbol{0}_{r\times d} \end{bmatrix}, \tag{82}$$

which can be directly used for calculation of the observer matrices $\boldsymbol{T}$ and $\boldsymbol{Q}$. The matrix dimensions are added as subscripts for clarity. It is easy to show that the transpose of Eq. (82):

$$\boldsymbol{S}^T \begin{bmatrix} \boldsymbol{T}^T \\ \boldsymbol{Q}^T \end{bmatrix} = \begin{bmatrix} \boldsymbol{I}_{n+m} \\ \boldsymbol{0} \end{bmatrix} \tag{83}$$

is equivalent to the normal linear system of the form $\boldsymbol{AX} = \boldsymbol{B}$, where $\boldsymbol{X}$ is an unknown variable matrix, $\boldsymbol{A}$ and $\boldsymbol{B}$ are known matrices. The only condition for the solution existence is that the system in Eq. (83) should be a consistent[1] system , *i.e.* $\text{rank}(\boldsymbol{A}) = \text{rank}\left(\begin{bmatrix} \boldsymbol{A} & \boldsymbol{B} \end{bmatrix}\right)$ or, equivalently:

$$\text{rank}\left(\boldsymbol{S}^T\right) = \text{rank}\left(\begin{bmatrix} \boldsymbol{S}^T & \begin{matrix} \boldsymbol{I}_{n+m} \\ \boldsymbol{0} \end{matrix} \end{bmatrix}\right). \tag{84}$$

[1] If the system has at least one solution it is called a consistent system

Expression (84) can be rewritten as, see (Klinkov, 2011):

$$\operatorname{rank}(\boldsymbol{S}) = n + \operatorname{rank}\left(\begin{bmatrix}\boldsymbol{D}_{r\times m} & -\boldsymbol{C}_{r\times n}\boldsymbol{W}_{n\times d}\end{bmatrix}\right). \tag{85}$$

Taking into account the definition of consistency (i.e. a solution for $\boldsymbol{T}$ and Q) which requires that the rank of matrix $\boldsymbol{S}$ be equal to $n+m+d$. This is only true for Eq. (85) if the following assumption is held:

Assumption 2. Matrix $\begin{bmatrix}\boldsymbol{D} & \boldsymbol{C}\boldsymbol{W}\end{bmatrix}$ has full column rank.

$$\operatorname{rank}\left(\begin{bmatrix}\boldsymbol{D} & \boldsymbol{C}\boldsymbol{W}\end{bmatrix}_{r\times(m+d)}\right) = m + d. \tag{86}$$

Corollary, this assumption requires that the number of outputs should be at least equal to the number of unknown inputs plus the size of nonlinear terms $\boldsymbol{f}_U$ which do not satisfy the Lipschitz condition: $\operatorname{rank}(r\times(m+d)) = min(r, m+d) \Rightarrow r \geq m+d$. In addition, the $\boldsymbol{D}$ matrix should not be zero or empty to fulfill this condition. That makes the usage of acceleration sensors obligatory in this method. If the equality in Eq. (84) is not held then the system is inconsistent (*i.e.* the measurements are not independent). However, it is still possible to find $\boldsymbol{S}^-$, a generalized inverse of $\boldsymbol{S}$, and use it for the computation of the general solution of matrices $\boldsymbol{T}$ and $\boldsymbol{Q}$. The solution for the consistent system is only a part of the overall class of general solutions which can be obtained according to (Rao and Mitra, 1971) so that Eq. 83 can be solved as:

$$\boldsymbol{T} = \begin{bmatrix}\boldsymbol{I}_{n+m} & 0\end{bmatrix}\boldsymbol{S}^- \begin{bmatrix}\boldsymbol{I}_n \\ \boldsymbol{0}\end{bmatrix} + \boldsymbol{Z}\left(\boldsymbol{I}_{n+r} - \boldsymbol{S}\boldsymbol{S}^-\right)\begin{bmatrix}\boldsymbol{I}_n \\ \boldsymbol{0}\end{bmatrix} \tag{87}$$

$$\boldsymbol{Q} = \begin{bmatrix}\boldsymbol{I}_{n+m} & 0\end{bmatrix}\boldsymbol{S}^- \begin{bmatrix}\boldsymbol{0} \\ \boldsymbol{I}_r\end{bmatrix} + \boldsymbol{Z}\left(\boldsymbol{I}_{n+r} - \boldsymbol{S}\boldsymbol{S}^-\right)\begin{bmatrix}\boldsymbol{0} \\ \boldsymbol{I}_r\end{bmatrix} \tag{88}$$

Combining some parts of Eqs. (87) and (88) into new matrices:

$$\boldsymbol{G}_{(n+r)\times n} = \left(\boldsymbol{I}_{n+r} - \boldsymbol{S}\boldsymbol{S}^-\right)\begin{bmatrix}\boldsymbol{I}_n \\ \boldsymbol{0}\end{bmatrix} \qquad \boldsymbol{J}_{(n+m)\times n} = \begin{bmatrix}\boldsymbol{I}_{n+m} & 0\end{bmatrix}\boldsymbol{S}^- \begin{bmatrix}\boldsymbol{I}_n \\ \boldsymbol{0}\end{bmatrix} \tag{89}$$

$$\boldsymbol{K}_{(n+r)\times r} = \left(\boldsymbol{I}_{n+r} - \boldsymbol{S}\boldsymbol{S}^-\right)\begin{bmatrix}\boldsymbol{0} \\ \boldsymbol{I}_r\end{bmatrix} \qquad \boldsymbol{V}_{(n+m)\times r} = \begin{bmatrix}\boldsymbol{I}_{n+m} & 0\end{bmatrix}\boldsymbol{S}^- \begin{bmatrix}\boldsymbol{0} \\ \boldsymbol{I}_r\end{bmatrix} \tag{90}$$

thus Eqs. (87) and (88) are now:

$$\boldsymbol{T} = \boldsymbol{J} + \boldsymbol{Z}\boldsymbol{G} \tag{91}$$

$$\boldsymbol{Q} = \boldsymbol{V} + \boldsymbol{Z}\boldsymbol{K}. \tag{92}$$

Computation of observer matrices $\boldsymbol{N}$ and $\boldsymbol{L}$. The first part of the observer has been determined now and can be used for the computation of the remaining observer matrices $\boldsymbol{N}$ and $\boldsymbol{L}$, where two methods can be applied. The first one assumes general nonlinear systems and uses an LMI algorithm (Boyd et al., 1994) to assure the asymptotic stability of the error dynamic system via a Lyapunov function. The second approach considers solely linear systems and uses the standard pole placement procedure (Niu et al., 2010).

$\boldsymbol{N}$ and $\boldsymbol{L}$ computation via LMI algorithm: Recalling the error dynamic Eq. (77) and substituting solution (91) for matrix $\boldsymbol{T}$ into the first part of condition 2, Eq. (81) of the error dynamics is:

$$\dot{e}(t) = \left(\bar{\boldsymbol{\Phi}} + \boldsymbol{Z}\bar{\boldsymbol{\Psi}} - \boldsymbol{FH}\right) e(t) + \\ + (\boldsymbol{J} + \boldsymbol{ZG}) \left[\boldsymbol{f}_L\left(\boldsymbol{e}(t) + \boldsymbol{\xi}(t), \boldsymbol{y}(t)\right) - \boldsymbol{f}_L\left(\boldsymbol{\xi}(t), \boldsymbol{y}(t)\right)\right] \quad (93)$$

where, $\boldsymbol{N}$ has been substituted by:

$$\boldsymbol{N} = \boldsymbol{TM} - \boldsymbol{FH} = \boldsymbol{JM} + \boldsymbol{ZGM} - \boldsymbol{FH} = \bar{\boldsymbol{\Phi}} + \boldsymbol{Z}\bar{\boldsymbol{\Psi}} - \boldsymbol{FH}. \quad (94)$$

Matrices $\bar{\boldsymbol{\Phi}} = \boldsymbol{JM}$ and $\bar{\boldsymbol{\Psi}} = \boldsymbol{GM}$ are introduced for the sake of simplicity. The error dynamics, Eq. (93), incorporates the nonlinear terms; therefore, one way to ensure its asymptotic convergence to zero is to use the Lyapunov stability method. To this end a Lyapunov candidate summarizing function is introduced:

$$\boldsymbol{V}\left(\boldsymbol{e}(t)\right) = \boldsymbol{e}^T(t)\boldsymbol{P}\boldsymbol{e}(t) \quad (95)$$

where $\boldsymbol{P}$ is a symmetric positive definite matrix. If the first derivative of $\boldsymbol{V}$ is smaller than zero for all time, then the nonlinear error function in Eq. (93) is stable and converges to zero. Taking a derivative of $\boldsymbol{V}$ with respect to time and substituting Eq. (93) for $\dot{\boldsymbol{e}}(t)$ leads to the following expression:

$$\dot{\boldsymbol{V}}(\boldsymbol{e}) = \boldsymbol{e}^T(t)\left(\boldsymbol{P}\bar{\boldsymbol{\Phi}} + \boldsymbol{PZ}\bar{\boldsymbol{\Psi}} - \boldsymbol{PFH} + \bar{\boldsymbol{\Phi}}^T\boldsymbol{P} + \bar{\boldsymbol{\Psi}}^T\boldsymbol{Z}^T\boldsymbol{P} - \boldsymbol{H}^T\boldsymbol{F}^T\boldsymbol{P}\right)e(t) \\ + \boldsymbol{e}^T(t)\left((\boldsymbol{PJ} + \boldsymbol{PZG})\widetilde{\boldsymbol{f}}_L + \widetilde{\boldsymbol{f}}_L^T\left(\boldsymbol{J}^T\boldsymbol{P} + \boldsymbol{G}^T\boldsymbol{Z}^T\boldsymbol{P}\right)\right)e(t) \quad (96)$$

where, $\boldsymbol{f}_L\left(\boldsymbol{e}(t) + \boldsymbol{\xi}(t), \boldsymbol{y}(t)\right) - \boldsymbol{f}_L\left(\boldsymbol{\xi}(t), \boldsymbol{y}(t)\right)$ is referred to as $\widetilde{\boldsymbol{f}}_L$ for simplicity. Using the Yong's matrix inequality $\boldsymbol{ZY}^T + \boldsymbol{YZ}^T \leq \kappa\boldsymbol{YY}^T + \frac{1}{\kappa}\boldsymbol{ZZ}^T$ (Zhan, 2002), where κ is a positive scalar, the second part of Eq. (96) is modified as:

$$\boldsymbol{e}^T(t)\left(\left(\boldsymbol{PJ}+\boldsymbol{PZG}\right)\widetilde{\boldsymbol{f}}_L+\widetilde{\boldsymbol{f}}_L^T\left(\boldsymbol{J}^T\boldsymbol{P}+\boldsymbol{G}^T\boldsymbol{Z}^T\boldsymbol{P}\right)\right)\boldsymbol{e}(t)\leq$$
$$\frac{1}{\kappa_1}\boldsymbol{e}^T(t)\boldsymbol{PJJ}^T\boldsymbol{Pe}(t)+\kappa_1\widetilde{\boldsymbol{f}}_L^T\widetilde{\boldsymbol{f}}_L+\frac{1}{\kappa_2}\boldsymbol{e}^T(t)\boldsymbol{PZGG}^T\boldsymbol{Z}^T\boldsymbol{Pe}(t)+\kappa_2\widetilde{\boldsymbol{f}}_L^T\widetilde{\boldsymbol{f}}_L. \tag{97}$$

Substitution of Eq. (97) into (96) subject to Assumption 1 gives (Ha and Trinh, 2004):

$$\dot{\boldsymbol{V}}(\boldsymbol{e})=\boldsymbol{e}^T(t)\boldsymbol{R}_{LMI}\boldsymbol{e}(t)+(\kappa_1+\kappa_2)\widetilde{\boldsymbol{f}}_L^T\widetilde{\boldsymbol{f}}_L\leq$$
$$\boldsymbol{e}^T(t)\left(\boldsymbol{R}_{LMI}+\gamma\left(\kappa_1+\kappa_2\right)\boldsymbol{I}\right)\boldsymbol{e}(t) \tag{98}$$

where

$$\boldsymbol{R}_{LMI}=\boldsymbol{P}\bar{\boldsymbol{\Phi}}+\bar{\boldsymbol{\Phi}}^T\boldsymbol{P}+\boldsymbol{PZ}\bar{\boldsymbol{\Psi}}+\bar{\boldsymbol{\Psi}}^T\boldsymbol{Z}^T\boldsymbol{P}-\boldsymbol{PFH}-\boldsymbol{H}^T\boldsymbol{F}^T\boldsymbol{P}$$
$$+\frac{1}{\kappa_1}\boldsymbol{PJJ}^T\boldsymbol{P}+\frac{1}{\kappa_2}\boldsymbol{PZGG}^T\boldsymbol{Z}^T\boldsymbol{P} \tag{99}$$

and γ is a Lipschitz constant from Assumption 1. Therefore, the estimation error $\boldsymbol{e}(t)\to 0$ as the time $t\to\infty$ if the derivative of Lyapunov function $\dot{\boldsymbol{V}}(\boldsymbol{e})<0$. Hence,

$$\left(\boldsymbol{R}_{LMI}+\gamma\left(\kappa_1+\kappa_2\right)\boldsymbol{I}\right)<0 \tag{100}$$

must hold. Eq. (100) is a Riccati inequality which should be solved in a computationally efficient manner. For the sake of simplicity new variables that combine constant matrices are introduced:

$$\boldsymbol{X}=\boldsymbol{PF}\qquad\qquad \boldsymbol{Y}=\boldsymbol{PZ}. \tag{101}$$

Riccati inequality (100) thus becomes:

$$\boldsymbol{P}\bar{\boldsymbol{\Phi}}+\bar{\boldsymbol{\Phi}}^T\boldsymbol{P}+\boldsymbol{Y}\bar{\boldsymbol{\Psi}}+\bar{\boldsymbol{\Psi}}^T\boldsymbol{Y}^T-\boldsymbol{XH}-\boldsymbol{H}^T\boldsymbol{X}^T+$$
$$+\frac{1}{\kappa_1}\boldsymbol{PJJ}^T\boldsymbol{P}+\frac{1}{\kappa_2}\boldsymbol{YGG}^T\boldsymbol{Y}^T+\gamma\left(\kappa_1+\kappa_2\right)\boldsymbol{I}<0 \tag{102}$$

Using the Schur complement (Zhang, 2005) Eq. (102) can be transformed into a more compact form of LMI:

$$\begin{bmatrix} \boldsymbol{P}\bar{\boldsymbol{\Phi}}+\bar{\boldsymbol{\Phi}}^T\boldsymbol{P}+\boldsymbol{Y}\bar{\boldsymbol{\Psi}}+\bar{\boldsymbol{\Psi}}^T\boldsymbol{Y}^T-\ldots & \boldsymbol{PJ} & \boldsymbol{YG} \\ \ldots-\boldsymbol{XH}-\boldsymbol{H}^T\boldsymbol{X}^T+\gamma\left(\kappa_1+\kappa_2\right)\boldsymbol{I} & & \\ \boldsymbol{J}^T\boldsymbol{P} & -\kappa_1\boldsymbol{I} & \boldsymbol{0} \\ \boldsymbol{G}^T\boldsymbol{Y}^T & \boldsymbol{0} & -\kappa_2\boldsymbol{I} \end{bmatrix}<0 \tag{103}$$

and solved numerically by the interior-point polynomial algorithm (Boyd et al., 1994). The estimation error $e(t) = \hat{\xi}(t) - \xi(t)$ will converge to zero if the matrices $\boldsymbol{P} = \boldsymbol{P}^T > 0$, $\boldsymbol{X}$, $\boldsymbol{Y}$ and positive scalars κ_1 and κ_2 exist such the LMI (103) is satisfied. Finally the observer matrices $\boldsymbol{N}$ and $\boldsymbol{L}$ can be calculated according to the following algorithm:

1. Provide a mathematical model in state space representation (either by analytical modeling or by an identification procedure).
2. Solve the LMI problem (103) (*e.g.* using the Matlab LMI toolbox) (Nesterov and Nemirovski, 1994).
3. If $\boldsymbol{P}$, $\boldsymbol{X}$, $\boldsymbol{Y}$, κ_1 and κ_2 satisfying the LMI (103) are found then go to step 4, otherwise choose a smaller value of γ.
4. Obtain the matrices $\boldsymbol{Z} = \boldsymbol{P}^{-1}\boldsymbol{Y}$ and $\boldsymbol{F} = \boldsymbol{P}^{-1}\boldsymbol{X}$. The observer matrices $\boldsymbol{N}$ and $\boldsymbol{L}$ are then given as in Eq.s (94) and (81):

$$\boldsymbol{N} = \bar{\boldsymbol{\Phi}} + \boldsymbol{Z}\bar{\boldsymbol{\Psi}} - \boldsymbol{F}\boldsymbol{H} \tag{104}$$

$$\boldsymbol{L} = \boldsymbol{F} + \boldsymbol{N}\boldsymbol{Q}. \tag{105}$$

If the matrix $\boldsymbol{S}$ in Eq. (87) and (88) is square of a full rank, the matrices $\boldsymbol{G}$ and $\boldsymbol{K}$ which represent the error after generalized inverse are equal to zero. Hence $\bar{\boldsymbol{\Psi}} = 0$ and one can choose $\boldsymbol{Z}$ and $\boldsymbol{Y}$ equal to zero. In this case the LMI (103) is reduced to:

$$\begin{bmatrix} \boldsymbol{P}\bar{\boldsymbol{\Phi}} + \bar{\boldsymbol{\Phi}}^T\boldsymbol{P} - \boldsymbol{X}\boldsymbol{H} - \boldsymbol{H}^T\boldsymbol{X}^T + \gamma^2\kappa_1\boldsymbol{I} & \boldsymbol{P}\boldsymbol{J} \\ \boldsymbol{J}^T\boldsymbol{P} & -\kappa_1\boldsymbol{I} \end{bmatrix} < 0. \tag{106}$$

The procedure of $\boldsymbol{N}$ and $\boldsymbol{L}$ computation stays the same as before except for minor modifications in Eq. (104) where $\boldsymbol{Z}\bar{\boldsymbol{\Psi}}$ is equal to zero.

$\boldsymbol{N}$ and $\boldsymbol{L}$ computation via pole placement: To ensure the stable dynamics of the error function in (77) subject to Assumption 1 (78) the matrix $\boldsymbol{N}$ should be Hurwitz stable or in other words. On closer observation, Eq. (94) can be reformulated as:

$$\underbrace{\boldsymbol{N}^T}_{\boldsymbol{P}} = \underbrace{\bar{\boldsymbol{\Phi}}^T}_{\boldsymbol{A}} - \underbrace{\begin{bmatrix} -\bar{\boldsymbol{\Psi}}^T & \boldsymbol{H}^T \end{bmatrix}}_{\boldsymbol{B}} \underbrace{\begin{bmatrix} \boldsymbol{Z}^T \\ \boldsymbol{F}^T \end{bmatrix}}_{\boldsymbol{K}} = \bar{\boldsymbol{\Phi}}^T - \boldsymbol{\Lambda}^T \begin{bmatrix} \boldsymbol{Z}^T \\ \boldsymbol{F}^T \end{bmatrix} \tag{107}$$

where, matrices $\bar{\boldsymbol{\Phi}}$, $\boldsymbol{H}$ and $\bar{\boldsymbol{\Psi}}$ are known matrices and $\boldsymbol{\Lambda}^T = \begin{bmatrix} -\bar{\boldsymbol{\Psi}}^T & \boldsymbol{H}^T \end{bmatrix}$, opening the similarity to a general closed loop state space system with a state feedback input as:

$$\dot{\boldsymbol{x}}(t) = \boldsymbol{A}\boldsymbol{x}(t) + \boldsymbol{B}\boldsymbol{u}(t) \qquad \text{with:} \qquad \boldsymbol{u}(t) = -\boldsymbol{K}\boldsymbol{x}(t) \tag{108}$$

$$\Rightarrow \dot{\boldsymbol{x}}(t) = (\boldsymbol{A} - \boldsymbol{BK})\,\boldsymbol{x}(t) \tag{109}$$

with a desired closed loop dynamics $(\boldsymbol{A} - \boldsymbol{BK})$ characterized by predefined closed loop poles $\boldsymbol{p}$ which can be achieved by proper choice of a feedback gain matrix $\boldsymbol{K}$ (*e.g.* by means of Ackermann's formula (Kailath, 1980)). Therefore, another possible way to compute observer matrices $\boldsymbol{N}$ and $\boldsymbol{L}$ is to use the standard pole placement procedure (Niu et al., 2010), in particular by assigning predefined negative eigenvalues to matrix $\boldsymbol{N}$ and successive calculation of matrices $\boldsymbol{Z}$ and $\boldsymbol{F}$ by means of a pole placement procedure (Kautsky et al., 1985) if the matrix pair $(\bar{\boldsymbol{\Phi}}^T, \boldsymbol{\Lambda})$ is controllable (*e.g.* the matrix pair $(\boldsymbol{\Lambda}, \bar{\boldsymbol{\Phi}})$ is observable). The observer matrix $\boldsymbol{L}$ is then computed as before with the help of Eq. (105).

Effect of the model inaccuracies and measurement noise. It is valid to mention that in the observer design the absence of measurement and process noise has been assumed. Nevertheless, the measurement and model uncertainties are always present in real world applications. Adding both process noise $\boldsymbol{w}(t)$ and measurement errors $\boldsymbol{v}(t)$ to the system in Eq.s (65) and (66) leads to:

$$\boldsymbol{E}\dot{\boldsymbol{\xi}}(t) = \boldsymbol{M}\boldsymbol{\xi}(t) + \boldsymbol{f}_L\left(\boldsymbol{\xi}(t), \boldsymbol{y}(t)\right) + \boldsymbol{W}\boldsymbol{f}_U\left(\boldsymbol{\xi}(t), \boldsymbol{y}(t)\right) + \boldsymbol{w}(t) \tag{110}$$

$$\boldsymbol{y}(t) = \boldsymbol{H}\boldsymbol{\xi}(t) + \boldsymbol{v}(t). \tag{111}$$

Employing the previously defined observer model as in Eqs. (67) and (68), the error dynamic Eq. (77) under satisfaction of conditions 1 and 2 is of the following form:

$$\begin{aligned}\dot{\boldsymbol{e}}(t) = \boldsymbol{N}\boldsymbol{e}(t) + \boldsymbol{T}\left[\boldsymbol{f}_L\left(\hat{\boldsymbol{\xi}}(t), \boldsymbol{y}(t)\right) - \boldsymbol{f}_L\left(\boldsymbol{\xi}(t), \boldsymbol{y}(t)\right)\right] + \\ + (\boldsymbol{L} - \boldsymbol{NQ})\,\boldsymbol{v}(t) + \boldsymbol{Q}\dot{\boldsymbol{v}}(t) + \boldsymbol{T}\boldsymbol{w}(t).\end{aligned} \tag{112}$$

Examination of the error dynamic equation shows that both inaccuracies are amplified by the observer matrices. Consequently, they will appear as a reconstruction error in the estimated variables or even lead to complete observer failure if the model and measurement noise are prevailing. This effect will be apparent when shown in the examples section.

5.3 Application Examples

The wind energy plant M5000-2. The Areva M5000-2 is a 5 MW wind turbine (Areva, 2012) with a tripod tower structure as shown in Figure 21(a). The equipment installation and data collection for the vibration tests were carried out by the Federal Institute for Material Research and

Testing, Germany. 167 sensors were installed on the M5000-2 onshore WEP in Bremerhaven. The external load reconstruction with the SS/IE method was performed offline as a sub-part of the overall SHM package for future offshore WEPs. For the observer design, an updated FE model was used and two sensor types (two 2-D accelerometers and two strain gauges) were needed, see Figure 21(a). Load estimation was then carried out for the section where the nacelle is connected to the tower. This section was chosen intentionally and included all possible forces that the tower should subtend (external wind forces, static nacelle orientation, blades dynamics, mass unbalance, etc.).

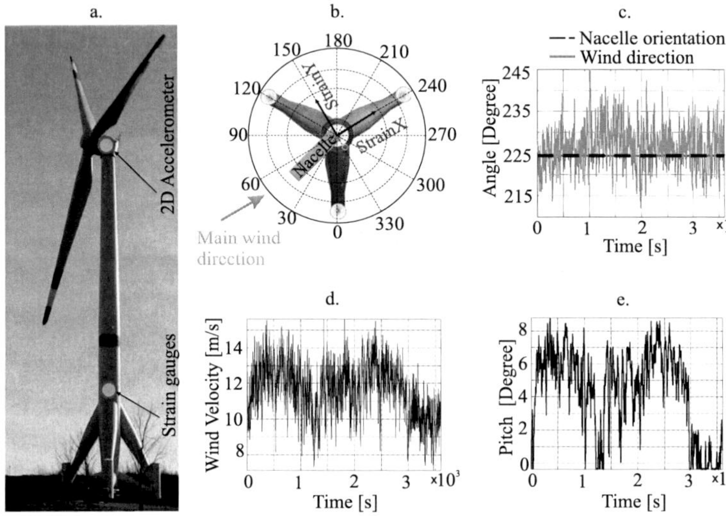

Figure 21: Operational state of wind turbine M5000-2 in Bremerhaven: (a) overall structure with schematic sensor positions; (b) and (c) nacelle orientation with respect to the wind together with strain gauges coordinates; (d) and (e) wind velocity and pitch angle of the rotor blades

In Figure 21(c-e), one of the working states of the WEP M5000-2, together with the operational conditions (wind velocity, pitch angle, etc.), is illustrated. In Figure 21(b), the WEP is shown from the top, with the thick line (under a direction of ca. 50°) representing the nacelle with the rotor. Two strain sensors are fixed to the tower under the 150°/330°-direction and have their own orientation in space which is displayed on the same subplot. The 2D-acceleration sensor is located in the nacelle which is always changing its orientation in space. Because the observer is based on a modal state space model which has fixed orientation, the measurements from the moving nacelle must be transformed.

The above operational conditions are very close to the ideal ones for the WEP M5000-2, see data sheet (Areva, 2012): a) the wind direction is almost orthogonal to the rotor area, with a nearly constant wind speed of approximately 12 m/s; b) the blades have a relatively small pitch angle. Such conditions allow using the Betz theory to validate the estimated load. According to Betz, the wind thrust load applied to the rotor area can be calculated as (Gasch and Twele, 2007):

$$F_{Betz} = \frac{1}{2} c \rho V_{wind}^2 A \tag{113}$$

where, c is an aerodynamic thrust coefficient V_{wind} the wind speed, ρ the air density and A the rotor area. The wind is assumed to blow orthogonal to the rotor area. The estimated and Betz forces are depicted in Figure 22(c) together with recorded measurements in Figure 22(a-b) for a time duration of 25 minutes and the above mentioned operational conditions.

Figure 22(c) shows a certain difference between estimated and Betz thrust force especially in the region where the wind velocity is increasing. This is due to the Betz theory as in Eq. (113), where the Betz thrust force follows the wind velocity behavior for a fixed coefficient c without accounting for pitch angles (see for a comparison the wind velocity in Figure 22(d) for the time interval 0-1500 seconds; the wind velocity increases and decreases as time evolves). As a result, the Betz force alone cannot be used for the validation; instead, one also needs to take the pitch angle of the rotor blades into consideration. A closer reading of Figure 22(c) shows that the estimated force is very close to the Betz force in the regions where the pitch angle is close to zero ($\approx$100 and $\approx$1300 seconds) and decreases as the rotor blades are pitched (see pitch and wind velocity in Figure 21(d-e) for the time interval 0-1500 seconds). A magnification of Figure 22(c) demonstrates that the estimated force has slower dynamics similar to the Betz thrust force, but also includes higher frequencies which are obviously the components of all interaction forces between the tower, nacelle and rotor blades. Observations of the estimated load spectrum in Figure 23(a) show that the external load lies in a low frequency range of up to 25 Hz. Moreover, a closer reading of the spectrum in Figure 23(b) demonstrates that the stall-effect (blade passage) and its multiples play a very important role in the dynamic excitation of the WEP. The rotor speed (Ω), blade passage frequencies (3Ω) etc. are plotted with dashed lines on both plots. The solid lines represent the first and the second natural frequency of the tower which are also present in the estimated load spectra. Moreover the identified spec-

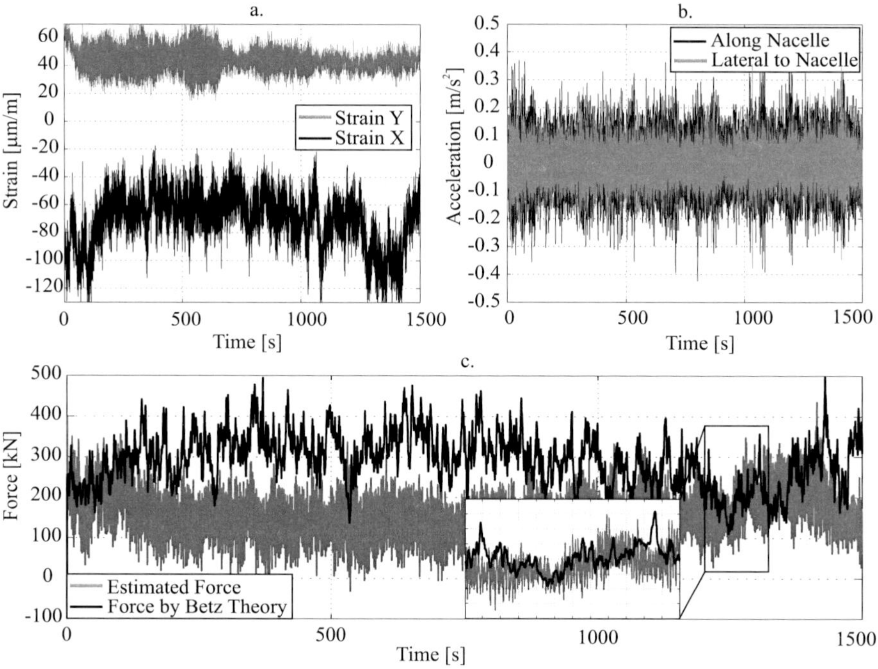

Figure 22: Estimated external force for M5000-2

trum of the wind velocity has a typical logarithmic decay and is shown in Figure 24. A comparison of the plots in Figure 23 and Figure 24 reveals the important fact that the load withstood by the WEP tower does not consist solely of the wind load but rather includes loads/forces that result from interactions of other components of the WEP. By means of the estimated force it is possible to calculate e.g. shear forces bending moments or stresses at any location of interest. Further investigations of the observer properties should be made together with a validation of the obtained result. Additionally, external moments that play a central role in the life expectancy and fatigue prediction of the tower structure should be estimated and validated in the same way as the external forces. Observations of the estimated load led to a conclusion that the WEP is experiencing a complete bench of forces that are arising throughout the WEP operation and need to be considered for a life expectancy prognosis. Among them are loads from pitching, yaw control, stalling and starting of the rotor.

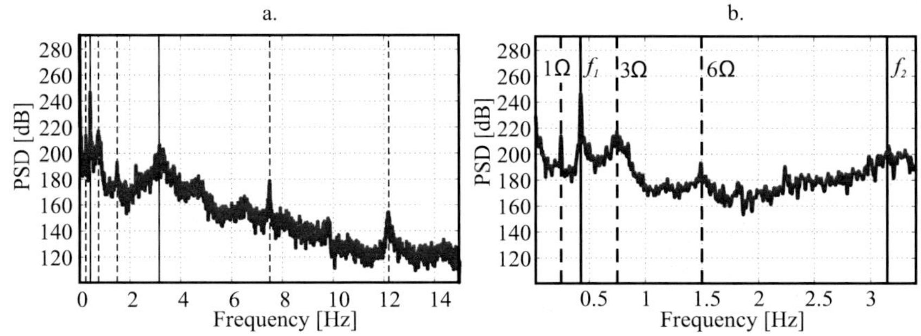

Figure 23: Estimated load spectrum from 0 to 15 Hz range (left); 0 to 3.5 Hz range (right)

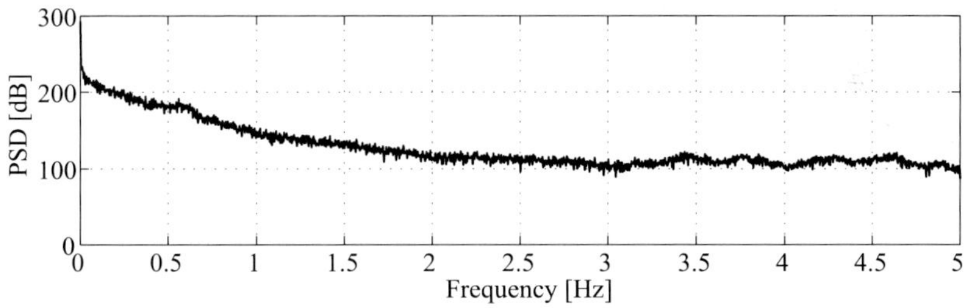

Figure 24: Wind velocity spectrum

Non-linear laboratory structure. A two storey laboratory structure with non-linear behavior, shown in Figure 25(a) and (b) was used to validate the SS&IE performance. The non-linearity is introduced by the addition of a stopper element to the supporting frame which constrains the movement of the lower mass in the positive x-direction, as shown in Figure 25. The structure without shaker in Figure 25(b) has been used as a basis for a simplified mathematical model, with the assumption that the system exhibited only planar motion in the z-x-plane). The nonlinear impact effect was taken into the model as an unknown external force which should be estimated together with the load applied from the shaker by the SS&IE algorithm. The stopper tip was fitted with a quartz load cell to register impacts in order to validate estimated nonlinearity. The nonlinearity was controlled by the distance b between the quartz sensor tip and m_1. The data for the observers has been acquired by two accelerometers and two strain gauges. The measured data were collected and fed online to the observers with the help of a dSpace controller board (DS1104). For structural excitation, a periodic load was generated by means of a shaker in the frequency range where the difference between the original structure which was connected to the shaker and the mathematical model which was used for the observer design was not large. Figure 26 shows an exemplary result of the load reconstruction for

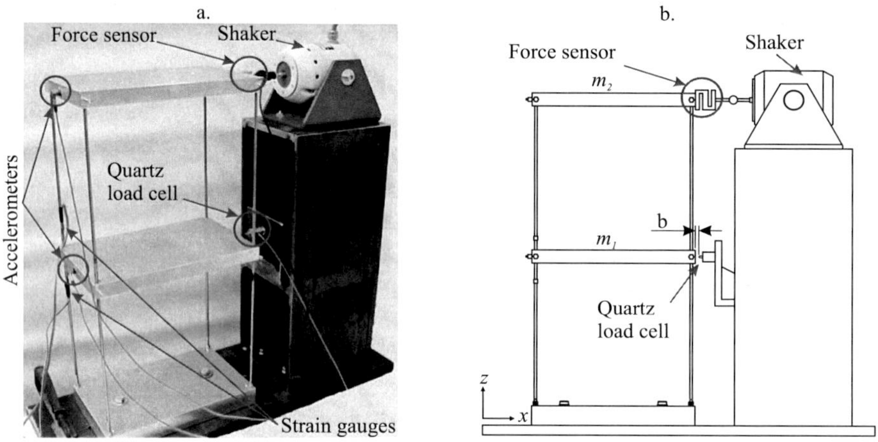

Figure 25: (a) A two storey lumped mass nonlinear structure, and (b) side view of same

a 7 Hz square wave excitation applied to the upper mass m_2. The distance b was held at 2 mm so that the lower mass impacted the holder as it was moving in the positive x-direction. Both upper and lower forces were reconstructed with a certain error; see zoomed parts of Figure 26. It can be seen that the contact forces have been reconstructed quite well. The analysis of the frequency content in the error signal led to the error source that due to model simplifications the higher modes were not taken into account. The above error occurs only if the impact duration is fast enough to excite the structure up to this frequency range and can be removed easily by the use of a more complete model for the observer design.

6 Conclusions

In this chapter different aspects of vibration-based methods have been presented. After a general discussion how to express the dynamics in terms of damage and environmental parameters, it appears to be necessary to split the environmental and operational influences from the damage effects for damage detection, because their influence might be of the same order of magnitude. The understanding of these environmental and the excitation

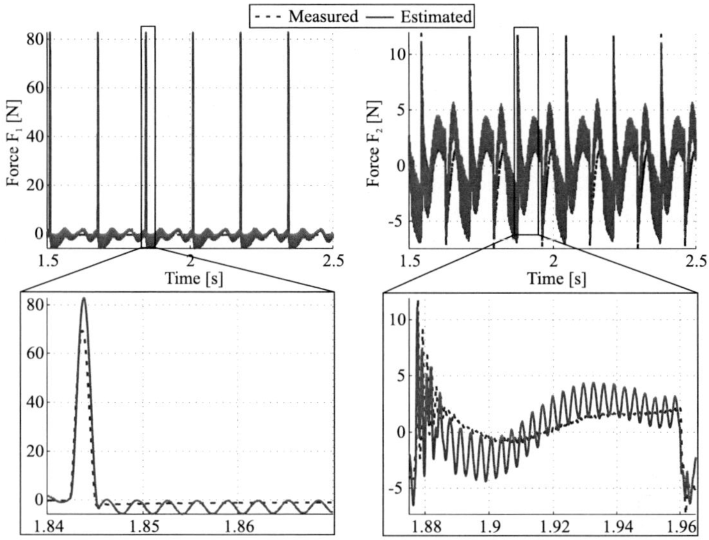

Figure 26: Load reconstruction on nonlinear system, square excitation $f_{ex} = 7Hz$ on m_2

sources of the structures is of big importance. In this context a long time monitoring of the dynamical properties over years has to be accompanied by proper approaches to separate the operational and environmental effects from the extracted features. Therefore, we need baseline measurements for the different combinations of the environmental and operational parameters. In a training phase when the structure is new and healthy, we collect data sets of different environmental and operational states and use them to find a reasonable number of representative classes. For these classes we calculate class centers to which we finally concentrate our knowledge of baseline states. New incoming data sets are then assigned to these classes using a weight which depends on the distance of the data set from the class center. Here, the NSFD method in combination with the classification techniques is proposed and validated using experimental and simulated vibration data. Further investigations should be concentrated on the probability of detection in dependency of the noise level on the signals and different number of reference states (class centers). It would be also interesting to apply the mentioned classification technique on different damage features e.g. modal parameters, statistical properties of signals, model or regression coefficients, etc. A further main topic of this chapter was the reconstruction of external loads acting on the structure, which is another kind of an inverse problem.

The solution of the problem is very important to update the lifetime consumption within damage prognosis. Having determined the external forces, it is possible to reconstruct also e.g. normal and shear forces, bending moments, local stresses or strains via a detailed finite element model. A comprehensive literature review was presented showing different concepts for the force identification. Especially, a combined state and input observer has been investigated which allows on-line reconstruction of the forces. This observer works for nonlinear structures, too. In the case that the structure is linear, it has been shown that the algorithm fades into a pole placement problem. Furthermore, it is desired that the method is robust against noise and tolerates modeling errors to a certain extent. The application to a real 5MW wind energy turbine and a laboratory structure with non-linear contact have shown that it is possible to reconstruct the unknown forces very well.

Bibliography

R. Adams and J.F. Doyle. Multiple force identification for complex structures. *Exp. Mech.*, 42(1):25–36, 2002.

Y. Aimin, J.-C. Golinval, and F. Marin. Fault detection algorithm based on null-space analysis for on-line structural health monitoring. In *Proc. 23th IMAC* [2] *, Orlando, FL, on CD-ROM*, 2005.

Areva. M5000 technical data, available from www.areva-wind.com, 2012.

D. Balageas, C.-P. Fritzen, and A. Guemes. *Structural Health Monitoring*. Hermes Science Publishing, London UK, 2006.

B. Balasko, J. Abonyi, and B. Feil. *Fuzzy Clustering and Data Analysis Toolbox*. www.fmt.vein.hu/softcomp, 2005.

M. Basseville, M. Abdelghani, and A. Benveniste. Subspace-based fault detection algorithms for vibration monitoring. *Automatica*, 36:101–109, 2000.

M. Basseville, A. Benveniste, M. Goursat, L. Hermans, L. Mevel, and H. van der Auweraer. Output-only modal subspace-based structural identification: From theory to industrial testing practice. *J. Dyn. Syst. Meas. Contr.*, 123:668–676, 2001.

M. Basseville, F. Bourquin, L. Mevel, H. Nasserand, and F. Treyssede. Merging sensor data from multiple temperature scenarios for vibration-based monitoring of civil structures. In *Proc. 3rd EWSHM* [3], pages 759–766, Granada, Spain, 2006. DEStech Publ., Lancaster, PA.

[2]IMAC: Int. Modal Analysis Conference

[3]EWSHM: European Workshop on Structural Health Monitoring

G. Bir. Multiblade coordinate transformation and its application to wind turbine analysis. *Wind Energy Symposium, Reno, Nevada*, 2008.

K. Bohle. *Sensitivitaetsbasierte Methoden zur modellgestuetzten Schadendiagnose mit Modaldaten.* PhD thesis (in German), University of Siegen, 2005.

S. Boyd, L. El Ghaoui, E. Feron, and V. Balakrishnan. *Linear Matrix Inequalities in Systems and Control Theory.* Philadelphia, 1994.

T.G. Carne, V.I. Bateman, and R.L. Mayes. Force reconstruction using a sum of weighted accelerometers technique. In *Proc. of 10th IMAC*, pages 291–298, San Diego, California, USA, 1992.

P. Cawley and R.D. Adams. The location of defects in structures from measurements of natural frequencies. *J. Strain Analysis*, 14:49–57, 1979.

M. Corless and J. Tu. State and input estimation for a class of uncertain systems. *Automatica*, 34(6):754–764, 1998.

S. Deng and T-Y. Heh. The study of structural system dynamic problems by recursive estimation method. *The Int. Journal of Advanced Manufacturing Technology*, 30(3-4):195–202, 2006.

A. Deramaeker, E. Reynders, G. DeRoeck, and J. Kullaa. Vibration-based structural health monitoring using output-only measurements under changing environment. *Mech. Syst. Sig. Process.*, 22:34–56, 2008.

K.-U. Dettmann and D. Soeffker. Adaptive modeling of reliability properties for control and supervision purposes. *Int. J. Appl. Math. Comput. Sci.*, 21(3):479486, 2011.

S.W. Doebling, C.R. Farrar, M.B. Prime, and D.W. Shevitz. Damage identification and health monitoring of structural systems from changes in their vibration characteristics: A literature review. Technical report, Report LA-12767-MS, Los Alamos National Laboratory, NM, 1996.

J.F. Doyle. An experimental method for determining the dynamic contact law. *Exp. Mech.*, 24(4):265–270, 1989.

J.F. Doyle. A wavelet deconvolution method for impact force identification. *Exp. Mech.*, 37:404–408, 1997.

J.F. Doyle. Force identification from dynamic response of a biomaterial beam. *Exp. Mech.*, 33:64–69, 1999.

J.F. Doyle and L.E. Kannal. Combining spectral super-elements, genetic algorithms and massive parallelism for computationally efficient flaw detection in complex structures. *Comput. Mech.*, 20:67–74, 1997.

J.F. Doyle and M.T. Martin. Impact force identification from wave propagation responses. *Int. J. Impact Eng.*, 18(1):65–77, 1996.

K. Elliott, R. Buehrle, and G. James. Space shuttle transportation loads diagnostics. In *Proc. 23rd IMAC*, Orlando, Florida, 2005.

C.R. Farrar and N. Lieven. Damage prognosis: the future of structural health monitoring. *Phil. Trans. R. Soc. A*, 365:623–632, 2007.

C.R. Farrar and K. Worden. An introduction to structural health monitoring. *Phil. Trans. R. Soc. A*, 365:303–315, 2007.

C.R. Farrar, H. Sohn, F.M. Hemez, M.C. Anderson, M.T. Bement, P.J. Corwell, S.W. Doebling, J.F. Schultze, N. Lieven, and A.N. Robertson. Damage prognosis: Current status and future needs. Technical report, Los Alamos National Laboratory, Los Alamos, NM, 2003.

S.D. Fassois and J.S. Sakellariou. Time series methods for fault detection and identification in vibrating structures. *Phil. Trans. R. Soc., 365*, pages 411–448, 2007.

M.I. Friswell and J.E. Mottershead. *Finite Element Model Updating in Structural Dynamics*. Kluwer Academic Publishers, 1995.

C.-P. Fritzen and K. Bohle. Damage identification using a modal kinetic energy criterion and output-only modal data-application to the z-24 bridge. In *Proc. 2nd EWSHM*, pages 185–194. DEStech, Publ. Lancaster, PA, 2004.

C.-P. Fritzen and K. Bohle. Identification of damage in large scale structures by means of measured frfs-procedure and application to the 140-highway-bridge. *Key Eng. Mater.*, 167-168:310–319, 1999a.

C.-P. Fritzen and K. Bohle. Model-based health monitoring of structures-application to the i40-highway-bridge. In *Proc. 2nd Conf. Identification in Eng. Systems IES99, Swansea, UK*, pages 492–505, 1999b.

C.-P. Fritzen and K. Bohle. Application of model-based damage identification to a seismically loaded structure. *Smart Mater. Struct.*, 10:452458, 2001.

C.-P. Fritzen and K. Bohle. Global damage identification of the steelquake structure using modal data. *Mech. Syst. Sig. Process.*, 17(1):111–117, 2003.

C.-P. Fritzen and K. Bohle. Parameter selection strategies in model-based damage detection. In *Proc. 2nd IWSHM*, pages 901–911. Technomic Publishing., Lancaster, PA, USA, 1999c.

C.-P. Fritzen and P. Kraemer. Self-diagnosis of smart structures based on dynamical properties. *Mech. Syst. Sig. Process.*, 23(6):18301845, 2009.

C.-P. Fritzen and P. Kraemer. Vibration based damage detection for structures of offshore wind energy plants. In *Proc. 8th IWSHM* [4], volume 2, pages 1656–1663. DEStech, Publ. Lancaster, PA, 2011.

C.-P. Fritzen and G. Mengelkamp. A kalman filter approach to the detection of structural damage. In *Proc. 4th IWSHM*, pages 1275–1284. DEStech Publ., Lancaster, PA, 2003.

[4]IWSHM: Int. Workshop on Structural Health Monitoring, Stanford University, Stanford, CA

C.-P. Fritzen, D. Jennewein, and T. Kiefer. Damage detection based on model-updating methods,mech. syst. sig. process. *Mech. Syst. Sig. Process.*, 12(1):163–186, 1998.

C.-P. Fritzen, G. Mengelkamp, and A. Guemes. A cfrp plate with piezoelectric actuators and sensors as self-diagnostic intelligent structure. In *Proc. ISMA 2002, KU Leuven, Belgium*, pages 185–191, 2002.

C.-P. Fritzen, G. Mengelkamp, and A. Guemes. Elimination of temperature effects on damage detection within a smart structure concept. In *4th IWSHM*, pages 1530–1538. DEStech Publ., Lancaster, PA, 2003a.

C.-P. Fritzen, G. Mengelkamp, and A. Guemes. Elimination of temperature effects on damage detection within a smart structure concept. In *Proc. 4th IWSHM*, pages 1530–1538, 2003b.

C.-P. Fritzen, P. Kraemer, and K.J. Xing. Damage detection and localization using time domain residuals. In *World Forum on Smart Mat. and Smart Struct. Techn., Chongqing and Nanjing, P.R. China*, 2007.

R. Gasch and J. Twele. *Windkraftanlagen.* Teubner, Wiesbaden, Germany, 5 edition, 2007.

G.G. Genaro and D.A. Rade. Input force identification in time domain. In *Proc. of the 16th Int. Modal Analysis Conference*, pages 124–129, 1998.

J.J. Gertler. *Fault Detection and Diagnosis in Engineering Systems.* Marcel Dekker, Inc., New York, Basel, Hong Kong, 1998.

V. Giurgiutiu. *Structural Health Monitoring with Piezoelectric Wafer Active Sensors.* Elsevier, 2008.

V. Giurgiutiu and S.E. Lyshewski. *Micromechatronics.* CRC Press, Boca Raton, 2004.

S. Gopalakrishnan and J.F. Doyle. Wave propagation in connected waveguides of varying cross-section. *J. Sound Vibr.*, 175(3):347–363, 1994.

G. Goral, K. Bydon, and T. Uhl. Intelligent transducers of in-operational loads in construction fatigue monitoring. *Mach. Dyn. Problems*, 2-3: 73–78, 2002.

Q.P. Ha and H. Trinh. State and input simultaneous estimation for a class of nonlinear systems. *Automatica*, 40:1779–1785, 2004.

Q.P. Ha, A.D. Nguyen, and H. Trinh. Simultaneous state and input estimation with application to a two-link robotic system. In *Proc. ASCC of the 5th Asian Control Conference*, pages 322–328, 2004.

P.C. Hansen, G.H. Golub, and D.P. O'Leary. Tikhonov regularization and total least squares. *SIAM J. Matrix Anal. Appl.*, 21(1):185–194, 1999.

W. Heylen, S. Lammens, and P. Sas. *Modal Analysis Theory and Testing.* Kath. Universiteit Leuven, Leuven, Belgium, 1998.

B. Hillary and D.J. Ewins. The use of strain gauges in force determination and frequency response function measurements. In *Proceeding of 2nd IMAC*, Orlando, 1984.

D.J. Inman, C.R. Farrar, V. Lopes Jr., and Steffen Jr. *Damage Prognosis.* Wiley, Chichester UK, 2005.

H. Inoue, J.J. Harrigan, and S.R. Reid. Review of inverse analysis for indirect measurement of impact force. *Appl. Mech. Rev.*, 54(6):503–524, 2001.

E. Jacquelin, A. Bennani, and P. Hamelin. Force reconstruction: analysis and regularization of a deconvolution problem. *J. Sound Vibr.*, 265: 81–107, 2003.

L. Jankowski. Off-line identification of dynamic loads. *Struct. Multidisc. Optim.*, 37(6):609–623, 2009.

P. Johannesson. *Rainflow analysis of switching Markov loads.* PhD thesis, Lund Inst. of Technology, Lund University, Sweden, 1999.

T. Kailath. *Linear Systems.* Prentice-Hall, 1980.

D.C. Kammer and A.D. Steltzner. Structural identification of mir using inverse system dynamics and mir/shuttle docking data. *J. Vib. Acoust.*, 132(2):230–237, 2001.

J. Kautsky, N. K. Nichols, and P. Van Dooren. Robust pole assignment in linear state feedback. *Int. Journal of Control*, 41:1129–1155, 1985.

M. Klinkov. *Identification of Unknown Structural Loads from Dynamic Measurements Using Robust Observers.* Phd thesis, University of Siegen, 2011.

P. Kraemer. *Damage diagnosis approaches for structural health and condition monitoring of offshore wind energy plants.* PhD thesis (in German), University of Siegen, 2011.

P. Kraemer and C.-P. Fritzen. Vibration analysis for structures of offshore wind energy plants. *Deutsche Windenergie Konferenz, Bremen, Germany*, 2010a.

P. Kraemer and C.-P. Fritzen. Aspects of operational modal analysis for structures of offshore wind energy plants. In *Structural Dynamics and Renewable Energy, Volume 1, Conf. Proc. of the Soc. Exp. Mech. Series, 2011, Volume 10*, pages 145–152, 2010b.

P. Kraemer, I. Buethe, and C.-P. Fritzen. Damage detection under changing operational and environmental conditions using self organizing maps. *Proc. of SMART 2011, Saarbrcken*, 2011.

K.B. Krishna and K. Pousga. Disturbance attenuation using proportional integral observer. *Int. J. Syst. Sci.*, 74(6):618–627, 2001.

J. Kullaa. Elimination of environmental influences from damage-sensitive features in a structural health monitoring system. *Proc. of 1st EWSHM*, pages 742–749, 2002.

J. Kullaa. Damage detection under a varying environment using the missing data concept. *Proc. 5th IWSHM*, pages 565–573, 2005.

P. Ladeveze and M. Reynier. A localization method of stiffness errors for the adjustment of fe models. In *Proc. 12th ASME Conf. on Mech. Vibr. and Noise, Montreal, Canada (ASME)*, pages 355–361, 1989.

M. Link, S. Stoehr, and M. Weiland. Identification of structural property degradations by computational model updating. *Key Eng. Mater.*, 347: 19–34, 2007.

J.-J. Liu, C.-K. Ma, I.-K. Kung, and D.-C. Lin. Input force estimation of a cantilever plate by using a system identification technique. *Computer Methods in Applied Mechanics and Engineering*, 190(11-12):1309–1322, 2000.

E. Lourens, G. Lombaert, G. DeRoeck, and G. Degrande. Identification of dynamic axle loads from bridge responses by means of an extended dynamic programming algorithm. In *Proc. WIW(Weight in Motion), Load Capacity and Bridge Performance*, pages 59–69, 2008.

E. Lourens, G. Lombaert, G. DeRoeck, and G. Degrande. Reconstructing dynamic moving loads using an extended dynamic programming algorithm and eigenvalue reduction. In *2nd Int. Conference on Computational Methods in Structural Dynamics and Earthquake Engineering*, volume CD-ROM. COMPDYN 2009, Rhodes, 2009.

D. G. Luenberger. An introduction to observers. *IEEE Trans. Autom. Control*, 16:596–602, 1971.

C.-K . Ma, J.-M. Chang, and D.-C. Lin. Input forces estimation of beam structures by an inverse method. *J. Sound Vibr.*, 259(2):387–407, 2003.

C.-K. Ma and C.-C. Ho. An inverse method for the estimation of input forces acting on non-linear structural systems. *J. Sound Vibr.*, 275:953–971, 2004.

C.-K. Ma and D.-C. Lin. Input forces estimation of a cantilever beam. *Inverse Prob. Eng.*, 00:1–18, 2000.

J. Maeck. *Damage Assessment of Civil Engineering Structures by Vibration Monitoring*. PhD thesis, KU Leuven, Leuven, Belgium, 2003.

R.K. Mehra and J. Peschon. An innovations approach to fault detection and diagnosis in dynamic systems. *Automatica*, 7:637–640, 1971.

J. Moll, P. Kraemer, and C.-P. Fritzen. Compensation of environmental influences for damage detection using classification techniques. *Proc. 4th EWSHM*, pages 1080–1087, 2008.

L. E. Mujica, J. Vehi, and J. Rodellar. Compensation of environmental influences for damage detection using classification techniques. *Lecture Notes in Computer Science: Artificial Neural Networks*, 4131:982–991, 2006.

H.G. Natke. *Einfuehrung in Theorie und Praxis der Zeitreihen- und Modalanalyse.* Vieweg, Germany, 1992.

H.G. Natke and C. Cempel. *Model-Aided Diagnosis of Mechanical Systems.* Springer, Germany, 1997.

Y. Nesterov and A. Nemirovski. *Interior Point Polynomial Methods in Convex Programming: Theory and Applications.* SIAM, Philadelphia, 1994.

Y. Niu, M. Klinkov, and C. P. Fritzen. Online reconstruction of concentrated and continuously distributed force using a robust observer technique. In *Proc. 5th EWSHM*, Sorrento, Italy, 2010.

L.J.L. Nordstroem. *Input estimation in structural dynamics.* PhD thesis, Chalmers University of Technology Sweden, 2005.

W. Ostachowicz and M. Krawczuk. On modelling of structural stiffness loss due to damage. In *Key Eng. Mater.*, volume 204-205, pages 185–200, 2001.

S. Pan, H. Su, H. Wang, and J. Chu. The study of joint input and state estimation with kalman filtering. *Transactions of the Institute of Measurement and Control, published online*, 00(0):1–18, 2010.

G. Park, H. Sohn, C.F. Farrar, and D.J. Inman. Overview of piezoelectric impedance-based health monitoring and path forward. *The Shock and Vibration Digest*, 35(6):451–463, 2003.

G. Park, C.R. Farrar, F. Lanza di Scalea, and S. Coccia. Performance assessment and validation of piezoelectric active-sensors in structural health monitoring. *Smart Mater. Struct.*, 15:1673–1683, 2006.

D.M. Peairs, G. Park, and D.J. Inman. Improving accessibility of the impedance-based structural health monitoring method. *J. Intell. Mater. Syst. Struct.*, 15:129–139, 2004.

D.M. Peairs, G. Park, and D.J. Inman. Simplified modeling for impedance-based health monitoring. *Key Eng. Mater.*, 293-294:643–652, 2005.

B. Peeters and G. DeRoeck. Stochastic system identification for operational modal analysis: A review. *J. Dynamic Systems, Measurement and Control*, 123:659–667, 2001.

B. Peeters and G. DeRoeck. Reference-based stochastic subspace identification for output-only modal analysis. *Mech. Syst. Sig. Process.*, 19(6): 855–878, 1999.

B. Peeters, J. Maeck, and G. DeRoeck. Vibration-based damage detection in civil engineering: excitation sources and temperature effects. *Smart Mater. Struct.*, 10:518–527, 2001.

C. R. Rao and S. K. Mitra. *Generalized inverse of matrices and it is applications.* John Wiley & Sons, New York, 1971.

A. Rytter. *Vibration Based Inspection of Civil Engineering Structures.* PhD thesis, Aalborg University, Denmark, 1993.

S. Seibold, C.-P. Fritzen, and D. Wagner. Employing identification procedures for the detection of cracks in rotors. *Modal Analysis*, 11(3,4): 204–215, 1996.

N. Sepehry, M. Shamshirsaz, and F. Abdollahi. Temperature variation effect compensation in impedance-based structural health monitoring using neural networks. *J. Intell. Mater. Syst. Struct.*, 22(17):1975–1982, 2011.

R. Seydel and F.K. Chang. Impact identification of stiffened composite panels: I system development. *Smart Mater. Struct.*, 10:354–369, 2001a.

R. Seydel and F.K. Chang. Impact identification of stiffened composite panels: Ii implementation studies. *Smart Mater. Struct.*, 10:370–379, 2001b.

D. Soeffker and I. Krajcin. Modified pio design for robust unknown input estimation. In *ASME DETC Conferences*, Chicago, Illinois, USA, 2003.

D. Soeffker, T-J. Yu, and P.C. Mueller. State estimation of dynamical systems with nonlinearities by using proportional-integral observer. *Int. J. Syst. Sci.*, 26(9):1571–1582, 1995.

D. Soeffker, J. Ahrens, H. Ulbrich, and I. Krajcin. Modellgestuetzte schaetzung von kontakt-kraeften und verschiebungen an rotierenden wellen. In *Schwingungen in rotierenden Maschinen.* Vieweg, 2003.

H. Sohn. Effects of environmental and operational variability on structural health monitoring. *Philos. Trans. R. Soc. London, Ser. A*, 365:539–560, 2007.

H. Sohn, C.R. Farrar, N.F. Hunter, and K. Worden. Structural health monitoring using statistical pattern recognition techniques. *J. Dyn. Syst. Meas. Contr.*, 123:706–711, 2001.

H. Sohn, K. Worden, and C. R. Farrar. Statistical damage classification under changing environmental and operational conditions. *J. Intell. Mater. Syst. Struct.*, 13:561–574, 2002.

H. Sohn, C.F. Farrar, F.M. Hemez, D.D. Shunk, D.W. Stinemates, B.R. Nadler, and J.J. Czarnecki. A review of structural health monitoring literature: 1996-2001. Technical report, Report LA-13976-MS, Los Alamos National Laboratory, Los Alamos, NM, 2004.

W. Staszewski, C. Boller, and G. Tomlinson (Eds.). *Health Monitoring of Aerospace Structures-Smart Sensor Technologies and Signal Processing.* Wiley, Chichester, 2004.

A.D. Steltzner and D.C. Kammer. Input force estimation using an inverse structural filter. In *Proc. 17th Int. Modal Analysis Conf.*, pages 954–960, 1999.

K. Stevens. Force identification problems - an overview. In *Proc. of SEM*, pages 838–844, Houston, 1987.

N. Stubbs, J.-T. Kim, and C.R. Farrar. Field verification of a nondestructive damage localization and severity estimation algorithm. *Proc. 13th IMAC, Nashville, TN, USA*, pages 210–218, 1995.

M. Tracy and F.-K. Chang. Identifying impacts in composite plates with piezoelectric strain sensors, part i: Theory. *J. Intell. Mater. Syst. Struct.*, 9(11):920–928, 1998a.

M. Tracy and F.-K. Chang. Identifying impacts in composite plates with piezoelectric strain sensors, part ii: Experiment. *J. Intell. Mater. Syst. Struct.*, 9(11):929–937, 1998b.

D.M. Trujillo and H.R. Busby. *Practical Inverse Analysis in Engineering.* CRC Press, London, 1997.

P.-C. Tuan, C.-C. Ji, L.-W. Fong, and W.-T. Huang. An input estimation approach to on-line two-dimensional inverse heat conduction problems. *Numer. Heat Transfer*, 29,3:345–364, 1996.

T. Uhl. Identification of loads in mechanical structures-helicopter case study. *Comp. Assisted Mechanics and Eng. Sc.*, 9:151–160, 2002.

T. Uhl. The inverse identification problem and its technical applications. *Archive Appl. Mechanics*, 77:325–327, 2007.

T. Uhl and A. Chudzikiewicz. Analytical and experimental investigation of low floor tram dynamics. *Veh. Syst. Dyn.*, 25(12):702–713, 2002.

T. Uhl and M. Petko. Load identification with the use of neural networks a design and application. *Mech. Syst. Sig. Process.*, 2006.

T. Uhl and J. Pieczara. Identification of operational loading forces for mechanical structures. *Arch. Transp.*, 16(2):109–126, 2003.

G. Vachtsevanos, F.L. Lewis, M. Roemer, A. Hess, and B. Wu. *Intelligent Fault Diagnosis and Prognosis for Engineering Systems.* Wiley, Hoboken, NJ, 2006.

G.S. Whiston. Remote impact analysis by use of propagated acceleration signals, i: Theoretical methods. *J. Sound Vibr.*, 97(1):35–51, 1984.

B. Wojciechowski. *Analysis and synthesis of proportional-integral observers for single input-output time-invariant continuous systems.* Phd thesis, Technical University of Gliwice, Poland, 1978.

K. Worden and J.M. Dulieu-Barton. An overview of intelligent fault detection in systems and structures, structural health monitoring. *Structural Health Monitoring*, 3(1):85–98, 2004.

A.-M. Yan, G. Kerschen, P. De Boe, and J.-C. Golinval. Structural damage diagnosis under varying environmental conditions - part 2: local pca for non-linear cases. *Mech. Syst. Sig. Process.*, 19(2):865–880, 2005.

X. Zhan. *Matrix Inequalities*, volume 1790 of *Lecture Notes in Mathematics.* Springer-Verlag, Heidelberg, 2002.

F. Zhang. *The Schur Complement and Its Applications*, volume 4 of *Numerical Methods and Algorithms.* Springer Verlag, New York, 2005.

Statistical Time Series Methods for Vibration Based Structural Health Monitoring

Spilios D. Fassois and Fotis P. Kopsaftopoulos
Stochastic Mechanical Systems & Automation (SMSA) Laboratory
Department of Mechanical & Aeronautical Engineering
University of Patras, 26504 Patras, Greece
E-mail: {fassois,fkopsaf}@mech.upatras.gr
Internet: www.smsa.upatras.gr

Abstract Statistical time series methods for vibration based structural health monitoring utilize random excitation and/or vibration response signals, statistical model building, and statistical decision making for inferring the health state of a structure. This includes damage detection, identification (including localization) and quantification. The principles and operation of methods that utilize the time or frequency domains are explained, and they are classified into various categories under the broad non-parametric and parametric classes. Representative methods from each category are outlined and their use is illustrated via their application to a laboratory truss structure.

1 Introduction

Structural Health Monitoring (SHM) involves the *continual* or *continuous* over time monitoring of a structure based on proper sensors which provide dynamic structural responses and other related data (such as environmental conditions), signal/data processing and analysis, as well as proper decision making for inferring the current health state of the structure. Once set up, an SHM procedure is ideally intended to be *global* (in the sense of covering the whole structure or a large part of it), *automated*, without necessitating human interaction, cost-effective, and capable of effectively treating the level I, II and III subproblems (Rytter, 1993), that is *damage detection* (simply detecting damage presence), *damage identification* (identifying the damage type/nature and location) and *damage quantification* (estimating the damage extent) – see section 2.2 for details.

Historically, SHM may be thought of as an evolution of classical Non-Destructive Testing (NDT) procedures (commonly based on ultrasound, acoustic, radiography, eddy current, and thermal field principles – Doherty 1987; Doebling et al. 1996, 1998; Farrar et al. 2001; Balageas et al. 2006). NDT is however different, in that it is exercised on demand – usually on a periodic basis – without permanent sensors mounted on the structure and not necessarily in an automated fashion. NDT typically works locally, requiring access to the vicinity of the suspected damage location, while the procedure is often time consuming and costly. On the other hand, the SHM philosophy and principles are much closer to the general theory of fault diagnosis (see Basseville and Nikiforov 1993; Rytter 1993; Doebling et al. 1996, 1998; Natke and Cempel 1997; Salawu 1997; Farrar et al. 2001).

Vibration based SHM is quite popular, as vibration is naturally available for many structures (aircraft, railway vehicles, bridges, and so on), while the technology for the precise measurement and processing of vibration signals has been available for a long time. For overviews of general vibration based methods see Doebling et al. (1996, 1998); Salawu (1997); Zou et al. (2000); Farrar et al. (2001); Sohn et al. (2003b); DeRoeck (2003); Carden and Fanning (2004); Montalvão et al. (2006). Also see Staszewski et al. (2004); Inman et al. (2005); Balageas et al. (2006); Fritzen (2006); Adams (2007).

Statistical time series SHM methods form an important and rapidly evolving class within the broader context of vibration based SHM. Their three *main elements* are: (i) random excitation and/or vibration response signals (referred to as *time series*), (ii) statistical model building, and (iii) statistical decision making for inferring the health state of a structure. As with all vibration based methods, the *fundamental principle* upon which they are founded is that small changes (damage) in a structure causes corresponding changes in the structural dynamics. These changes are reflected – though often in a very subtle form – in the measured vibration responses. Hence, the main idea is on the use of measured vibration signals in order to detect, identify, and quantify changes in the *underlying structural dynamics* which are attributed to damage. It could be argued that the effects of damage might be evident simply on the vibration response level, without the need for elaborate analysis. This was indeed used in the early days, and it might still be adequate in certain applications. The drawback of such a simple approach however is that the increased vibration level may be due to alternative reasons, such as increased excitation. Furthermore, modern SHM methods which are based on the dynamics aim at detecting damage at an *early stage*, well before increases in the vibration level are noticed. Statistical time series SHM methods are discussed in references such as Natke and Cempel (1997); Basseville et al. (2000); Sohn and Farrar (2000); Fugate et al. (2001); Yan et al. (2004); Carden and Brownjohn (2008) – see Fassois and Sakellariou (2007, 2009) for recent overviews.

From an *operational viewpoint*, statistical time series SHM methods involve two distinct phases: In an initial *baseline (training) phase*, the methods employ random excitation and/or vibration response (displacement, velocity or acceleration) signals obtained from the structure under its healthy state, as well as from a number of potential damage states, for identifying suitable non-parametric or parametric statistical time series models that describe the structural dynamics in each considered state. A statistical quantity, referred to as *characteristic quantity* and characterizing each structural state, is then extracted.

In the continually or continuously implemented *inspection phase*, the procedure is repeated under the current conditions using freshly measured signals. The current characteristic quantity is obtained, and damage detection, identification (including localization) and quantification (magnitude estimation) are accomplished via statistical decision making procedures consisting of "comparing", in a statistical sense, the current characteristic quantity with that of each potential structural state determined in the baseline phase.

Statistical time series SHM methods thus involve inverse type procedures, as the models employed are *data based* rather than physics based. In addition to the features of the general vibration based methods, statistical time series methods offer further unique *advantages* such as (Fassois and Sakellariou, 2007, 2009; Basseville et al., 2004; Sakellariou and Fassois, 2008; Kopsaftopoulos and Fassois, 2010):

(i) No need for physics based or analytical, such as Finite Element (FE), models.

(ii) No need for complete structural models. In fact models describing only part of the dynamics and based on a very limited number (even a single pair) of excitation and/or vibration response signals are sometimes adequate.

(iii) Inherent accounting for uncertainties (measurement, environmental, operational and so on) through statistical tools.

(iv) Statistical decision making with specified performance characteristics.

(v) Effective operation even in the "low" frequency range.

(iv) Effective use of naturally obtained random vibration signals. This is very important as it implies that there is no need for interrupting the normal operation of the structure.

Of course, like with any other family, statistical time series SHM methods have their limitations. Since only partial structural models are employed, they may identify (locate) damage only to the extent allowed by the type of model used. In addition, adequate "training" in the baseline phase is needed in order to tackle the damage identification and magnitude estimation subproblems, and sufficient user expertise is required.

The goal of this chapter is to provide an overview of the principles and main classes of statistical time series SHM methods. Simple *non-parametric*, as well as more advanced *parametric* methods are reviewed, while both the *response-only* and *excitation-response* measurement cases are considered. The methods are mainly presented for continual (periodic) inspection, although extensions to continuous (on-line) monitoring are possible (for instance through the use of a time window sliding over the measured signals).

For purposes of simplicity, the presentation is limited to the *single (scalar)* vibration signal case – extensions to the *multiple (vector)* signal case are available, and pertinent references are provided in the bibliographical remarks. Likewise, the presentation focuses on *time-invariant (stationary)* and *linear* structural dynamics in a *Gaussian* context, although – with proper modifications – the concepts may extend to the time-varying and non-linear cases. One important aspect, which due to space limitations is not covered in the chapter, is *environmental effects* or in broader terms the *effects of varying operating conditions*. These can be very important and – if not properly accounted for – they may have an adverse effect on SHM operation (Doebling et al., 1996; Sohn, 2007; Deraemaeker et al., 2008; Figueiredo et al., 2011).

The rest of this chapter is organized as follows: The general framework of the methods is presented in section 2. Non-parametric and parametric time series models for representing the structural dynamics are presented in section 3. Selected non-parametric and parametric time series SHM methods are presented in sections 4 and 5, respectively. The application of selected methods to damage diagnosis for a laboratory truss structure is presented in section 6. Concluding remarks are finally summarized in section 7.

2 The General Workframe

2.1 The Structural States and the Data Sets

Let S_o designate the structure under consideration in its *nominal* (healthy) state and S_u the structure in an unknown, to be determined, state. Furthermore, let $S_A, S_B, \ldots$ designate the structure under damage of type $A, B, \ldots$, respectively. Each damage type includes a continuum of damages of common nature or location but of any admissible damage magnitude.

Statistical time series SHM methods are typically based on discretized excitation $x[t]$ and/or response $y[t]$ (for $t = 1, 2, \ldots, N$) random vibration data records. The discrete time t corresponds to the actual time $(t-1) \cdot T_s$, with T_s designating sampling period. Let the complete excitation and

Table 1. Workframe setup: structural state, vibration signals used, and the characteristic quantity.

	Baseline Phase		
Structural state	S_o (healthy structure)	S_A (damage type A)†	...
Vibration signals	$z_o[t] = (x_o[t], y_o[t])$	$z_A[t] = (x_A[t], y_A[t])$	...
	$Z_o = (X_o, Y_o)$	$Z_A = (X_A, Y_A)$	...
Characteristic quantity	Q_o	Q_A	...
	Inspection Phase		
Structural state	S_u (current structure in unknown state)		
Vibration signals	$z_u[t] = (x_u[t], y_u[t])$		
	$Z_u = (X_u, Y_u)$		
Characteristic quantity	Q_u		

†Various damage magnitudes may be considered within each damage type.

response signals be designated as X and Y, respectively, or, collectively as $Z = (X, Y)$. The subscript $(o, A, B, \ldots, u)$ is used for designating the state of the structure that provided the signals.

All signals measured under the various structural states need to be suitably preprocessed (Doebling et al., 1996; Ljung, 1999; Fassois, 2001; Sohn, 2007). This may include low or bandpass filtering within the frequency range of interest, signal subsampling (in case the original sampling frequency is too high) and normalization, which includes sample mean subtraction and division by its sample standard deviation. Normalization is used in the linear case only, for improving numerical accuracy, but also counteracting different operating or environmental conditions or excitation levels.

2.2 The Baseline and Inspection Phases

As indicated, the methods consist of an initial *baseline phase*, while normal operation takes place under the continually or continuously implemented *inspection phase* (see Table 1). In the *baseline (training) phase* the data records $Z_o, Z_A, Z_B, \ldots$ are obtained and analyzed, while the data record Z_u, corresponding to an unknown (to be determined) structural state, is obtained and analyzed in the *inspection* phase. Based on each data record a non-parametric or parametric time series model representing part of the structural dynamics is identified and validated. From each estimated model, the corresponding estimate[1] of a *characteristic quantity* Q is extracted ($\widehat{Q}_o, \widehat{Q}_A, \widehat{Q}_B, \ldots$ in the baseline phase, and $\widehat{Q}_u$ in the inspection phase).

[1] Estimators/estimates are designated by a hat.

Table 2. The damage detection, identification, and estimation subproblems.

		Damage detection
H_o :	$Q_u \sim Q_o$	Null hypothesis (healthy structure)
H_1 :	$Q_u \nsim Q_o$	Alternative hypothesis (damaged structure)
		Damage identification
H_A :	$Q_u \sim Q_A$	Hypothesis A (damage type A)
H_B :	$Q_u \sim Q_B$	Hypothesis B (damage type B)
	$\vdots$	$\vdots$
		Damage estimation
Estimate the damage magnitude given the damage type.		

$\sim$ indicates a proper relationship, such as equality, inequality and so on.

Damage detection, identification and quantification. *Damage detection* is then based on proper comparison of the true (but not precisely known) current characteristic quantity Q_u to the true (but also not precisely known) characteristic quantity Q_o of the healthy structure (Table 2). This is accomplished via a binary composite statistical hypothesis test that employs the corresponding estimates $\widehat{Q}_u$ and $\widehat{Q}_o$.

Damage identification is similarly based on proper comparison of the current characteristic quantity Q_u to each of $Q_A, Q_B, \ldots$ (Table 2) via statistical hypothesis testing that also employs the corresponding estimates. *Damage estimation (quantification)* generally is a more complicated task that requires proper formulation and the use of interval estimation techniques (Table 2). The workframe of a general statistical time series SHM method is presented in Figure 1.

The design of a binary statistical hypothesis test is generally based on two error probabilities:

(i) The *type I error* or *false alarm* probability, designated as α, which is the probability of rejecting the null hypothesis H_o when it is true.

(ii) The *type II error* or *missed damage* probability, designated as β, which is the probability of accepting the null hypothesis H_o when it is not true.

The majority of the designs presented herein are based on the type I error probability α. In selecting α it should be noted that as it decreases (resp. increases), β increases (resp. decreases). The reader is referred to Lehmann and Romano (2008), Basseville and Nikiforov (1993, section 4.2), and Montgomery (1991, section 3.3) for details on statistical hypothesis testing.

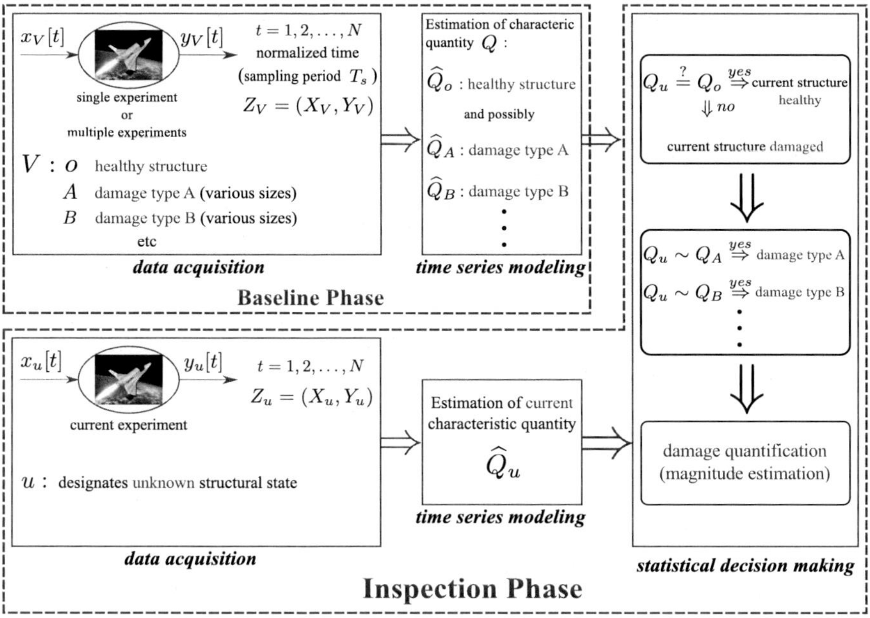

Figure 1. Workframe for statistical time series SHM methods.

2.3 Classes of Statistical Time Series SHM Methods

An important classification of the methods follows the precise nature of the problem in terms of the available signals. *Response-only methods* are based on response signals only, and tackle a generally more difficult problem, while *excitation-response methods* are based on both types of signals. Response-only and excitation-response methods are both treated in sections 4 and 5.

An additional important classification follows the *non-parametric* or *parametric* nature of the statistical time series model used, on which the characteristic quantity Q is based. Non-parametric methods are thus based on corresponding models (see section 3.1), such as non-parametric Power Spectral Density (PSD) or Frequency Response Function (FRF) representations (Söderström and Stoica, 1989; Ljung, 1999; Bendat and Piersol, 2000), and have received limited attention in the literature (Fassois and Sakellariou, 2007, 2009; Kopsaftopoulos and Fassois, 2010, 2011; Benedetti et al., 2011). Parametric methods are likewise based on corresponding models (see section 3.2), such as AutoRegressive Moving Average (ARMA) or State Space (SS) representations (Söderström and Stoica, 1989; Ljung, 1999; Fassois, 2001).

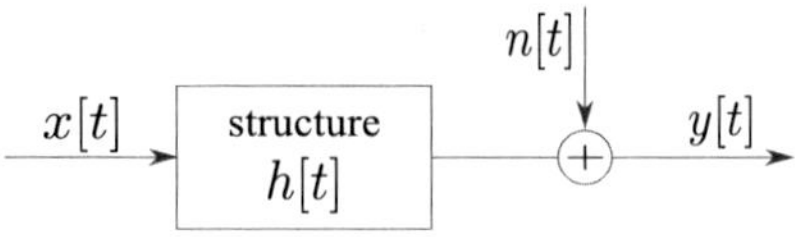

Figure 2. Excitation-response representation of linear time-invariant structural dynamics with additive response noise.

This type of methods has attracted considerable attention and their principles have been used in a number of studies (Fassois and Sakellariou, 2007, 2009; Kopsaftopoulos and Fassois, 2010, 2011). Non-parametric methods offer simplicity and the computational efficiency – selected non-parametric methods are presented in section 4. Parametric methods typically are more elaborate and are characterized by higher computational complexity, but are generally capable of achieving better performance – selected parametric methods are presented in section 5.

Various other classifications are possible, for instance *scalar (univariate)* versus *vector (multivariate)* methods, *time-invariant (stationary)* versus *time-varying (non-stationary)* methods, *linear dynamics* versus *nonlinear dynamics* methods, and so forth. Due to space limitations only the scalar, time-invariant and linear dynamics case is treated in this chapter.

3 Statistical Time Series Models of the Structural Dynamics

Statistical time series models for the representation of linear and time-invariant structural dynamics are presented in this section for the univariate (scalar) case. Let $h[t]$ designate the time-discretized impulse response function describing the causality relation between an excitation and vibration response signal, as illustrated in Figure 2. Let $x[t]$ represent the excitation and $y[t]$ the noise-corrupted response signal, with the additive noise $n[t]$ being a stationary Gaussian process, mutually uncorrelated with the excitation, and with zero mean but unknown autocovariance (and thus auto power spectral density).

The excitation and response signals are related through the convolution summation plus noise expression:

$$y[t] = h[t] * x[t] + n[t] = \sum_{\tau=0}^{\infty} h[\tau] \cdot x[t-\tau] + n[t] \tag{1}$$

with $*$ designating discrete convolution. Alternatively, using the backshift operator $\mathcal{B}$ (defined such that[2] $\mathcal{B}^i \cdot y[t] = y[t-i]$), the above may be expressed as:

$$y[t] = H(\mathcal{B}) \cdot x[t] + n[t], \quad \text{with} \quad H(\mathcal{B}) = \sum_{\tau=0}^{\infty} h[\tau] \cdot \mathcal{B}^{\tau} \tag{2}$$

where $H(\mathcal{B})$ stands for the discrete-time structural dynamics transfer function.

3.1 Non-Parametric Models

Assuming that $x[t]$ is a random stationary excitation, then the response $y[t]$ will also be stationary in the steady state. Furthermore, $y[t]$ will be Gaussian if $x[t]$ and $n[t]$ are jointly Gaussian. In this case each signal is fully characterized by its first two moments, *mean* and *AutoCovariance Function (ACF)*. Instead of the autocovariance function $\gamma_{yy}[\tau]$ (or its normalized version $\rho_{yy}[\tau]$), its Fourier transform which is the *auto Power Spectral Density (PSD)* $S_{yy}(\omega)$ (Kay 1988, p. 3, Box et al. 1994, pp. 39–40) may be employed. Thus:

$$\mu_y = E\{y[t]\} \tag{3}$$

$$\gamma_{yy}[\tau] = E\{\tilde{y}[t] \cdot \tilde{y}[t-\tau]\} \tag{4}$$

$$\rho_{yy}[\tau] = \frac{\gamma_{yy}[\tau]}{\gamma_{yy}[0]} \in [-1, 1] \tag{5}$$

$$S_{yy}(\omega) = \sum_{\tau=-\infty}^{\infty} \gamma_{yy}[\tau] \cdot e^{-j\omega\tau T_s}. \tag{6}$$

In these expressions $E\{\cdot\}$ designates statistical expectation, j the imaginary unit, τ time lag, $\omega \in [0, 2\pi/T_s]$ frequency (rad/s), T_s the sampling period, and $\tilde{y}[t] = y[t] - \mu_y$. Moreover, note that $\gamma_{yy}[0]$ is equal to the variance σ_y^2 of the response $y[t]$. The first order moment, along with any suitable form of the second order moment, constitute a *response-only* time series model (summary in Table 3).

In the *excitation-response* case, a complete joint description of the excitation and response signals is given in terms of the means μ_x and μ_y, and the ACFs $\gamma_{xx}[\tau]$, $\gamma_{yy}[\tau]$ and *Cross Covariance Function (CCF)* $\gamma_{yx}[\tau]$ (or their normalized equivalents). An equivalent representation may be obtained by using the auto PSDs $S_{xx}(\omega)$, $S_{yy}(\omega)$ and *Cross Spectral Density (CSD)* $S_{yx}(j\omega)$.

[2] The backshift operator is alternatively designated as q^{-1}, or as z^{-1} in the z domain – yet the designation $\mathcal{B}$ is exclusively used in this chapter.

Table 3. The elements of non-parametric response-only models.

Mean:	$\mu_y = E\{y[t]\}$
ACF:	$\gamma_{yy}[\tau] = E\{\tilde{y}[t] \cdot \tilde{y}[t-\tau]\}$
normalized ACF:	$\rho_{yy}[\tau] = \gamma_{yy}[\tau]/\gamma_{yy}[0] \in [-1,1]$
auto PSD:	$S_{yy}(\omega) = \sum_{\tau=-\infty}^{\infty} \gamma_{yy}[\tau] \cdot e^{-j\omega\tau T_s}$

$\tilde{y}[t] = y[t] - \mu_y$

The response characteristics are related to those of the excitation and the noise through the expressions (Box et al., 1994, pp. 455–456):

$$\mu_y = H(j\omega)|_{\omega=0} \cdot \mu_x \tag{7a}$$

$$\gamma_{yx}[\tau] = \gamma_{xx}[\tau] * h[\tau] \tag{7b}$$

$$\gamma_{yy}[\tau] = \gamma_{yx}[\tau] * h[-\tau] + \gamma_{nn}[\tau] \tag{7c}$$

$$S_{yx}(j\omega) = H(j\omega) \cdot S_{xx}(\omega) \tag{8a}$$

$$S_{yy}(\omega) = H^{\star}(j\omega) \cdot S_{yx}(j\omega) + S_{nn}(\omega) = |H(j\omega)|^2 \cdot S_{xx}(\omega) + S_{nn}(\omega) \tag{8b}$$

with $*$ designating discrete convolution, the superscript * complex conjugation, and $|\cdot|$ complex magnitude. $H(j\omega) = H(\mathcal{B})|_{\mathcal{B}=e^{-j\omega T_s}}$ designates the structural *Frequency Response Function (FRF)*, $S_{nn}(\omega)$ the noise auto PSD, and $S_{yx}(j\omega)$ the CSD between the designated signals defined as the Fourier transform of the corresponding CCF.

In addition, the squared *coherence function* is defined as (Bendat and Piersol, 2000, p. 196):

$$\gamma^2(\omega) = \frac{|S_{yx}(j\omega)|^2}{S_{xx}(\omega) \cdot S_{yy}(\omega)} = \frac{1}{1 + \frac{S_{nn}(\omega)}{|H(j\omega)|^2 \cdot S_{xx}(\omega)}} \in [0,1]. \tag{9}$$

The elements of non-parametric excitation-response time series models are summarized in Table 4.

3.2 Parametric Models

Parametric time series models may be obtained via proper parametrizations of Equation (1). In the response-only case it is assumed, without loss of generality, that the excitation is white (uncorrelated), that is $\gamma_{xx}[\tau] = 0$ for $\tau \neq 0$, in which case the signal is often designated as $w[t]$, while $n[t] \equiv 0$. Moreover, the signals are assumed zero mean, thus $\tilde{y}[t] = y[t] - \mu_y$ is typically used.

Table 4. The elements of non-parametric excitation-response models.

Means:	μ_x	μ_y
ACFs:	$\gamma_{xx}[\tau]$	$\gamma_{yy}[\tau]$
normalized ACFs:	$\rho_{xx}[\tau]$	$\rho_{yy}[\tau]$
CCF:	$\gamma_{yx}[\tau] = E\{\tilde{y}[t] \cdot \tilde{x}[t-\tau]\}$	
normalized CCF:	$\rho_{yx}[\tau] = \gamma_{yx}[\tau]/\sqrt{\gamma_{xx}[0] \cdot \gamma_{yy}[0]} \in [-1, 1]$	
auto PSDs:	$S_{xx}(\omega)$	$S_{yy}(\omega)$
CSD:	$S_{yx}(j\omega) = \sum_{\tau=-\infty}^{\infty} \gamma_{yx}[\tau] \cdot e^{-j\omega\tau T_s}$	

$\tilde{x}[t] = x[t] - \mu_x$, $\tilde{y}[t] = y[t] - \mu_y$

The parametrization of Equation (1) in the response-only case leads to the celebrated *AutoRegressive Moving Average (ARMA)* model (Box et al., 1994, pp. 52–53):

$$y[t] + \sum_{i=1}^{na} a_i \cdot y[t-i] = w[t] + \sum_{i=1}^{nc} c_i \cdot w[t-i] \tag{10a}$$

which, using the backshift operator, is also written as:

$$\left(1 + \sum_{i=1}^{na} a_i \cdot \mathcal{B}^i\right) \cdot y[t] = \left(1 + \sum_{i=1}^{nc} c_i \cdot \mathcal{B}^i\right) \cdot w[t] \Longleftrightarrow$$
$$\Longleftrightarrow A(\mathcal{B}) \cdot y[t] = C(\mathcal{B}) \cdot w[t], \quad w[t] \sim \text{iid}\, \mathcal{N}(0, \sigma_w^2) \tag{10b}$$

with a_i, c_i, $A(\mathcal{B})$, and $C(\mathcal{B})$ designating the AR and MA parameters and corresponding polynomials, respectively, iid stands for identically independently distributed, $\mathcal{N}(\cdot, \cdot)$ designates normal distribution with the indicated mean and variance, while na, nc are the model's AR, MA orders, respectively. The model parameter vector is $\boldsymbol{\theta} = [\text{coef}(A)\ \text{coef}(B)]^T$. It should be noted that $w[t]$ coincides with the *one-step-ahead-prediction error* and is also referred to as the *model residual* or *innovations* (Box et al. 1994, p. 134, Ljung 1999, p. 70). Corresponding *Vector AutoRegressive Moving Average (VARMA)* models are available for the multivariate case (see Söderström and Stoica, 1989; Lütkepohl, 2005).

The ARMA representation of Equations (10a) – (10b) may be equivalently set into *State Space (SS)* form, consisting of a first order state equation plus an output equation (Söderström and Stoica 1989, p. 157, Box et al. 1994, pp. 163–164):

$$\boldsymbol{\psi}[t+1] = \boldsymbol{A} \cdot \boldsymbol{\psi}[t] + \boldsymbol{K} \cdot \boldsymbol{v}[t], \quad \boldsymbol{v}[t] \sim \text{iid}\, \mathcal{N}(\boldsymbol{0}, \boldsymbol{\Sigma_v}) \tag{11a}$$
$$y[t] = \boldsymbol{C} \cdot \boldsymbol{\psi}[t] + \boldsymbol{v}[t] \tag{11b}$$

Table 5. Parametric response-only models.

ARMA:	$A(\mathcal{B}) \cdot y[t] = C(\mathcal{B}) \cdot w[t]$	$A(\mathcal{B}) = 1 + \sum_{i=1}^{na} a_i \mathcal{B}^i$
		$C(\mathcal{B}) = 1 + \sum_{i=1}^{nc} c_i \mathcal{B}^i$
	$w[t] \sim \text{iid } \mathcal{N}(0, \sigma_w^2)$	na, nc : AR, MA orders
State Space (SS):	$\boldsymbol{\psi}[t+1] = \boldsymbol{A} \cdot \boldsymbol{\psi}[t] + \boldsymbol{K} \cdot \boldsymbol{v}[t]$	$\boldsymbol{\psi}[t]$: state vector
	$y[t] = \boldsymbol{C} \cdot \boldsymbol{\psi}[t] + \boldsymbol{v}[t]$	$\boldsymbol{v}[t] \sim \text{iid } \mathcal{N}(\boldsymbol{0}, \boldsymbol{\Sigma}_v)$
		$\boldsymbol{A}$: system matrix
		$\boldsymbol{C}$: output matrix
		$\boldsymbol{K}$: Kalman gain matrix

with $\boldsymbol{\psi}[t]$ designating the model's state vector and $\boldsymbol{v}[t]$ a zero mean uncorrelated (white) vector sequence with covariance $\boldsymbol{\Sigma_v}$. $\boldsymbol{A}$, $\boldsymbol{C}$, $\boldsymbol{K}$ designate proper matrices and the parameter vector is $\boldsymbol{\theta} = [\text{vec}(\boldsymbol{A}) \ \text{vec}(\boldsymbol{C}) \ \text{vec}(\boldsymbol{K})]$, with vec$(\cdot)$ designating the column vector operator. These parametric response-only time series models are summarized in Table 5.

In the excitation-response case different parametrizations of Equation (1) lead to different models. The simplest one is the *AutoRegressive with eXogenous excitation (ARX)* model (Söderström and Stoica 1989, pp. 149–151, Ljung 1999, p. 81, Fassois 2001):

$$y[t] + \sum_{i=1}^{na} a_i \cdot y[t-i] = \sum_{i=0}^{nb} b_i \cdot x[t-i] + w[t] \tag{12a}$$

or using the backshift operator:

$$A(\mathcal{B}) \cdot y[t] = B(\mathcal{B}) \cdot x[t] + w[t] \quad \Longleftrightarrow$$

$$\Longleftrightarrow \quad y[t] = \frac{B(\mathcal{B})}{A(\mathcal{B})} \cdot x[t] + \frac{1}{A(\mathcal{B})} \cdot w[t], \quad w[t] \sim \text{iid } \mathcal{N}(0, \sigma_w^2) \tag{12b}$$

with $A(\mathcal{B})$ and $B(\mathcal{B})$ designating the AR and X polynomials, respectively, and na, nb the corresponding orders. The parameter vector is defined as $\boldsymbol{\theta} = [\text{coef}(A) \ \text{coef}(B)]^T$, while $w[t]$ is a zero mean, uncorrelated (white) signal, which coincides with the model based one-step-ahead prediction error. In the special case of $na = 0$, a Finite Impulse Response (FIR) model is obtained. The ARX model block diagram is depicted in Figure 3.

A more general and flexible representation is the *AutoRegressive Moving Average with eXogenous excitation (ARMAX)*, which additionally involves a Moving Average (MA) polynomial $C(\mathcal{B})$ for describing the noise dynamics

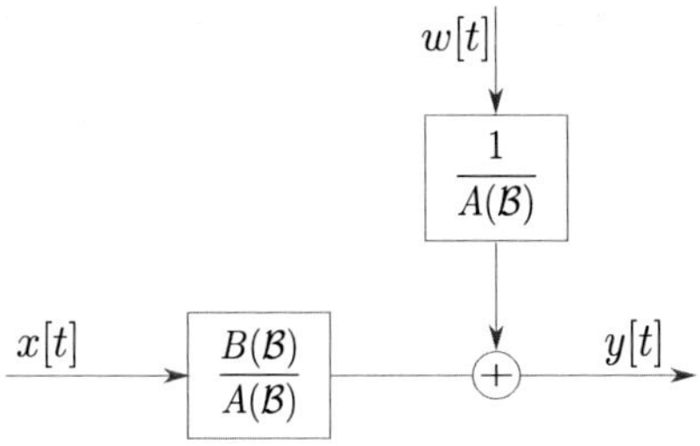

Figure 3. The ARX model structure.

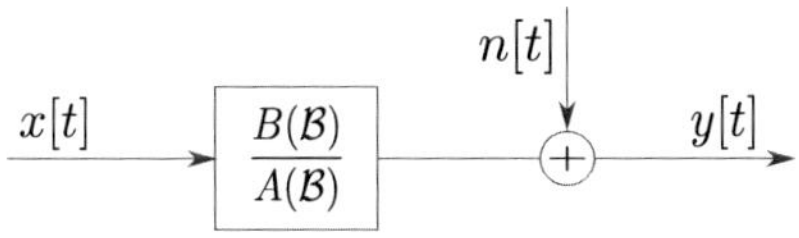

Figure 4. The Output Error (OE) model structure.

(Söderström and Stoica 1989, pp. 149–151, Ljung 1999, p. 83, Fassois 2001):

$$y[t] + \sum_{i=1}^{na} a_i \cdot y[t-i] = \sum_{i=0}^{nb} b_i \cdot x[t-i] + w[t] + \sum_{i=1}^{nc} c_i \cdot w[t-i] \qquad (13a)$$

or using the backshift operator:

$$A(\mathcal{B}) \cdot y[t] = B(\mathcal{B}) \cdot x[t] + C(\mathcal{B}) \cdot w[t] \quad \Longleftrightarrow$$

$$\Longleftrightarrow \quad y[t] = \frac{B(\mathcal{B})}{A(\mathcal{B})} \cdot x[t] + \frac{C(\mathcal{B})}{A(\mathcal{B})} \cdot w[t], \quad w[t] \sim \text{iid}\, \mathcal{N}(0, \sigma_w^2) \qquad (13b)$$

with nc designating the MA order. The parameter vector is defined as $\boldsymbol{\theta} = [\text{coef}(A)\ \text{coef}(B)\ \text{coef}(C)]^T$.

The *Output Error (OE)* representation models the excitation-response dynamics while avoiding noise modelling (Söderström and Stoica 1989, p. 153, Ljung 1999, p. 85, Fassois 2001):

$$y[t] = \frac{B(\mathcal{B})}{A(\mathcal{B})} \cdot x[t] + n[t]. \qquad (14)$$

The parameter vector in this case is $\boldsymbol{\theta} = [\text{coef}(A)\ \text{coef}(B)]^T$ and the model's block diagram is depicted in Figure 4.

Table 6. Parametric excitation-response models.

ARX:	$A(\mathcal{B}) \cdot y[t] = B(\mathcal{B}) \cdot x[t] + w[t]$	na, nb : AR, X orders $A(\mathcal{B}) = 1 + \sum_{i=1}^{na} a_i \mathcal{B}^i$ $B(\mathcal{B}) = b_0 + \sum_{i=1}^{nb} b_i \mathcal{B}^i$
ARMAX:	$A(\mathcal{B}) \cdot y[t] = B(\mathcal{B}) \cdot x[t] + C(\mathcal{B}) \cdot w[t]$	nc : MA order $C(\mathcal{B}) = 1 + \sum_{i=1}^{nc} c_i \mathcal{B}^i$
Output Error:	$y[t] = \frac{B(\mathcal{B})}{A(\mathcal{B})} \cdot x[t] + n[t]$	$n[t]$: autocorrelated zero mean
Box-Jenkins:	$y[t] = \frac{B(\mathcal{B})}{A(\mathcal{B})} \cdot x[t] + \frac{C(\mathcal{B})}{D(\mathcal{B})} \cdot w[t]$	$D(\mathcal{B}) = 1 + \sum_{i=1}^{nd} d_i \mathcal{B}^i$
State Space:	$\boldsymbol{\psi}[t+1] = \boldsymbol{A} \cdot \boldsymbol{\psi}[t] + \boldsymbol{B} \cdot x[t] + \boldsymbol{K} \cdot \boldsymbol{v}[t]$ $y[t] = \boldsymbol{C} \cdot \boldsymbol{\psi}[t] + \boldsymbol{D} \cdot x[t] + \boldsymbol{v}[t]$	$\boldsymbol{\psi}[t]$: state vector $\boldsymbol{v}[t] \sim \text{iid}\, \mathcal{N}(\mathbf{0}, \boldsymbol{\Sigma}_v)$
	$\boldsymbol{A}$: system matrix, $\boldsymbol{B}$: input matrix, $\boldsymbol{K}$: Kalman gain matrix $\boldsymbol{C}$: output matrix, $\boldsymbol{D}$: dir. transm. matrix	

$w[t] \sim \text{iid}\, \mathcal{N}(0, \sigma_w^2)$

Alternatively, the *Box-Jenkins (BJ)* representation independently models the structural and noise dynamics (Söderström and Stoica 1989, pp. 148–154, Ljung 1999, p. 87):

$$y[t] = \frac{B(\mathcal{B})}{A(\mathcal{B})} \cdot x[t] + \frac{C(\mathcal{B})}{D(\mathcal{B})} \cdot w[t], \quad w[t] \sim \text{iid}\, \mathcal{N}(0, \sigma_w^2) \tag{15}$$

with parameter vector $\boldsymbol{\theta} = [\text{coef}(A)\ \text{coef}(B)\ \text{coef}(C)\ \text{coef}(D)]^T$.

Finally, a State Space (SS) representation in this case assumes the form (Ljung, 1999, pp. 97–101)):

$$\boldsymbol{\psi}[t+1] = \boldsymbol{A} \cdot \boldsymbol{\psi}[t] + \boldsymbol{B} \cdot x[t] + \boldsymbol{K} \cdot \boldsymbol{v}[t], \ \boldsymbol{v}[t] \sim \text{iid}\, \mathcal{N}(\mathbf{0}, \boldsymbol{\Sigma_v}) \tag{16a}$$

$$y[t] = \boldsymbol{C} \cdot \boldsymbol{\psi}[t] + \boldsymbol{D} \cdot x[t] + \boldsymbol{v}[t] \tag{16b}$$

with $\boldsymbol{\psi}[t]$ designating the model's state vector and $\boldsymbol{v}[t]$ a zero mean uncorrelated (white) vector sequence with covariance $\boldsymbol{\Sigma_v}$ and the model parameter vector being $\boldsymbol{\theta} = [\text{vec}(\boldsymbol{A})\ \text{vec}(\boldsymbol{B})\ \text{vec}(\boldsymbol{C})\ \text{vec}(\boldsymbol{D})\ \text{vec}(\boldsymbol{K})]$. The parametric excitation-response time series models are summarized in Table 6.

3.3 Identification of Time Series Models

Model identification refers to the estimation of statistical time series models based on excitation $x[t]$ and/or response $y[t]$ random vibration data records (for $t = 1, 2, \ldots, N$) that are properly preprocessed and are collectively designated as $Z = (X, Y)$. This is achieved via *estimators*, which operate on the obtained data records to provide *estimates* of the quantities of interest.

The estimator of a quantity Q is designated as $\widehat{Q}$ and is a function of the random data Z, thus $\widehat{Q} = g(Z)$. Viewing each data point as the realization (observed value) of an underlying random variable, the estimator is a function of several random variables, and thus a random variable as well with probability density function (pdf) $f_{\widehat{Q}}$. For Gaussian distribution this is expressed as $\widehat{Q} \sim \mathcal{N}(\mu_{\widehat{Q}}, \text{var}[\widehat{Q}])$, with the arguments $\mu_{\widehat{Q}}$ and $\text{var}[\widehat{Q}]$ designating the estimator mean and variance, respectively.

Estimators may be characterized by several important properties. One of them is unbiasedness, implying that its mean $\mu_{\widehat{Q}}$ coincides with the true value of the quantity being estimated, that is $E\{\widehat{Q}\} = Q$; when this is true the estimator is referred to as *unbiased.* Oftentimes, unbiasedness or other estimator properties (such as Gaussianity or minimum variance which is referred to as *efficiency*) are only valid *asymptotically*, as the data record length (in samples) tends to infinity ($N \to \infty$). Some other properties such as *consistency* by definition refer to the estimator asymptotic behaviour in a proper sense ($\lim_{N\to\infty} \widehat{Q}(N) \to Q$). Certain properties of non-parametric statistical time series estimators are summarized in Table 7.

The identification of parametric time series models is divided into two main tasks: parameter estimation and model structure selection. Typical methods for parameter estimation include the Least Squares (LS), the Prediction Error (PE), the Maximum Likelihood (ML), and subspace methods (Söderström and Stoica 1989, Chapters 7–8, Ljung 1999, Chapter 7).

Model structure selection, which refers to the determination of the model orders, is generally a more complicated procedure and is typically achieved by identifying increasingly higher order models until no further "improvement" is observed. "Improvement" may be judged via a variety of criteria, such as the model Residual Sum of Squares (RSS – often normalized by the Signal Sum of Squares, SSS), or the negative likelihood, or those criteria that include a penalty term for high model dimensionality, such as the Akaike Information Criterion (AIC) and the Bayesian Information Criterion (BIC) (Söderström and Stoica 1989, pp. 442–443, Ljung 1999, pp. 505–507, Fassois 2001). Other "practical" criteria, such as PSD or FRF or modal frequency stabilization diagrams are also used (Fassois, 2001). In all cases it is advised to simultaneously monitor the Signal Per (estimated) Parameter (SPP) ratio which must be maintained sufficiently high (say above 15), as well as the numerical accuracy (for instance by monitoring the condition

Table 7. Estimation of non-parametric statistical time series model characteristics.

Quantity	Power Spectral Density (PSD)	Frequency Response Function (FRF)
Estimator	$\widehat{S}_{yy}(\omega) = \frac{1}{K}\sum_{i=1}^{K} \widehat{Y}_L^i(j\omega)\widehat{Y}_L^i(-j\omega)$	$\widehat{H}(j\omega) = \widehat{S}_{yx}(j\omega)/\widehat{S}_{xx}(\omega)$
	$\widehat{Y}_L^i(j\omega) = \frac{1}{\sqrt{L}}\sum_{t=1}^{L} a[t]\widehat{y}^i[t]e^{-j\omega T_s}$	
	$\widehat{y}^i[t] = y^i[t] - \widehat{\mu}_y$	
	(i-th segment of length L)	
Properties	$2K\widehat{S}_{yy}(\omega)/S_{yy}(\omega) \sim \chi^2(2K)$	$E\{\lvert\widehat{H}(j\omega)\rvert\} \approx \lvert H(j\omega)\rvert$
		$\text{var}[\lvert\widehat{H}(j\omega)\rvert] \approx \frac{1-\gamma^2(\omega)}{\gamma^2(\omega)2K}\lvert H(j\omega)\rvert^2$
Comments	Welch method (no overlap)	
	K : number of data segments	For $N \to \infty$, $a[t] = 1$
	$a[t]$: time window	$\gamma^2(\omega) \to 1$ or $K \to \infty$

Remarks:
$\omega \in [0, 2\pi/T_s]$ stands for frequency (rad/s); j stands for the imaginary unit.
K stands for the number of segments used in Welch spectral estimation.
$\gamma^2(\omega)$ stands for the coherence function (Bendat and Piersol, 2000, p. 196).
The FRF magnitude estimator distribution may be approximated as Gaussian for small relative errors (Bendat and Piersol, 2000, pp. 274–275).
MATLAB functions: *pwelch.m* for $\widehat{S}_{yy}$, *tfestimate.m* for $\widehat{H}$.

number of the matrices inverted during estimation).

4 Selected Non-Parametric Time Series SHM Methods

Non-parametric methods are those in which the characteristic quantity Q is constructed based on non-parametric time series models (Tables 3 and 4). A response-only *Power Spectral Density (PSD)* method and an excitation-response *Frequency Response Function (FRF)* method are outlined in the sequel. For alternative methods that employ a novelty measure see Worden (1997); Worden et al. (2000); Worden and Manson (2003); Manson et al. (2003).

4.1 PSD Based Method

This method tackles damage detection and identification via changes in the auto Power Spectral Density (PSD) of the measured vibration response signal when the excitation is not available (response-only case). The method's characteristic quantity thus is $Q = S_{yy}(\omega) = S(\omega)$. The main idea is based on the comparison of the current structure's response PSD, $S_u(\omega)$, to that of the healthy structure's, $S_o(\omega)$ (or, in fact, to that corresponding to any other structural state). The response signals must be normalized in order to properly account for potentially different levels of excitation.

Damage detection is based on the following hypothesis testing problem:

$$\begin{array}{ll} H_o : S_u(\omega) = S_o(\omega) & \text{(null hypothesis – healthy structure)} \\ H_1 : S_u(\omega) \neq S_o(\omega) & \text{(alternative hypothesis – damaged structure).} \end{array} \tag{17}$$

As the true PSDs, $S_u(\omega)$ and $S_o(\omega)$, are unknown, their estimates, $\widehat{S}_u(\omega)$ and $\widehat{S}_o(\omega)$, obtained via the Welch method (with K non-overlapping segments; refer to Table 7) are used (Kay, 1988, pp. 3 and 76). Then, the quantity F, below, follows $\mathcal{F}$ distribution with $(2K, 2K)$ degrees of freedom for each frequency ω (as the ratio of two random variables each following a normalized χ^2 distribution with $2K$ degrees of freedom; see Appendix):

$$F = \frac{\widehat{S}_o(\omega)/S_o(\omega)}{\widehat{S}_u(\omega)/S_u(\omega)} \sim \mathcal{F}(2K, 2K). \tag{18}$$

Under the null hypothesis H_o of a healthy structure, the true PSDs coincide, $S_u(\omega) = S_o(\omega)$, thus:

$$\text{Under } H_o: \quad F = \frac{\widehat{S}_o(\omega)}{\widehat{S}_u(\omega)} \sim \mathcal{F}(2K, 2K). \tag{19}$$

Then F should be in the range $[f_{\alpha/2}, f_{1-\alpha/2}]$ with probability $1 - \alpha$, and decision making is as follows for a selected α (false alarm) risk level – see Figure 5:

$$\begin{array}{ll} f_{\frac{\alpha}{2}}(2K, 2K) \leq F \leq f_{1-\frac{\alpha}{2}}(2K, 2K) \quad (\forall\, \omega) & \\ \qquad \Longrightarrow H_o \text{ is accepted} & \text{(healthy structure)} \\ \text{Else} \quad \Longrightarrow H_1 \text{ is accepted} & \text{(damaged structure),} \end{array} \tag{20}$$

with $f_{\alpha/2}, f_{1-\alpha/2}$ designating the $\mathcal{F}$ distribution's $\alpha/2$ and $1-(\alpha/2)$ critical points (f_α is defined such that $\text{Prob}(F \leq f_\alpha) = \alpha$).

Damage identification may be achieved by performing hypotheses testing similar to the above for damages from each potential damage type (see Table 2). Damage quantification may be achieved by possibly associating specific quantitative changes in the PSD with specific damage magnitudes.

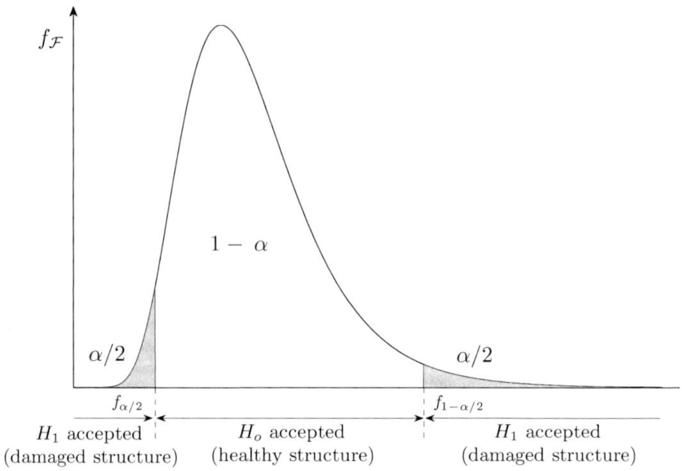

Figure 5. Statistical hypothesis testing based on an $\mathcal{F}$ distributed statistic (two-tail test).

Bibliographical remarks. Sakellariou et al. (2001) present the application of a variant of the PSD based method to fault detection on a railway vehicle suspension. Liberatore and Carman (2004) present the application of a simplified version (not using a statistical framework) to a simply supported aluminum beam. In the non-stationary or non-linear cases, time-frequency, polyspectra, or wavelet-based models may be used (Farrar and Doebling, 1997; Staszewski, 1998, 2000; Hou et al., 2000; Hera and Hou, 2004; Peng and Chu, 2004; Staszewski and Robertson, 2007).

4.2 FRF Magnitude Based Method

The FRF magnitude based method is similar to the PSD based method, but refers to the the excitation-response case and employs the FRF magnitude as the characteristic quantity $Q = |H(j\omega)|$. A somewhat similar approach may be used in case the excitation is unavailable but several responses are available (Sakellariou and Fassois, 2006; Mao and Todd, 2011). The main idea is the comparison of the FRF magnitude $|H_u(j\omega)|$ of the current state of the structure to that of the healthy structure $|H_o(j\omega)|$ (or, in fact, to that corresponding to any other structural state).

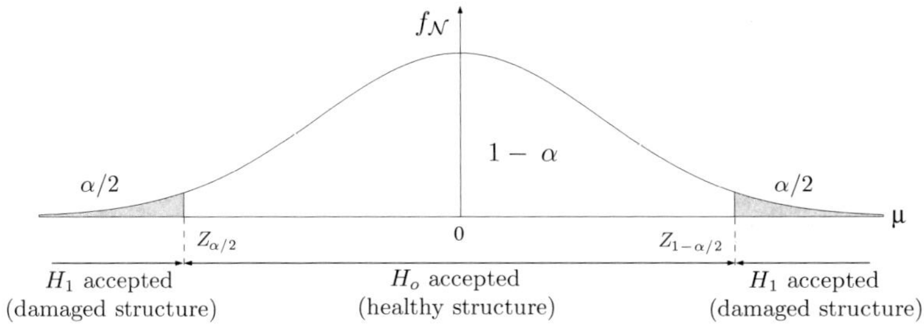

Figure 6. Statistical hypothesis testing based on a Gaussian distributed statistic (two-tail test).

Damage detection is based on the following hypothesis testing problem:

$$\begin{array}{l} H_o: \quad \delta|H(j\omega)| = |H_o(j\omega)| - |H_u(j\omega)| = 0 \\ \text{(null hypothesis -- healthy structure)} \\ H_1: \quad \delta|H(j\omega)| = |H_o(j\omega)| - |H_u(j\omega)| \neq 0 \\ \text{(alternative hypothesis -- damaged structure).} \end{array} \tag{21}$$

As the true FRFs $H_u(j\omega)$ and $H_o(j\omega)$ are unknown, their corresponding estimates, $\widehat{H}_u(j\omega)$ and $\widehat{H}_o(j\omega)$, obtained as indicated in Table 7, are employed. As indicated there, the FRF magnitude estimator may, asymptotically ($N \to \infty$), be considered to approximately follow Gaussian distribution (Bendat and Piersol, 2000, p. 338). As the data records Z_u and Z_o are mutually independent, the two FRF magnitude estimators are also mutually independent, implying that their difference $|\widehat{H}_o(j\omega)| - |\widehat{H}_u(j\omega)|$ is Gaussian with mean equal to the true magnitude difference and variance equal to the sum of the two variances.

Under the null hypothesis H_o (healthy structure), the true FRF magnitudes coincide ($|H_u(j\omega)| = |H_o(j\omega)|$), hence:

$$\text{Under } H_o: \quad \delta|\widehat{H}(j\omega)| = |\widehat{H}_o(j\omega)| - |\widehat{H}_u(j\omega)| \sim \mathcal{N}(0, 2\sigma_o^2(\omega)). \tag{22}$$

The variance $\sigma_o^2(\omega) = \text{var}[|\widehat{H}_o(j\omega)|]$ is generally unknown, but may be estimated in the baseline phase (Table 7). Assuming negligible variability of this estimate (which is true for "large" N), equality of the two FRF magnitudes may be examined at the selected α (false alarm) risk level through the following statistical test (Figure 6):

$$\begin{array}{lll} Z = \big|\delta|\widehat{H}(j\omega)|\big| / \sqrt{2\widehat{\sigma}_o^2(\omega)} \leq Z_{1-\frac{\alpha}{2}} & (\forall\, \omega) & \Longrightarrow H_o \text{ is accepted} \\ \qquad\qquad \text{Else} & & \Longrightarrow H_1 \text{ is accepted,} \end{array} \tag{23}$$

with $Z_{1-\alpha/2}$ designating the standard normal distribution's $1-\alpha/2$ critical point.

Damage identification may be similarly achieved by performing hypotheses testing similar to the above for damages from each potential damage type (see Table 2). Damage quantification may be achieved by possibly associating specific quantitative changes in the FRF magnitude with specific damage magnitudes.

Bibliographical remarks. Hwang and Kim (2004) present an FRF based method (though not in a statistical context), whose effectiveness is numerically demonstrated via simulation examples based on Finite Element (FE) models of a simple cantilever and a helicopter rotor blade. The method is reported to achieve satisfactory damage diagnosis. Rizos et al. (2008) employ the method for skin damage detection in stiffened aircraft panels. Changes in the FRF magnitude estimates due to damage are shown to exceed their normal variability bounds.

5 Selected Parametric Time Series SHM Methods

Parametric time series methods are those in which the characteristic quantity Q is constructed based on parametric time series representations. The response-only or excitation-response cases may be dealt with through the use of corresponding models (Basseville and Nikiforov 1993, section 6.2, Natke and Cempel 1997, Gertler 1998, section 4.2). Although parametric methods may operate on either the time or frequency domains, the former case is more widely used and is in the focus of this section.

Parametric methods may be further classified into three main categories:

(i) *Model parameter based methods*, which tackle damage detection and identification via a characteristic quantity Q that is a function of the estimated model parameters. These methods require that a model is re-estimated during the inspection phase based on the current signals Z_u.

(ii) *Model residual based methods*, which tackle damage detection and identification via characteristic quantities Q that are functions of the model residuals generated by driving the current signals Z_u through predetermined, in the baseline phase, models corresponding to the considered structural states. An advantage of these methods is that no model re-estimation is required in the inspection phase.

(iii) *Functional model (FM) based methods* constitute a family of advanced schemes which are capable of properly treating the subproblems of damage detection, identification (localization) and magnitude estimation within a unified framework. Damages characterized by a *double continuum* of locations and magnitudes on a structural topology may be considered. Model estimation is required in the inspection phase.

5.1 Model Parameter Based Methods

These methods perform damage detection and identification based on a characteristic quantity $Q = f(\boldsymbol{\theta})$, which is a function of the parameter vector $\boldsymbol{\theta}$ of a parametric time series model ($Q = \boldsymbol{\theta}$ in the typical case). Transformed forms of the parameter vector $\boldsymbol{\theta}$ may be also used, with the most common and historically important case being that in which the vector of the model natural frequencies is employed (Doebling et al., 1996; Salawu, 1997; Sohn et al., 2003a; Uhl and Mendrok, 2004; Rizos et al., 2008; Hios and Fassois, 2009b).

Let $\widehat{\boldsymbol{\theta}}$ designate a proper estimator of the parameter vector $\boldsymbol{\theta}$ (Söderström and Stoica, 1989, pp. 198–199, Ljung, 1999, pp. 212–213). For sufficiently long signals ("large" N) the estimator is (under mild assumptions) Gaussian distributed with mean equal to its true value $\boldsymbol{\theta}$ and a certain covariance $\boldsymbol{P_\theta}$ (Söderström and Stoica, 1989, pp. 205–207, Ljung, 1999, p. 303), that is:

$$\widehat{\boldsymbol{\theta}} \sim \mathcal{N}(\boldsymbol{\theta}, \boldsymbol{P_\theta}). \tag{24}$$

Damage detection is based on testing for statistically significant changes in $\boldsymbol{\theta}$ between the nominal and current structural states through the hypothesis testing problem:

$$\begin{array}{l} H_o : \delta\boldsymbol{\theta} = \boldsymbol{\theta}_o - \boldsymbol{\theta}_u = \mathbf{0} \\ \text{(null hypothesis – healthy structure)} \\ H_1 : \delta\boldsymbol{\theta} = \boldsymbol{\theta}_o - \boldsymbol{\theta}_u \neq \mathbf{0} \\ \text{(alternative hypothesis – damaged structure).} \end{array} \tag{25}$$

Due to the mutual independence of the Z_u and Z_o data records, the difference between the two parameter vector estimators also follows Gaussian distribution:

$$\delta\widehat{\boldsymbol{\theta}} = \widehat{\boldsymbol{\theta}}_o - \widehat{\boldsymbol{\theta}}_u \sim \mathcal{N}(\delta\boldsymbol{\theta}, \delta\boldsymbol{P}) \tag{26}$$

with

$$\delta\boldsymbol{\theta} = \boldsymbol{\theta}_o - \boldsymbol{\theta}_u, \quad \delta\boldsymbol{P} = \boldsymbol{P}_o + \boldsymbol{P}_u, \tag{27}$$

where $\boldsymbol{P}_o, \boldsymbol{P}_u$ designate the corresponding covariance matrices.

Under the null (H_o) hypothesis, $\delta\widehat{\boldsymbol{\theta}} = \widehat{\boldsymbol{\theta}}_o - \widehat{\boldsymbol{\theta}}_u \sim \mathcal{N}(\mathbf{0}, 2\boldsymbol{P}_o)$ and the quantity $\chi^2_{\boldsymbol{\theta}}$, below, follows χ^2 distribution with d (parameter vector dimensionality) degrees of freedom (as the sum of squares of independent standardized Gaussian variables – see Ljung 1999, p. 558 and Appendix):

$$\text{Under } H_o: \quad \chi^2_{\boldsymbol{\theta}} = \delta\widehat{\boldsymbol{\theta}}^T \cdot \delta \boldsymbol{P}^{-1} \cdot \delta\widehat{\boldsymbol{\theta}} \sim \chi^2(d) \tag{28}$$

with $\delta \boldsymbol{P} = 2\boldsymbol{P}_o$.

As the covariance matrix $\boldsymbol{P}_o$ corresponding to the healthy structure is unavailable, the estimated covariance $\widehat{\boldsymbol{P}}_o$ is used. Treating this as a quantity characterized by negligible variability (reasonable for large N), leads to the following test constructed at the α (false alarm) risk level (Figure 7):

$$\begin{array}{lcll} \chi^2_{\boldsymbol{\theta}} \leq \chi^2_{1-\alpha}(d) & \Longrightarrow & H_o \text{ is accepted} & \text{(healthy structure)} \\ \text{Else} & \Longrightarrow & H_1 \text{ is accepted} & \text{(damaged structure),} \end{array} \tag{29}$$

with $\chi^2_{1-\alpha}(d)$ designating the χ^2 distribution's $1-\alpha$ critical point. See Ljung 1999, p. 559 for an alternative approach based on the $\mathcal{F}$ distribution.

Damage identification may be based on the multiple hypotheses testing problem of Table 2 comparing the parameter vector $\widehat{\boldsymbol{\theta}}_u$ belonging to the current state of the structure to those corresponding to different damage types $\widehat{\boldsymbol{\theta}}_A, \widehat{\boldsymbol{\theta}}_B, \ldots$. Nevertheless, this is expected to work only for damages of specific magnitudes and cannot generally account for the continuum of damage magnitudes within each damage type. A *geometric method* aiming at circumventing this difficulty and also being suitable for damage estimation is presented in Sakellariou and Fassois (2006). Essentially this is a predecessor of Functional Model (FM) based methods and the reader is directed to subsection 5.3.

Bibliographical remarks. The principles of the model parameter based methods have been used in a number of studies. Sohn and Farrar (2000) use the parameters of an AR model and statistical process control charts for damage detection in a concrete bridge column, as it is progressively damaged. Adams and Farrar (2002) use frequency domain ARX models for damage detection in a simulated structural system and a three-story building model. Nair et al. (2006) employ the first three autoregressive parameters of an ARMA model to tackle damage detection. The postulated method is applied to analytical and experimental data from the ASCE benchmark structure. Sakellariou and Fassois (2006) employ the parameter vector of an Output Error (OE) model and statistical hypothesis testing procedures in order to tackle damage diagnosis in a six-story building model under earthquake excitation. Nair and Kiremidjian (2007) employ Gaussian mixture modelling of the parameter vector of an ARMA model to tackle damage detection on the ASCE benchmark structure. Zheng and Mita (2007) apply a two-stage damage diagnosis method (though not in a statistical context)

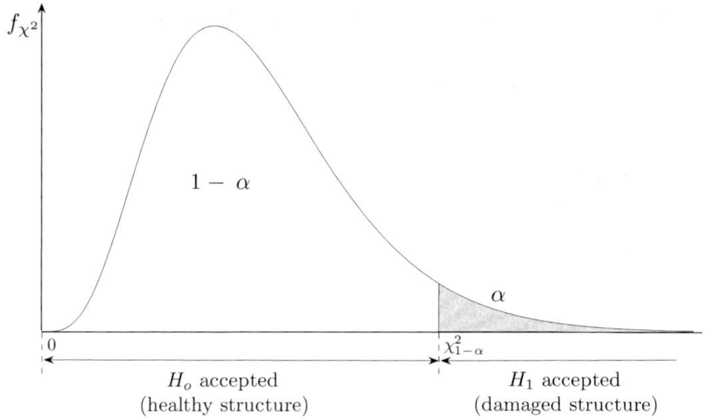

Figure 7. Statistical hypothesis testing based on a χ^2 distributed statistic (one-tail test).

on a five-storey steel structure. Damage detection is achieved in the first stage using the distance between two ARMA models, while damage localization is achieved in the second stage via pre-whitening filters. Carden and Brownjohn (2008) use ARMA model parameters for damage detection and identification in the IASC-ASCE benchmark four-story frame structure, in the Z24 bridge, and in the Malaysia-Singapore Second Link bridge. Hios and Fassois (2009b) employ the model parameter vector, or, alternatively, the modal parameters, of a global Functionally Pooled VAR model for damage detection in a smart composite beam under varying temperature. Mosavi et al. (2012) employ the distances between the parameters of Vector AR (VAR) models in order to detect and localize damage in a two-span continuous steel beam subject to ambient vibrations. In the non-linear model case Wei et al. (2005) use Non-linear AutoRegressive Moving Average with eXogenous excitation (NARMAX) models for fault detection and identification in carbon fiber-reinforced epoxy plates based on a deterministic index, which mirrors changes incurred in the modal parameters.

5.2 Model Residual Based Methods

Model residual based methods tackle damage detection and identification using characteristic quantities that are functions of the residual sequences obtained by driving the current signal(s) Z_u through suitable and predetermined (in the baseline phase) models $\mathcal{M}_o, \mathcal{M}_A, \mathcal{M}_B, \ldots$, each one corresponding to a particular structural state. The general idea is that

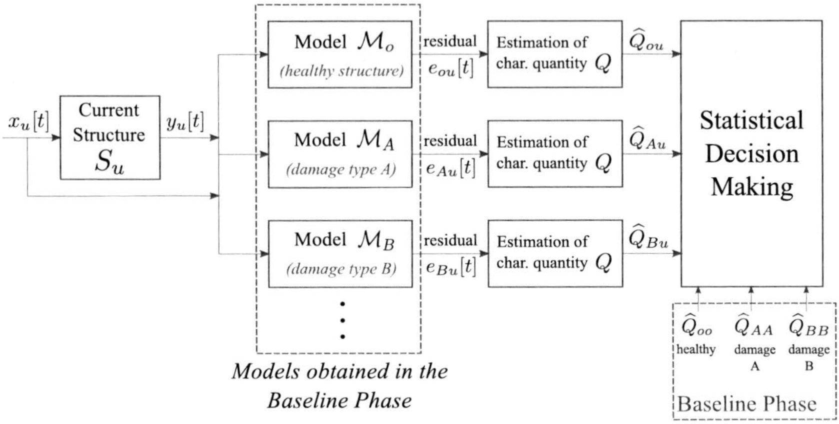

Figure 8. Schematic representation of a "general" residual based statistical time series SHM method (the operations associated with the baseline phase are within the dashed boxes).

the residual sequence obtained by a model that truly reflects the current structural state possesses certain distinct properties, and is thus possible to distinguish. An advantage of the methods is that no model re-identification is required in the inspection phase. The methods have a long history of development and application, mainly within the more general context of engineering systems (Basseville and Nikiforov, 1993; Natke and Cempel, 1997; Gertler, 1998).

Let $\mathcal{M}_V$ designate the model representing the structure in its V state ($V = o$ or $V = A, B, \ldots$). The residual series obtained by driving the current signal(s) Z_u through each one of the aforementioned models are designated as $e_{ou}[t], e_{Au}[t], e_{Bu}[t], \ldots$ and are characterized by variances $\sigma^2_{ou}, \sigma^2_{Au}, \sigma^2_{Bu}, \ldots$, respectively. Notice that the first subscript designates the model employed, while the second the structural state corresponding to the currently used excitation and/or response signal(s). The characteristic quantities obtained from the corresponding residual series are designated as $Q_{ou}, Q_{Au}, Q_{Bu}, \ldots$. On the other hand, the characteristic quantities obtained using the baseline data records are designated as Q_{VV} ($V = o$ or $V = A, B, \ldots$). A schematic representation for a "general" (generic) residual based SHM method is illustrated in Figure 8.

A residual variance based method. In this case the characteristic quantity is the variance of the model residual sequence. Damage detection is based on the fact that the residual series $e_{ou}[t]$, obtained by driving the current signals Z_u through the model $\mathcal{M}_o$ corresponding to the nominal (healthy) structure, should be characterized by variance $\sigma_{ou}^2 = \sigma_{oo}^2$ which becomes minimal if and only if the current structure is healthy ($S_u = S_o$).

Damage detection is thus based on the following hypothesis testing problem:

$$\begin{array}{ll} H_o: & \sigma_{ou}^2 \leq \sigma_{oo}^2 \text{ (null hypothesis – healthy structure)} \\ H_1: & \sigma_{ou}^2 > \sigma_{oo}^2 \text{ (alternative hypothesis – damaged structure).} \end{array} \tag{30}$$

Under the null (H_o) hypothesis, the residuals $e_{ou}[t]$ are (just like the residuals $e_{oo}[t]$) iid Gaussian with zero mean and variance σ_{oo}^2. Hence the quantities $N_u\widehat{\sigma}_{ou}^2/\sigma_{oo}^2$ and $(N_o - d)\widehat{\sigma}_{oo}^2/\sigma_{oo}^2$ follow central χ^2 distributions with N_u and $N_o - d$ degrees of freedom, respectively (as sums of squares of independent standardized Gaussian random variables; see Appendix). Note that N_o and N_u designate the number of samples used in estimating the residual variance in the healthy and current cases, respectively (typically $N_o = N_u = N$), and d designates the dimensionality of the estimated model parameter vector. N_u and N_o should be adjusted to $N_u - 1$ and $N_o - 1$, respectively, in case each estimated mean is subtracted from each residual sequence. Consequently, the following statistic follows $\mathcal{F}$ distribution with $(N_u, N_o - d)$ degrees of freedom (as the ratio of two independent and normalized χ^2 random variables; see Appendix):

$$\text{Under } H_o: \quad F = \frac{\frac{N_u\widehat{\sigma}_{ou}^2}{\sigma_{oo}^2 N_u}}{\frac{(N_o-d)\widehat{\sigma}_{oo}^2}{\sigma_{oo}^2(N_o-d)}} = \frac{\widehat{\sigma}_{ou}^2}{\widehat{\sigma}_{oo}^2} \sim \mathcal{F}(N_u, N_o - d). \tag{31}$$

The following test is thus constructed at the α (false alarm) risk level (Figure 9):

$$\begin{array}{lll} F \leq f_{1-\alpha}(N_u, N_o - d) & \Longrightarrow H_o \text{ is accepted} & \text{(healthy structure)} \\ \text{Else} & \Longrightarrow H_1 \text{ is accepted} & \text{(damaged structure)} \end{array} \tag{32}$$

with $f_{1-\alpha}(N_u, N_o - d)$ designating the corresponding $\mathcal{F}$ distribution's $1 - \alpha$ critical point.

Damage identification may be similarly achieved via pairwise tests of the form:

$$\begin{array}{ll} H_o: & \sigma_{Xu}^2 \leq \sigma_{XX}^2 \text{ (structure under damage type } X) \\ H_1: & \sigma_{Xu}^2 > \sigma_{XX}^2 \text{ (structure not under damage type } X). \end{array} \tag{33}$$

An alternative possibility could be based on obtaining the residual series $e_{Au}[t]$, $e_{Bu}[t]$, ... Then, the current damage type is determined as that

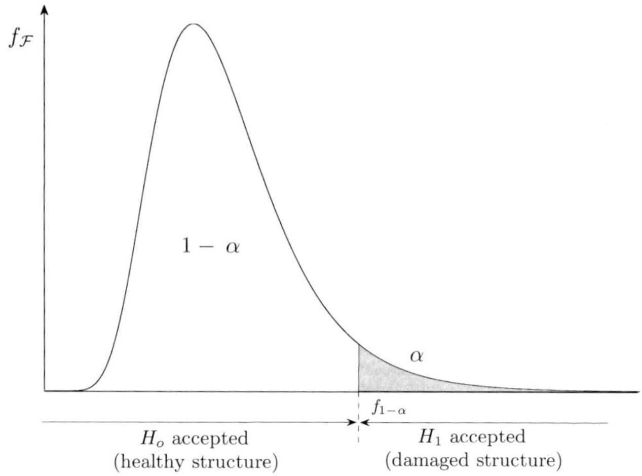

Figure 9. Statistical hypothesis testing based on an $\mathcal{F}$ distributed statistic (one-tail test).

one for which the residual series is characterized by minimal variance – notice that by including the residual sequence $e_{ou}[t]$ in the test, both damage detection and identification may be achieved. On the other hand, damage quantification may be possibly achieved in the limited case of a single damage type by associating specific values of the residual variance with specific damage magnitudes.

Bibliographical remarks. Sohn et al. (2001) use the prediction errors of a so-called AutoRegressive and AutoRegressive with eXogenous inputs (AR-ARX) model. The method is assessed via numerical simulations and its application to an eight degree-of-freedom mass-spring system, data obtained from a patrol boat, and a three-storey building model. In a related work, Sohn and Farrar (2001) employ the standard deviation ratio of the residuals of a AR-ARX model as the damage sensitive feature to infer the structural health state of an eight degree-of-freedom mass-spring system.

Fugate et al. (2001) use the AR model residuals, along with statistical process control methods, to monitor their mean and variance in order to detect damage on a concrete bridge column. Basseville et al. (2004) present a method based on subspace identification and state space model residuals in order to treat damage detection and localization. Yan et al. (2004) use an

X-bar control chart on state space model residuals for damage detection and identification in an aircraft skeleton and the Z24 bridge benchmark. Lu and Gao (2005) use an ARX model and the standard deviation of its residuals to treat damage detection and localization in a two and eight degree-of-freedom simulated mass-spring system. Sohn et al. (2005) explore the use of extreme value statistics on model residuals in order to classify damage.

An estimate of the standard deviation along with higher-order moments of the residuals obtained from vector AR models are used to detect damage by Mattson and Pandit (2006). A damage detection threshold level is identified from available training data, while the method is assessed via data obtained from an eight degree-of-freedom test bed. Zhang (2007) explores data normalization procedures and a probability based measure expressing changes in the ARX residual variance for damage detection and identification in a three-span continuous girder bridge simulation model. Gao and Lu (2009) present a formulation that enables the construction of residual generators via state-space representations. Damage detection is demonstrated via numerical results and experimental examples on a laboratory test frame. Rao and Ratnam (2012) employ the AR model residuals along with Shewhart and weighted moving average control charts for damage detection and identification in welded structures.

In the time-varying (non-stationary) case Poulimenos and Fassois (2004) employ a Time-varying AutoRegressive with eXogenous excitation (TARX) model along with statistical hypothesis testing employing its residual variance in order to detect faults in a bridge-like structure with moving mass. Spiridonakos and Fassois (2009) tackle fault detection in a time-varying extendable prismatic link structure via Functional Series Vector Time Dependent AutoRegressive (FS-VTAR) model residuals.

Methods employing Neural Network (NN) type non-linear models and deterministic decision making based on the response error (residual) are presented in Masri et al. (2000) and Huang et al. (2003).

A likelihood based method. In this case damage detection is based on the likelihood function under the null hypothesis H_o of a healthy structure (Gertler, 1998, pp. 119–120). The hypothesis testing problem thus is:

$$\begin{aligned} H_o: &\quad \boldsymbol{\theta}_o = \boldsymbol{\theta}_u \text{ (null hypothesis – healthy structure)} \\ H_1: &\quad \boldsymbol{\theta}_o \neq \boldsymbol{\theta}_u \text{ (alternative hypothesis – damaged structure)}, \end{aligned} \tag{34}$$

with $\boldsymbol{\theta}_o, \boldsymbol{\theta}_u$ designating the parameter vectors corresponding to the healthy and current structure, respectively. Assuming serial independence of the residual sequence, the Gaussian likelihood function $L_y(Y, \boldsymbol{\theta}/X)$ for the data

Y given X is obtained as (Box et al., 1994, p. 226):

$$L_y(Y, \boldsymbol{\theta}/X) = \frac{1}{\left(\sqrt{2\pi\sigma^2}\right)^N} \cdot \exp\left\{-\frac{1}{2\sigma^2}\sum_{t=1}^{N} e^2[t, \boldsymbol{\theta}]\right\} \tag{35}$$

with $e[t, \boldsymbol{\theta}]$ designating the model residual (one-step-ahead prediction error) characterized by zero mean and variance σ^2.

Under the null (H_o) hypothesis, the residual series $e_{ou}[t]$ generated by driving the current signal(s) through the nominal model is (just like $e_{oo}[t]$) iid Gaussian with zero mean and variance σ_{oo}^2. Damage detection is then based on the likelihood function under H_o evaluated for the current data, by requiring it to be larger or equal to a (properly selected) threshold l in order for the null (H_o) hypothesis to be accepted:

$$\begin{array}{lcll} L_y(Y, \boldsymbol{\theta}_o/X) \geq l & \Longrightarrow & H_o \text{ is accepted} & \text{(healthy structure)} \\ \text{Else} & \Longrightarrow & H_1 \text{ is accepted} & \text{(damaged structure).} \end{array} \tag{36}$$

Obviously, the evaluation of $L_y(Y, \boldsymbol{\theta}_o/X)$ requires knowledge of the true innovations variance σ_{oo}^2. This is typically a-priori unknown, but it may be estimated quite accurately for long data records ("large" N), in which case its variability may be also neglected. Then, under the null (H_o) hypothesis the statistic $N\widehat{\sigma}_{ou}^2/\widehat{\sigma}_{oo}^2$ follows χ^2 distribution with N degrees of freedom (as the sum of squares of mutually independent standardized Gaussian variables; see Appendix). This leads to the re-expression of the previous decision making rule as:

$$\begin{array}{ll} \chi_N^2 = \frac{N\widehat{\sigma}_{ou}^2}{\widehat{\sigma}_{oo}^2} \leq \chi_{1-\alpha}^2(N) & \Longrightarrow H_o \text{ is accepted (healthy structure)} \\ \qquad \text{Else} & \Longrightarrow H_1 \text{ is accepted (damaged structure),} \end{array} \tag{37}$$

with $\chi_{1-\alpha}^2(N)$ designating the χ^2 distribution's $1-\alpha$ critical point. Note that N should be adjusted to $N-1$ in case the estimated mean is subtracted from the residual series $e_{ou}[t]$.

Damage identification may be achieved by computing the likelihood function for the current signal(s) for the various values of $\boldsymbol{\theta}$ ($\boldsymbol{\theta}_A, \boldsymbol{\theta}_B, \ldots$) and accepting the hypothesis that corresponds to the maximum value of the likelihood:

$$\max_V L_y(Y, \boldsymbol{\theta}_V/X) \Longrightarrow H_V \text{ is accepted (damage type } V). \tag{38}$$

It should be mentioned that damage detection may be also achieved by including $\boldsymbol{\theta}_o$ in the hypothesis testing procedure. Damage quantification may be possibly achieved in the limited case of a single type of damage, by associating specific values of the likelihood with specific damage magnitudes.

A residual uncorrelatedness based method. This method is based on the fact that the residual sequence $e_{ou}[t]$ obtained by driving the current signal(s) Z_u through the nominal model $\mathcal{M}_o$ (resp. $\mathcal{M}_V$) is *uncorrelated* (white) if and only if the current structure is in its nominal (healthy) S_o (resp. S_V). Damage detection may be then based on the following hypothesis testing problem:

$$\begin{array}{ll} H_o: & \rho[\tau]=0 \quad \tau=1,2,\ldots,r \\ & \text{(null hypothesis – healthy structure)} \\ H_1: & \rho[\tau]\neq 0 \quad \text{for some } \tau \\ & \text{(alternative hypothesis – damaged structure)} \end{array} \tag{39}$$

with $\rho[\tau]$ being the normalized autocovariance (ACF) function (see Table 7) of the $e_{ou}[t]$ residual sequence. Thus, the method's characteristic quantity is $Q = \big[\rho[1]\ \rho[2]\ \ldots\ \rho[\tau]\big]^T$, with r being a design variable.

Under the null (H_o) hypothesis, the residual sequence $e_{ou}[t]$ is iid Gaussian with zero mean, and the statistic χ^2_ρ, below, follows χ^2 distribution with r degrees of freedom (Box et al., 1994, p. 314):

$$\text{Under } H_o: \quad \chi^2_\rho = N(N+2)\cdot\sum_{\tau=1}^{r}(N-\tau)^{-1}\cdot\widehat{\rho}^2[\tau] \ \sim\ \chi^2(r) \tag{40}$$

with $\widehat{\rho}[\tau]$ designating the estimator of $\rho[\tau]$.

Decision making is then based on the following test at the α (false alarm) risk level:

$$\begin{array}{lll} \chi^2_\rho \leq \chi^2_{1-\alpha}(r) & \Longrightarrow & H_o \text{ is accepted} \quad \text{(healthy structure)} \\ \text{Else} & \Longrightarrow & H_1 \text{ is accepted} \quad \text{(damaged structure)} \end{array} \tag{41}$$

with $\chi^2_{1-\alpha}(r)$ designating the χ^2 distribution's $1-\alpha$ critical point.

Damage identification may be achieved by similarly examining which one of the $e_{Vu}[t]$ (for $V=A,B,\ldots$) residual series is uncorrelated. As with the previous methods, only damages of specific magnitudes (but not the continuum of damage magnitudes) may be considered.

A Sequential Probability Ratio Test (SPRT) based method. This method employs the Sequential Probability Ratio Test (SPRT – Wald 1947; Ghosh and Sen 1991) in order to detect a change in the standard deviation σ_{ou} of the model residual sequence $e_{ou}[t]$. The SPRT based method employs both α (false alarm) and β (missed damage) error probabilities in its design. Damage detection is based on the SPRT of strength (α,β) for the following hypothesis testing problem:

$$\begin{array}{ll} H_o: & \sigma_{ou}\leq\sigma_o \text{ (null hypothesis – healthy structure)} \\ H_1: & \sigma_{ou}\geq\sigma_1 \text{ (alternative hypothesis – damaged structure)}, \end{array} \tag{42}$$

with σ_{ou} designating the standard deviation of the residual signal $e_{ou}[t]$ obtained by driving the current signal(s) through the healthy structural model, and σ_o, σ_1 user defined values. The basis of the SPRT is the logarithm of the likelihood ratio function based on n ($n \leq N$) samples:

$$\mathcal{L}(n) = \sum_{t=1}^{n} \ln \frac{f(e_{ou}[t]|H_1)}{f(e_{ou}[t]|H_o)} = n \cdot \ln \frac{\sigma_o}{\sigma_1} + \frac{\sigma_1^2 - \sigma_o^2}{2\sigma_o^2\sigma_1^2} \cdot \sum_{t=1}^{n} e_{ou}^2[t] \tag{43}$$

with $f(e_{ou}[t]|H_i)$ designating the probability density function of the residual sequence under hypothesis H_i ($i = 0, 1$).

Decision making is then based on the following test at the (α, β) risk levels:

$$\begin{array}{rcll} \mathcal{L}(n) \leq B & \Longrightarrow & H_o \text{ is accepted} & \text{(healthy structure)} \\ \mathcal{L}(n) \geq A & \Longrightarrow & H_1 \text{ is accepted} & \text{(damaged structure)} \\ B < \mathcal{L}(n) < A & \Longrightarrow & \text{no decision is made} & \text{(continue the test)} \end{array} \tag{44}$$

with:

$$A = \ln \frac{1-\beta}{\alpha} \quad \text{and} \quad B = \ln \frac{\beta}{1-\alpha}. \tag{45}$$

Following a decision at a stopping time $\widehat{n}$, it is possible to continue the test by resetting $\mathcal{L}(\widehat{n}+1)$ to zero and using additional residual samples.

Damage identification may be achieved by performing SPRTs similar to the above separately for damages of each potential type.

Bibliographical remarks. Sohn et al. (2003a) combine the SPRT with extreme value statistics for treating statistical damage classification in a laboratory three-story building model, while Oh and Sohn (2009) use the SPRT to tackle damage diagnosis under environmental and operational variations. Kopsaftopoulos and Fassois (2011) use the SPRT in order to detect and identify damage in a scale aircraft skeleton structure.

5.3 Functional Model (FM) Based SHM Methods

The Functional Model (FM) based methods provide a *unified framework* for the combined treatment of the damage detection, identification (localization), and quantification (magnitude estimation) subproblems. An important asset of the methods is overcoming the limitation of treating damages occurring *only* at *specific and pre-specified* locations and of *specific magnitudes*. Indeed, FM methods allow for the full and complete treatment of damages, achieving precise *localization* over *continuous topologies* on a structure and over the *continuum* of all damage magnitudes.

The cornerstone of the methods is the new class of *Functional Models (FMs)*, which allows for the analytical inclusion of *both* damage location and magnitude information (Kopsaftopoulos and Fassois, 2006; Sakellariou and Fassois, 2007; Hios and Fassois, 2009a,c). FM models essentially permit the extension of the notion of *damage mode* to include damage not only of all possible magnitudes, but also of all possible *locations* on a specific (continuous) topology on a structure. FMs are based on the *pooling* of multiple data records, thus they are also referred to as *Functionally Pooled* (FP) models, and are explicitly parametrized in terms of both damage location and magnitude. In the simple, special, case where only a finite number (instead of a continuum) of damage locations is considered, the parametrization may be limited to damage magnitude alone (Sakellariou and Fassois, 2008). This case is treated in the sequel for the sake of simplicity and clarity. Information on the more general case is provided in the bibliographical remarks.

Let the damage magnitude, within a specific damage mode (type) V, be represented by a scalar variable $k \in \mathbb{R}$. The healthy structure typically corresponds to $k = 0$. Then, a simple model capable of representing the structural dynamics under mode V is the *Functionally Pooled AutoRegressive with eXogenous excitation (FP-ARX)* model:

$$y_k[t] + \sum_{i=1}^{na} a_i(k) \cdot y_k[t-i] = \sum_{i=0}^{nb} b_i(k) \cdot x_k[t-i] + w_k[t] \tag{46a}$$

$$w_k[t] \sim \text{iid}\, \mathcal{N}\big(0, \sigma_w^2(k)\big), \quad k \in \mathbb{R} \tag{46b}$$

$$a_i(k) = \sum_{j=1}^{p} a_{i,j} \cdot G_j(k), \quad b_i(k) = \sum_{j=1}^{p} b_{i,j} \cdot G_j(k). \tag{46c}$$

In this expression $x_k[t]$, $y_k[t]$, and $w_k[t]$ designate the excitation, response, and innovations (residual) signals, respectively, corresponding to a specific damage magnitude k. The form of the model resembles that of a conventional ARX. Yet, its AR and X parameters, as well as its innovations variance, are explicit functions of the damage magnitude k, belonging to p-dimensional functional subspaces spanned by the mutually independent basis functions $G_1(k)$, $G_2(k)$, …, $G_p(k)$ (*functional basis*). The constants $a_{i,j}$ and $b_{i,j}$ designate the AR and X, respectively, coefficients of projection. The parameter vector to be estimated from the measured signals is $\boldsymbol{\theta} = [\alpha_{i,j} \vdots b_{i,j}]^T$.

In the *baseline phase*, a suitable FP model, corresponding to each considered damage mode, is estimated using signals obtained under various damage magnitudes k. In the *inspection phase*, given the current signal(s)

Z_u, damage detection is based on the FP-ARX model of damage mode V (or in fact any other). This is now re-parametrized in terms of the currently unknown damage magnitude k and the innovations variance $\sigma_{e_u}^2$ ($e_u[t]$ designates the re-parametrized model's innovations), by replacing the coefficients of projection by their corresponding estimates available from the baseline phase:

$$\mathcal{M}_V(k, \sigma_{e_u}^2): \; y_u[t] + \sum_{i=1}^{na} a_i(k) \cdot y_u[t-i] = \sum_{i=0}^{nb} b_i(k) \cdot x_u[t-i] + e_u[t]. \quad (47)$$

Damage detection is based on the following hypothesis testing problem:

$$\begin{array}{lll} H_o: & k = 0 & \text{(null hypothesis – healthy structure)} \\ H_1: & k \neq 0 & \text{(alternative hypothesis – damaged structure).} \end{array} \quad (48)$$

Estimates of k and $\sigma_{e_u}^2$ are obtained based on the current data Z_u and the Nonlinear Least Squares (NLLS) estimator (refer to Ljung 1999, pp. 327–329 for details on NLLS estimation):

$$\widehat{k} = \arg\min_k \sum_{t=1}^{N} e_u^2[t], \qquad \widehat{\sigma}_{e_u}^2 = \frac{1}{N} \sum_{t=1}^{N} \widehat{e}_u^2[t]. \quad (49)$$

Assuming that the structure is indeed under a damage belonging to mode V (or in healthy condition which simply corresponds to $k = 0$), the above estimator may be shown to be asymptotically ($N \to \infty$) Gaussian, with mean equal to its true value k and variance $\sigma_{\widehat{k}}^2$ provided by the Cramer-Rao lower bound (Sakellariou and Fassois, 2008):

$$\widehat{k} \sim \mathcal{N}(k, \sigma_{\widehat{k}}^2). \quad (50)$$

Under the null hypothesis H_o, the t statistic, below, follows t distribution with $N-1$ degrees of freedom (to be adjusted to $N-2$ in case the estimated mean is subtracted from the residuals in the evaluation of $\widehat{\sigma}_k$; see Appendix):

$$t = \widehat{k}/\widehat{\sigma}_k \sim t(N-1) \quad (51)$$

which leads to the following test at the α (false alarm) risk level:

$$\begin{array}{lll} t_{\frac{\alpha}{2}}(N-1) \leq t \leq t_{1-\frac{\alpha}{2}}(N-1) & \Longrightarrow & H_o \text{ is accepted (healthy structure)} \\ \text{Else} & \Longrightarrow & H_1 \text{ is accepted (damaged structure)} \end{array} \quad (52)$$

with t_α designating the t distribution's α critical point.

Once damage is detected, *damage mode identification* is achieved through successive estimation and validation of the re-parametrized model $\mathcal{M}_V(k, \sigma_{e_u}^2)$ of Equation (47) of each damage mode ($V = A, B, \ldots$) using the current signal(s) Z_u. The procedure terminates as soon as a particular model is successfully validated, with the corresponding damage mode identified as current. Model validation may be based on statistical tests examining the hypothesis of residual uncorrelatedness (see section 5.2).

Following this, an *interval estimate* (at the α risk level) of the damage magnitude k is then obtained based on the point estimates $\widehat{k}$ and $\widehat{\sigma}_k$:

$$k \text{ interval estimate:} \quad \left[\widehat{k} + t_{\frac{\alpha}{2}}(N-1) \cdot \widehat{\sigma}_k,\ \widehat{k} + t_{1-\frac{\alpha}{2}}(N-1) \cdot \widehat{\sigma}_k\right]. \quad (53)$$

Bibliographical remarks. Sakellariou et al. (2002) and Sakellariou and Fassois (2008) present the application of an FM based method which considers a finite number of fault locations for on-board fault detection in railway vehicle systems, and damage detection, localization and magnitude estimation in a scale aircraft skeleton structure, respectively. Kopsaftopoulos and Fassois (2007, 2012) postulate the generalization of the FM based method in order to include damage not only of all possible magnitudes, but also of all possible locations on continuous topologies on a structure. The method's effectiveness is assessed via its application to damage detection, precise localization and magnitude estimation on a scale aircraft skeleton structure.

6 Application of the Methods to a Laboratory Truss Structure

6.1 The Laboratory Truss Structure and Problem Definition

The use of various methods is now illustrated through their application to damage diagnosis on a laboratory truss structure. The structure and part of the experimental set-up are shown in Figure 10. It consists of twenty eight elements with rectangular cross sections (15×15 mm) jointed together via steel elbow plates and bolts. All parts are constructed from standard aluminium with the overall dimensions being $1400 \times 700 \times 800 \times 700$ mm.

The damages considered correspond to complete loosening of various bolts at different joints of the structure. Five distinct types, each corresponding to the loosening of bolts joining together various horizontal, vertical and diagonal elements, are considered (Table 8 and Figure 10).

The structure is suspended through a set of cords and is excited vertically at Point X through an electromechanical shaker (MB Dynamics Modal 50A, max load 225 N) equipped with a stinger (Figure 10). The force excitation is

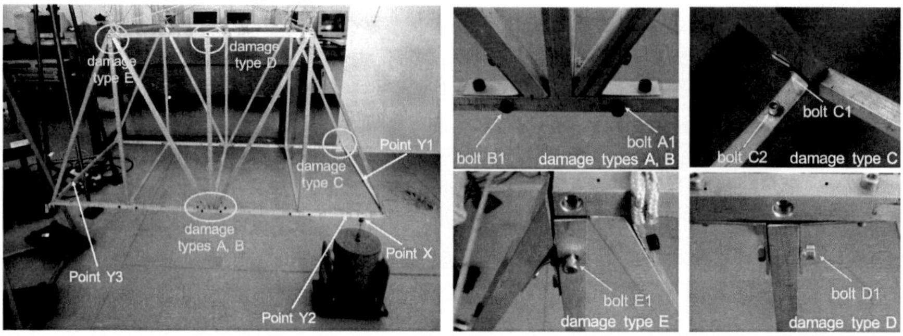

Figure 10. The aluminum truss structure and the experimental set-up: The force excitation (Point X), the vibration measurement positions (Points Y1, Y2, Y3), and the considered damage types (A, B, C, D, and E).

Table 8. The considered damage types and experimental details.

Structural State	Description	No of Experiments
Healthy	—	40 (1 baseline)
Damage type A	loosening of bolt A1	32 (1 baseline)
Damage type B	loosening of bolts A1 and B1	32 (1 baseline)
Damage type C	loosening of bolts C1 and C2	32 (1 baseline)
Damage type D	loosening of bolt D1	32 (1 baseline)
Damage type E	loosening of bolt E1	32 (1 baseline)

Sampling frequency: $f_s = 256$ Hz, Signal bandwidth: $[0.5 - 100]$ Hz
Signal length N in samples (s): Non-parametric methods: $N = 30\ 720$ (120 s)
Parametric methods: $N = 10\ 000$ (39 s)

a random Gaussian signal measured via an impedance head (PCB 288D01, sensitivity 98.41 mV/lb), while the resulting strain responses are measured at different points via dynamic strain gauges (PCB ICP 740B02, 0.005 − 100 kHz, 50 mV/$\mu\varepsilon$). The analysis bandwidth is 0.5 − 100 Hz and the sampling frequency $f_s = 256$ Hz. The measured signals are driven through a signal conditioning device (PCB 481A02) into the data acquisition system (SigLab 20–42). In this study damage detection and identification is based on one of the three vibration response signals (Points Y1, Y2 and Y3 – Figure 10) at a time, so that *scalar* (univariate) versions of the methods are used.

Experimental details and the number of experiments for each damage type are presented in Table 8. One experiment per damage type is executed in the baseline phase, while several are executed in the inspection phase. The excitation and response signals are in all cases pre-processed, so that the sample mean is subtracted and each signal is normalized by its sample standard deviation.

Both response-only and excitation-response methods are used. In all cases, the fact that only a single response is used and also in a very limited bandwidth, renders the damage diagnosis problem challenging. This allows for the exploration of the capabilities and limitations of the methods when very limited information is available.

6.2 Baseline Phase: Structural Identification

Non-parametric identification. Non-parametric identification is based on $N = 30\ 720$ (≈ 120 s) sample-long excitation and/or response signals.

Sample response PSD and FRF magnitude estimates for the healthy and various damage states of the structure are depicted in Figure 11 for response Y3 (MATLAB functions *pwelch.m* and *tfestimate.m*, respectively – details in Table 9). It is evident that the healthy and damage estimates are quite similar in the 0.5 – 30 Hz range, where the first twelve modes are included. Significant discrepancies between the healthy and damage type C, D and E curves are observed in the 30 – 58 Hz range, where the next three modes are included. These discrepancies become even more evident for damage types C and E in the 58 – 100 Hz range, where the next eight modes are included. Figure 12 depicts the Welch based FRF magnitude estimates for the healthy and damage type A structural states, along with their corresponding 95% confidence intervals (refer to Table 7). Although both FRF magnitude curves are quite similar, small discrepancies are evident at specific frequencies. Yet the statistical significance of these discrepancies has to be confirmed.

Parametric identification. In the parametric case only excitation-response representations are presently employed. Estimation is based on $N = 10\ 000$ (≈ 39 s) sample-long excitation and response signals which are used for estimating AutoRegressive with eXogenous excitation (ARX) models (MAT-

Table 9. Non-parametric estimation details.

Data length	$N = 30\ 720$ samples (≈ 120 s)
Method	Welch, zero overlap
Segment length (samples)	$L = 2\ 048$ samples
No of non-overlapping segments	$K = 15$ segments
Window type	Hamming
Frequency resolution	$\Delta f = 0.125$ Hz

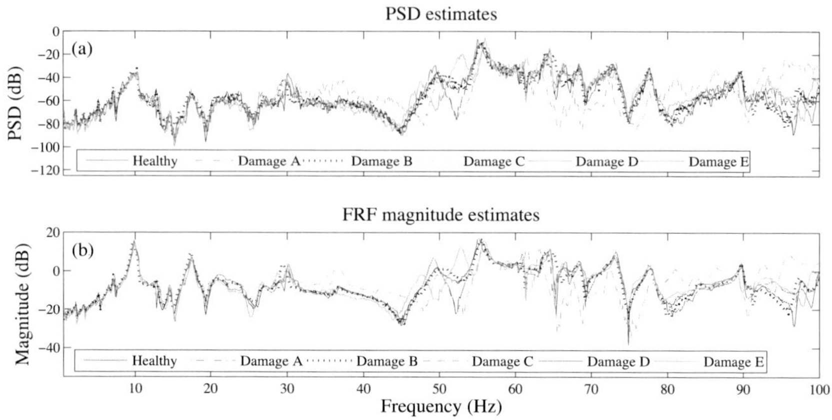

Figure 11. Indicative non-parametric (Welch based) estimates for the healthy and damaged structural states (response Y3): (a) Power Spectral Density (PSD) and (b) Frequency Response Function (FRF) magnitude estimates.

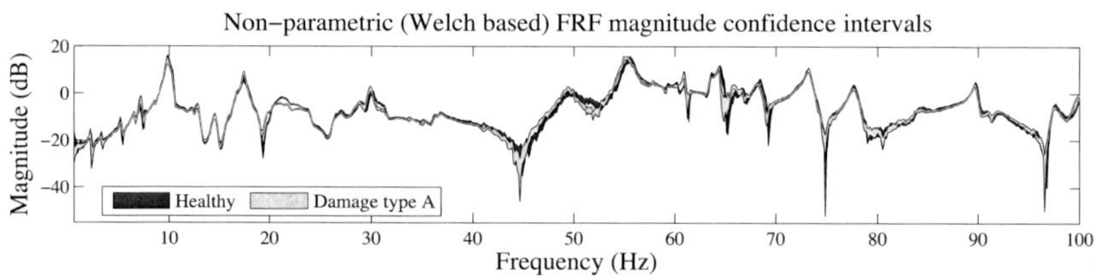

Figure 12. Non-Parametric (Welch based) FRF magnitude estimates for the healthy and damage type A structural states along with their 95% confidence intervals (response Y3).

LAB function *arx.m*). The modelling strategy consists of the successive fitting of ARX(na, nb) models ($na = nb = n$ is presently used) until a suitable model is selected. Model parameter estimation is based on minimization of a quadratic Prediction Error (PE) criterion leading to a Least Squares (LS) estimator (Ljung 1999, p. 206, Fassois 2001). Model order selection, which is crucial for successful identification, is based on a combination of tools, including the Bayesian Information Criterion (BIC – Figure 13a), the RSS/SSS (Residual Sum of Squares over Signal Sum of Squares – Figure 13b), residual series whiteness, and "stabilization diagrams".

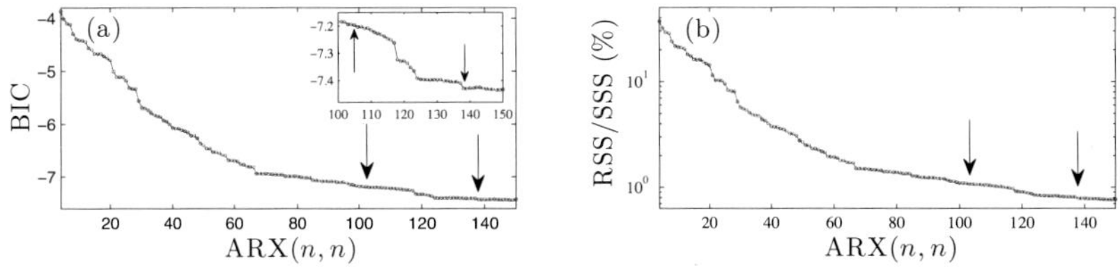

Figure 13. Order selection criteria for ARX(n,n) type parametric models in the healthy case (response Y3): (a) BIC and (b) RSS/SSS.

Table 10. Selected models and estimation details.

Response	Selected Model	No parameters	SPP	BIC	RSS/SSS (%)
Y1	ARX(112, 112)	225 parameters	44.4	−5.19	0.43
Y2	ARX(136, 136)	273 parameters	36.6	−5.83	0.22
Y3	ARX(103, 103)	207 parameters	48.3	−4.31	1.07
Parameter estimation method: Weighted Least Squares (WLS)					
Signal length: $N = 10\ 000$ samples (≈ 39 s)					

The identification procedure leads to the selection of an ARX(112, 112), an ARX(136, 136) and an ARX(103, 103) model for vibration responses Y1, Y2 and Y3, respectively. The selected ARX models, as well as their estimation details, numbers of estimated parameters, signal Samples Per Parameter (SPP), BIC, and RSS/SSS values are summarized in Table 10. Figure 14 presents a comparison between the parametric (response Y3 – ARX(103, 103) based) FRF magnitude estimate and its non-parametric (Welch based) counterpart; the agreement is excellent.

6.3 Inspection Phase: SHM via Selected Non-Parametric Methods

PSD based method. Typical PSD based damage detection results are presented in Figure 15 based on response Y3. Evidently, correct detection at the $\alpha = 10^{-4}$ risk level is achieved in each case, as the test statistic is shown not to exceed the critical points (dashed horizontal lines) in the healthy case, while it exceeds them in each damage case. In each case the sensor location with respect to the damage location is characterized as "local" or "remote". Observe that damage type C (two bolts loosened) is the easiest to detect (note the logarithmic scale on the vertical axis of Figure 15), while damage type A (one bolt loosened) is the hardest (the test statistic is within the

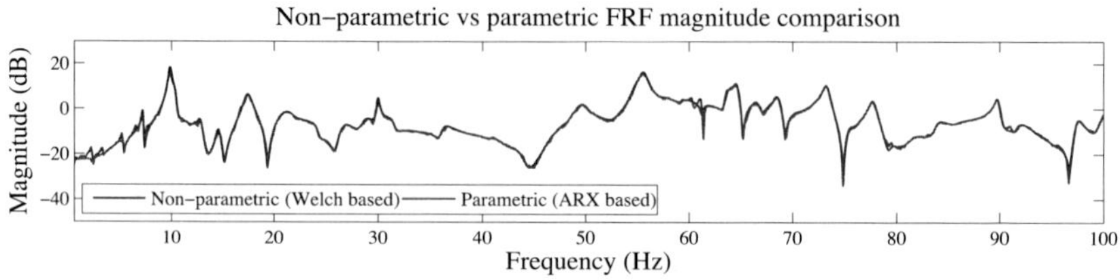

Figure 14. Parametric (ARX based) and non-parametric (Welch based) FRF magnitude estimates for the healthy structure (response Y3).

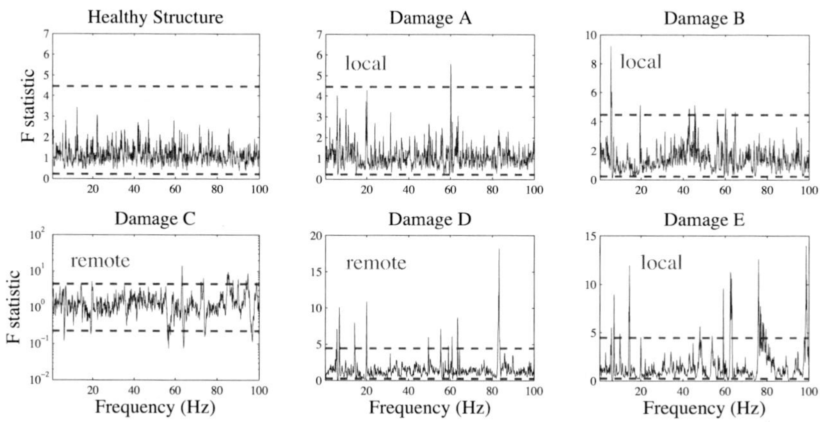

Figure 15. PSD based damage detection (response Y3): Indicative results at the $\alpha = 10^{-4}$ risk level. "Local" and "remote" damages are considered with the actual structural state indicated above each plot box. A damage is detected if the test statistic exceeds the critical points (dashed horizontal lines).

critical points for most frequencies).

Representative damage identification results are, at the $\alpha = 10^{-4}$ risk level, presented in Figure 16 based on response Y1. The actual damage is of type A. When testing the hypothesis of damage type A, the test statistic does not exceed the critical points, while it clearly exceeds them when testing any other hypothesis.

FRF based method. Figure 17 presents typical FRF based damage detection results based on response Y2. In each case the sensor location with

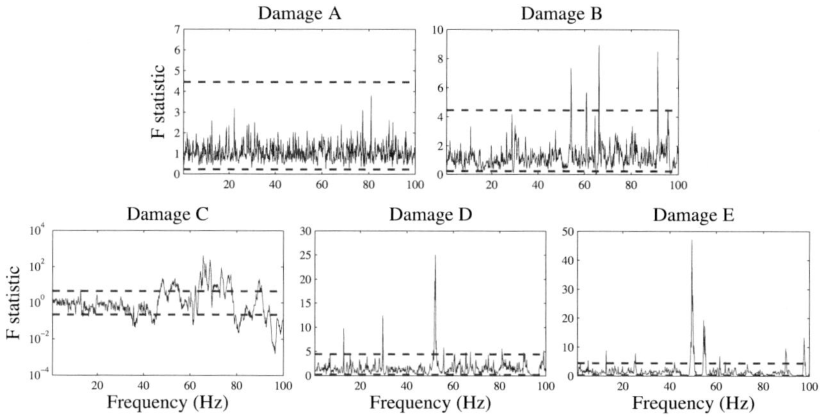

Figure 16. PSD based damage identification (response Y1): Indicative results at the $\alpha = 10^{-4}$ risk level, with the actual damage being of type A ("remote" to the sensor location). Each considered hypothesis is shown above each plot box. A hypothesis is accepted as true if the corresponding test statistic does not exceed the critical points (dashed horizontal lines).

respect to the damage location is characterized as "local" or "remote". Correct detection at the $\alpha = 10^{-5}$ risk level is achieved in each case, as the test statistic is shown not to exceed the critical points (dashed horizontal lines) in the healthy case, while it exceeds them in each damage case. Like before, damage type C is the easiest to detect, while damage types A and B are the hardest.

Indicative damage identification results at the $\alpha = 10^{-5}$ risk level are presented in Figure 18 based on response Y1. The actual damage is of type C. When testing the damage type C hypothesis the statistic does not exceed the critical points, while it clearly does so when testing any other hypothesis.

6.4 Inspection Phase: SHM via Selected Parametric Methods

Model parameter based method. The model parameter based method (excitation-response case) is based on the ARX models (Table 10) obtained in the baseline phase, as well as on their counterparts obtained in the inspection phase using the current data records Z_u.

Figure 19 presents typical parametric damage detection results based on the excitation–Y2 response pair. Correct detection is achieved at the $\alpha = 10^{-12}$ risk level, as the test statistic does not exceed the critical point

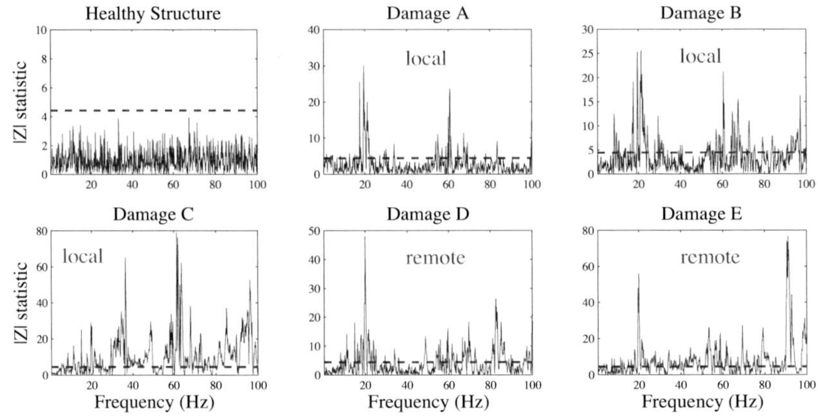

Figure 17. FRF based damage detection (response Y2): Indicative results at the $\alpha = 10^{-5}$ risk level. "Local" and "remote" damages are considered with the actual structural state indicated above each plot box. A damage is detected if the test statistic exceeds the critical point (dashed horizontal line).

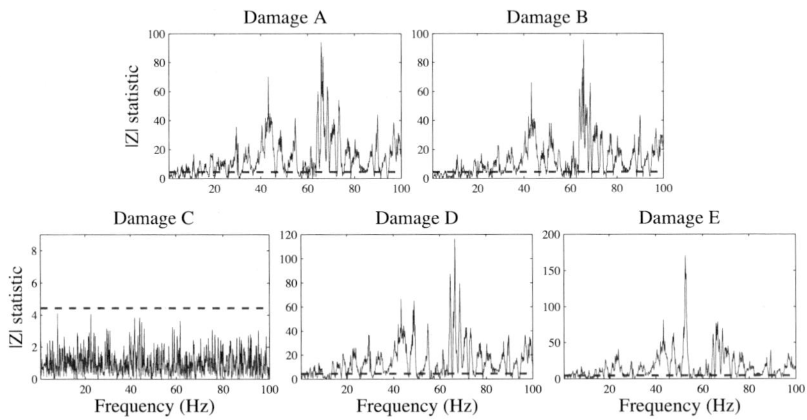

Figure 18. FRF based damage identification (response Y1): Indicative results at the $\alpha = 10^{-5}$ risk level, with the actual damage being of type C ("local" to the sensor location). Each considered hypothesis is shown above each plot box. A hypothesis is accepted as true if the corresponding test statistic does not exceed the critical point (dashed horizontal line).

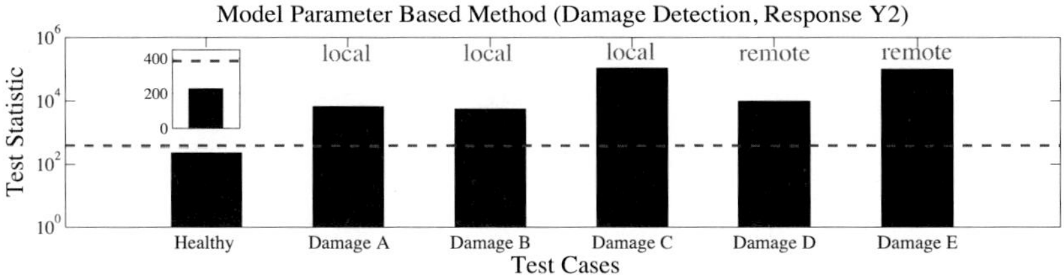

Figure 19. Model parameter based damage detection (excitation–Y2 response): Indicative results at the $\alpha = 10^{-12}$ risk level. "Local" and "remote" damages are considered and characterized above each bar. A damage is detected if the test statistic (bar) exceeds the critical point (dashed horizontal line).

in the healthy case, while it exceeds it in each damage case. Note the logarithmic scale on the vertical axis, indicating significant differences in the statistic between the healthy and the damage cases. The ability of the method to properly identify the damage type is demonstrated in Figure 20 at the $\alpha = 10^{-12}$ risk level using the excitation–Y3 response pair (two test cases are shown).

Residual likelihood based method. This method is based on the ARX models identified in the baseline phase (Table 10). No identification is required in the inspection phase.

Typical damage detection results are presented in Figure 21 based on the excitation–Y1 response pair. Correct detection is achieved at the $\alpha = 10^{-12}$ risk level as the test statistic does not exceed the critical point in the healthy case while exceeding it in each damage case. Damage identification results (two test cases) are presented in Figure 22 based on the excitation–Y3 response pair. In each test case correct identification is achieved at the $\alpha = 10^{-12}$ risk level – it is worth noting the logarithmic scale on the vertical axis.

SPRT based method. Like in the previous method the ARX models identified in the baseline phase (Table 10) are used, with no identification required in the inspection phase.

Typical damage detection results are presented in Figure 23 based on the excitation–Y2 response pair. The characterization of each damage as "local" or "remote" to the response sensor is indicated in each subplot. A

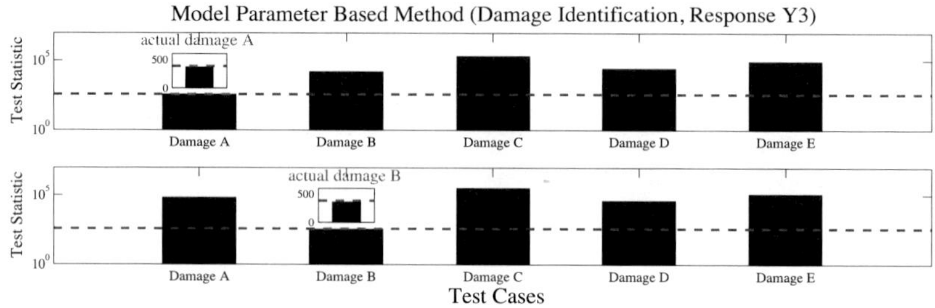

Figure 20. Model parameter based damage identification (excitation–Y3 response): Two indicative test cases at the $\alpha = 10^{-12}$ risk level. The actual damage is indicated in each test case (subplot). A hypothesis is accepted as true if the corresponding test statistic (bar) does not exceed the critical point (dashed horizontal line).

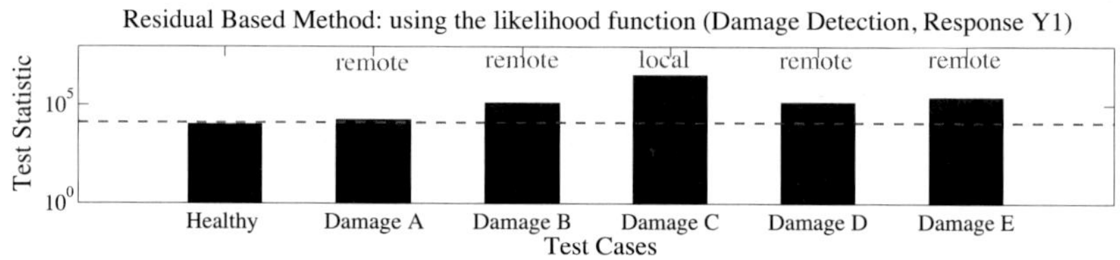

Figure 21. Residual likelihood based damage detection (excitation–Y1 response): Indicative results at the $\alpha = 10^{-12}$ risk level. “Local” and “remote” damages are considered above each bar. A damage is detected if the test statistic (bar) exceeds the critical point (dashed horizontal line).

damage is detected when the test statistic exceeds the *upper* critical point (upper dashed line), while the structure is detected as healthy when the test statistic exceeds the *lower* critical point (lower dashed line). After a critical point is exceeded and a decision is made, the test statistic is reset to zero and the test continues. Evidently, correct detection, at the $\alpha = \beta = 0.01$ risk levels ($q = 1.1$), is obtained in each test case, as the test statistic is shown to exceed multiple times the lower critical point in the healthy case, while it repeatedly exceeds the upper critical point in each damage test case. Observe that damage type A appears harder to detect, as the number of detections in this case is lowest among those of the various damage types, while damage types C and E appear easiest to detect. This is in agreement

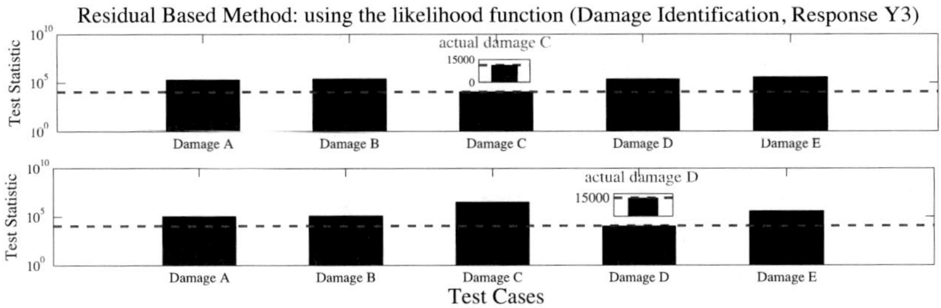

Figure 22. Residual likelihood based damage identification (excitation–Y3 response): Two indicative results at the $\alpha = 10^{-12}$ risk level. The actual damage is indicated in each case (subplot). A hypothesis is accepted as true if the corresponding test statistic (bar) does not exceed the critical point (dashed horizontal line).

with the remarks made in subsection 6.2 and Figure 11.

The ability of the method to identify damage is demonstrated via the test case of Figure 24 based on the excitation–Y3 response pair. The actual damage type (type B) is properly identified as the test statistic exceeds (for the corresponding hypothesis) the lower critical point.

6.5 Discussion

Summary results for all test cases considered and for each one of the three vibration responses (Y1, Y2 and Y3) are presented in Table 11. Evidently, both non-parametric and parametric methods achieve accurate damage detection with mostly zero false alarms at the selected risk levels. Only the FRF based method exhibits one and two false alarms for the vibration responses Y1 and Y3, respectively. The ability of the methods to effectively detect damage is demonstrated by the fact that no missed damage cases are observed. The damage identification results confirm the ability of the methods to accurately identify the damage type. No damage misclassifications are observed, except for the FRF based method where misclassification errors occur for damage type A (Table 11).

It is important to emphasize that these results are achieved using a *single* vibration response signal, or a *single* excitation-response pair, and also a particularly *low frequency range* (0.5 – 100 Hz). It is well known that the problem is more challenging in a "low" frequency range, yet the results demonstrate that it is properly handled without the need for higher frequency ranges. The fact that the methods are capable of detecting and

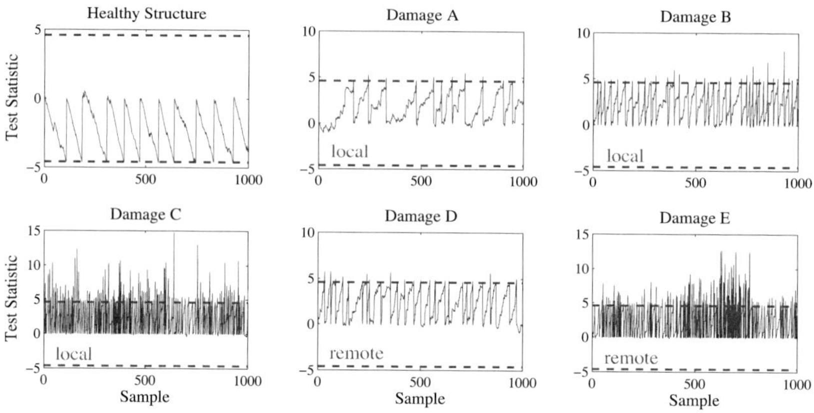

Figure 23. SPRT based damage detection (excitation–Y2 response): Indicative results at the $\alpha = \beta = 0.01$ risk levels ($q = 1.1$). "Local" and "remote" damages are considered with the actual structural state indicated above each plot box. A damage is detected if the test statistic exceeds the *upper* critical point (upper dashed line), while the structure is in its healthy state when the test statistic exceeds the *lower* critical point (lower dashed line).

identifying damage using response sensors that are relatively close ("local") or far ("remote") from the actual damage location should be also emphasized. Of course, performance is somewhat affected by distance; this is demonstrated for the damage type A case in conjunction with the FRF magnitude based method where the lowest misclassification rate occurs for sensor Y2 (Table 11) which is closest to the damage location. Yet, the ability of the methods in this respect is remarkable and underscores the fact that a few (or even a single) vibration response sensors may be adequate for proper detection and identification.

Nevertheless, in using statistical time series SHM methods, a number of issues require attention on part of the user. First, careful model identification, especially in the parametric case, is crucial for successful damage diagnosis. Parametric models require accurate parameter estimation and model structure selection in order to properly represent the structural dynamics and be effectively used for damage diagnosis. Therefore parametric methods require adequate user expertise and are somewhat more elaborate than their non-parametric counterparts.

Another issue of primary importance is the proper selection of the α

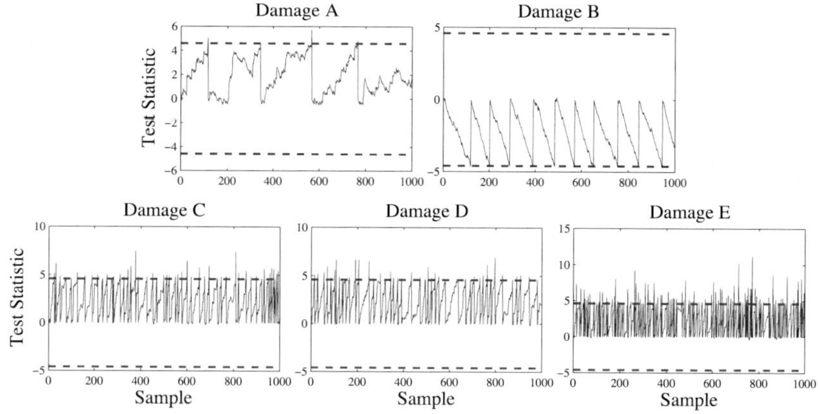

Figure 24. SPRT based damage identification (excitation–Y3 response): Indicative results at the $\alpha = \beta = 0.01$ risk levels ($q = 1.1$), with the actual damage being of type B ("local" to the sensor location). Each considered hypothesis is indicated above each plot box. A hypothesis is accepted as true if the corresponding test statistic exceeds the *lower* critical point (lower dashed line).

(false alarm) risk level. If not properly adjusted, false alarm, missed damage, and damage misclassification cases may occur. The user is advised to make an initial investigation on false alarm rates for different α levels using several healthy data sets. Afterwards, potential missed damage cases may be checked with data corresponding to various damaged structural states. When applying the model residual uncorrelatedness based method, the user should be aware of the fact that the selected max lag r value may affect performance. Thus, a tentative inquiry on this affects the false alarm rate should be undertaken.

Moreover, in order for most parametric methods to work effectively, a very small value of α is very often required. This is due to the fact that the current time series models are incapable of fully capturing the experimental, operational and environmental uncertainties based on just a *single* data record (Hios and Fassois, 2009a; Michaelides and Fassois, 2008). For this reason, a very small α is often selected in order to compensate for the lack of proper uncertainty modelling. This is the subject of on-going research efforts.

It should be noted that only detection is possible for damage types not modeled in the baseline phase, while essentially no work is reported on mul-

Table 11. Summary damage detection and identification results.

Method	Damage Detection					
	False alarms	Missed damage cases				
		dam A	dam B	dam C	dam D	dam E
PSD	0/0/0	0/0/0	0/0/0	0/0/0	0/0/0	0/0/0
FRF	**1/0/2**	0/0/0	0/0/0	0/0/0	0/0/0	0/0/0
Mod. parameter†	0/0/0	0/0/0	0/0/0	0/0/0	0/0/0	0/0/0
Res. variance†	0/0/0	0/0/0	0/0/0	0/0/0	0/0/0	0/0/0
Res. likelihood†	0/0/0	0/0/0	0/0/0	0/0/0	0/0/0	0/0/0
Res. uncorrelatedness†	0/0/0	0/0/0	0/0/0	0/0/0	0/0/0	0/0/0
SPRT	0/0/0	0/0/0	0/0/0	0/0/0	0/0/0	0/0/0
Method	**Damage Identification**					
	Damage misclassification cases					
		dam A	dam B	dam C	dam D	dam E
PSD		0/0/0	0/0/0	0/0/0	0/0/0	0/0/0
FRF		**2/1/2**	0/0/0	0/0/0	0/0/0	0/0/0
Mod. parameter†		0/0/0	0/0/0	0/0/0	0/0/0	0/0/0
Res. variance†		0/0/0	0/0/0	0/0/0	0/0/0	0/0/0
Res. likelihood†		0/0/0	0/0/0	0/0/0	0/0/0	0/0/0
Res. uncorrelatedness†		0/0/0	0/0/0	0/0/0	0/0/0	0/0/0
SPRT		0/0/0	0/0/0	0/0/0	0/0/0	0/0/0

False alarms for responses Y1/Y2/Y3 out of 39 test cases each; †adjusted α.
Missed damages for responses Y1/Y2/Y3 out of 31 test cases each.
Damage misclassification for responses Y1/Y2/Y3 out of 31 test cases each.

tiple damages. In its more complete form in which damage may occur at an infinite number of locations on a continuum, the damage identification problem requires precise localization which – in the context of statistical time series methods – is possible only through the new Functional Model (FM) methods. These are more elaborate but also allow for damage magnitude estimation.

7 Concluding Remarks and Future Research

- Statistical time series SHM methods are capable of achieving damage detection, identification (including precise localization), and quantification (magnitude estimation) in both the *response-only* and *excitation-response* cases based on (i) random excitation and/or response signals, (ii) statistical model building, and (iii) statistical decision making under uncertainties.
- The methods may be classified as non-parametric or parametric. *Non-parametric* methods are generally simpler, while they mainly focus on the damage detection subproblem, although simple forms of the

damage identification subproblem may be also tackled. *Parametric* methods are more elaborate, as they necessitate the use of proper parameter estimation and model structure selection techniques. Yet, they offer potentially high performance along with more effective and precise damage identification (including localization) and quantification.

- The use of random excitation and/or vibration response signals is important and implies that signals obtained under *normal operating conditions* may be potentially employed, which is practically very important. This is further enhanced by the ability of the methods to work in the *lower frequency range*, as vibration signals obtained under normal operating conditions are often characterized by low frequency content.
- Two additional and practically important advantages of the methods is (i) the use of *simple and partial* (both space and bandwidth wise) dynamical models and (ii) a *limited number* of measured signals.
- The handling of uncertainties via proper statistical techniques is an additional asset.
- Sequential methods require, on average, a substantially smaller number of observations than their fixed sample size counterparts (Wald, 1947; Ghosh and Sen, 1991; Lehmann and Romano, 2008) and warrant further investigation. An added bonus is their direct suitability for on-line implementation, although other methods may be adapted as well.
- On the other hand, statistical time series SHM methods are limited to the identification of damage only to the extent allowed by the specific type of model employed, while they also require adequate user expertise. Another limitation relates to the requirement for vibration data records corresponding to the potential damage states of the structure in the baseline phase and in case that damage identification and estimation are sought. Such data may be difficult to obtain, but either laboratory scale models or analytical (like tuned finite element) models may be alternatively used.
- Further research is necessary for exploring the limits and applicability of the methods in tackling the less studied damage identification and estimation (quantification) subproblems, including the important multiple damage case. Effectively handling environmental effects, and distinguishing them from those of damage, is a critical issue warranting further investigation, as is the more effective handling of uncertainties. More automated methods and methods suitable for the multivariate (multi sensor) case (especially in conjunction with large

structures where dense sensor arrays may be used) need to be developed. Similarly, methods suitable for structures exhibiting nonlinear and/or time-varying (non-stationary) dynamics have received only limited attention and warrant further consideration.

Bibliography

D. E. Adams. *Health Monitoring of Structural Materials and Components.* John Wiley & Sons Inc., 2007.

D. E. Adams and C. R. Farrar. Classifying linear and nonlinear structural damage using frequency domain arx models. *Structural Health Monitoring*, 1(2):185–201, 2002.

D. Balageas, C. P. Fritzen, and A. Guemes, editors. *Structural Health Monitoring.* ISTE Ltd, 2006.

M. Basseville and I. V. Nikiforov. *Detection of Abrupt Changes.* PTR Prentice-Hall, 1993.

M. Basseville, M. Abdelghani, and M. Benveniste. Subspace-based faul detection algorithms for vibration monitoring. *Automatica*, 36(1):101–109, 2000.

M. Basseville, L. Mevel, and M. Goursat. Statistical model–based damage detection and localization: subspace–based residuals and damage-to-noise sensitivity ratios. *Journal of Sound and Vibration*, 275:769–794, 2004.

J. S. Bendat and A. G. Piersol. *Random Data: Analysis and Measurement Procedures.* Wiley–Interscience: New York, 3rd edition, 2000.

M. Benedetti, V. Fontanari, and D. Zonta. Structural health monitoring of wind towers: remote damage detection using strain sensors. *Smart Materials and Structures*, 20:13pp, 2011.

G. E. P. Box, G. M. Jenkins, and G. C. Reinsel. *Time Series Analysis: Forecasting & Control.* Prentice Hall: Englewood Cliffs, NJ, third edition, 1994.

E. P. Carden and J. M. Brownjohn. Arma modelled time–series classification for structural health monitoring of civil infrastructure. *Mechanical Systems and Signal Processing*, 22(2):295–314, 2008.

E. P. Carden and P. Fanning. Vibration-based condition monitoring: A review. *Structural Health Monitoring*, 3(4):355–377, 2004.

A. Deraemaeker, E. Reynders, G. De Roeck, and J. Kullaa. Vibration-based structural health monitoring using output-only measurements under changing environment. *Mechanical Systems and Signal Processing*, 22:34–56, 2008.

G. DeRoeck. The state-of-the-art of damage detection by vibration monitoring: the SIMCES experience. *Journal of Structural Control*, 10:127–134, 2003.

S. W. Doebling, C. R. Farrar, M. B. Prime, and D. W. Shevitz. Damage identification and health monitoring of structural and mechanical systems from changes in their vibration characteristics: A literature review. Technical Report LA-13070-MS, Los Alamos National Laboratory, 1996.

S. W. Doebling, C. R. Farrar, and M. B. Prime. A summary review of vibration–based damage identification methods. *Shock and Vibration Digest*, 30(2):91–105, 1998.

J. E. Doherty. Nondestructive evaluation. In A. S. Kobayashi, editor, *Handbook on Experimental Mechanics*, chapter 12. Society for Experimental Mechanics, 1987.

C. R. Farrar and S. W. Doebling. Using statistical analysis to enhance modal based damage identification. In J. M. Dulieu, W. J. Staszewski, and K. Worden, editors, *Structural Damage Assessment Using Advanced Signal Processing Procedures*, pages 199–211. Academic Press, Sheffield, 1997.

C. R. Farrar, S. W. Doebling, and D. A. Nix. Vibration–based structural damage identification. *The Royal Society – Philosophical Transactions: Mathematical, Physical and Engineering Sciences*, 359:131–149, 2001.

S. D. Fassois. Parametric identification of vibrating structures. In S.G. Braun, D.J. Ewins, and S.S. Rao, editors, *Encyclopedia of Vibration*, pages 673–685. Academic Press, 2001.

S. D. Fassois and J. S. Sakellariou. Time series methods for fault detection and identification in vibrating structures. *The Royal Society – Philosophical Transactions: Mathematical, Physical and Engineering Sciences*, 365:411–448, 2007.

S. D. Fassois and J. S. Sakellariou. Statistical time series methods for structural health monitoring. In C. Boller, F. K. Chang, and Y. Fujino, editors, *Encyclopedia of Structural Health Monitoring*, pages 443–472. John Wiley & Sons Ltd., 2009.

E. Figueiredo, G. Park, C .R. Farrar, K. Worden, and J. Figueiras. Machine learning algorithms for damage detection under operational and environmental variability. *Structural Health Monitoring*, 10(6):559–572, 2011.

C. P. Fritzen. Vibration-based techniques for structural health monitoring. In D. Balageas, C. P. Fritzen, and A. Guemes, editors, *Structural Health Monitoring*, pages 45–224. ISTE, 2006.

M. L. Fugate, H. Sohn, and C. R. Farrar. Vibration–based damage detection using statistical process control. *Mechanical Systems and Signal Processing*, 15(4):103–119, 2001.

F. Gao and Y. Lu. An acceleration residual generation approach for structural damage identification. *Journal of Sound and Vibration*, 319:163–181, 2009.

J. J. Gertler. *Fault Detection and Diagnosis in Engineering Systems.* Marcel Dekker, 1998.

B. K. Ghosh and P. K. Sen, editors. *Handbook of Sequential Analysis.* Marcel Dekker, Inc., New York, 1991.

A. Hera and Z. Hou. Application of wavelet approach for asce structural health monitoring benchmark studies. *Journal of Engineering Mechanics*, 130(1):96–104, 2004.

J. D. Hios and S. D. Fassois. Stochastic identification of temperature effects on the dynamics of a smart composite beam: assessment of multi–model and global model approaches. *Smart Materials and Structures*, 18(3): 035011 (15pp), 2009a.

J. D. Hios and S. D. Fassois. Statistical damage detection in a smart structure under different temperatures via vibration testing: a global model based approach. *Key Engineering Materials*, Vols. 413–414:261–268, 2009b.

J. D. Hios and S. D. Fassois. Stochastic identification under multiple operating conditions: Functionally pooled varma methods. In *Proceedings of the 15th Symposium on System Identification (SYSID)*, Saint-Malo, France, July 2009c.

Z. Hou, M. Noori, and R. St. Amand. Wavelet-based approach for structural damage detection. *Journal of Engineering Mechanics*, 126(7):677–683, 2000.

C. S. Huang, S. L. Hung, C. M. Wen, and T. T. Tu. A neural network approach for structural identification and diagnosis of a building from seismic response data. *Earthquake Engineering and Structural Dynamics*, 32:187–206, 2003.

H. Y. Hwang and C. Kim. Damage detection in structures using a few frequency response measurements. *Journal of Sound and Vibration*, 270: 1–14, 2004.

D. J. Inman, C. R. Farrar, V. Lopez Jr., and V. Steffen Jr., editors. *Damage Prognosis for Aerospace, Civil and Mechanical Systems.* John Wiley & Sons, 2005.

S. M. Kay. *Modern Spectral Estimation: Theory and Application.* Prentice Hall: New Jersey, 1988.

F. P. Kopsaftopoulos and S. D. Fassois. Identification of stochastic systems under multiple operating conditions: The vector dependent FP–ARX parametrization. In *Proceedings of 14th Mediterranean Conference on Control and Automation*, Ancona, Italy, 2006.

F. P. Kopsaftopoulos and S. D. Fassois. Vibration–based structural damage detection and precise assessment via stochastic functionally pooled models. *Key Engineering Materials*, 347:127–132, 2007.

F. P. Kopsaftopoulos and S. D. Fassois. Vibration based health monitoring for a lightweight truss structure: experimental assessment of several statistical time series methods. *Mechanical Systems and Signal Processing*, 24:1977–1997, 2010.

F. P. Kopsaftopoulos and S. D. Fassois. Scalar and vector time series methods for vibration based damage diagnosis in a scale aircraft skeleton structure. *Journal of Theoretical and Applied Mechanics*, 49(3):727–756, 2011.

F. P. Kopsaftopoulos and S. D. Fassois. A stochastic functional model based method for vibration based damage detection, precise localization, and magnitude estimation. *Mechanical Systems and Signal Processing, to appear*, 2012.

E. L. Lehmann and J. P. Romano. *Testing Statistical Hypotheses*. Springer, 3rd edition, 2008.

S. Liberatore and G. P. Carman. Power spectral density analysis for damage identification and location. *Journal of Sound and Vibration*, 274(3–5): 761–776, 2004.

L. Ljung. *System Identification: Theory for the User*. Prentice–Hall, 2nd edition, 1999.

Y. Lu and F. Gao. A novel time–domain auto–regressive model for structural damage diagnosis. *Journal of Sound and Vibration*, 283:1031–1049, 2005.

H. Lütkepohl. *New Introduction to Multiple Time Series Analysis*. Springer-Verlag Berlin, 2005.

G. Manson, K. Worden, and D. Allman. Experimental validation of a structural health monitoring methodology: part II, novelty detection on a gnat aircraft. *Journal of Sound and Vibration*, 259(2):345–363, 2003.

Z. Mao and M. Todd. A model for quantifying uncertainty in the estimation of noise-contaminated measurements of transmissibility. *Mechanical Systems and Signal Processing*, 2011. doi: doi:10.1016/j.ymssp.2011.10.002.

S. F. Masri, A. W. Smyth, A. G. Chassiakos, T. K. Caughey, and N. F. Hunter. Application of neural networks for detection of changes in nonlinear systems. *American Society of Civil Engineers (ASCE) Journal of Engineering Mechanics*, 126(7):666–676, 2000.

S. G. Mattson and S. M. Pandit. Statistical moments of autoregressive model residuals for damage localization. *Mechanical Systems and Signal Processing*, 20:627–645, 2006.

P. G. Michaelides and S. D. Fassois. Stochastic identification of structural dynamics from multiple experiments – epxerimental variability analysis. In *Proceedings of the ISMA Conference on Noise and Vibration Engineering*, Leuven, Belgium, September 2008.

D. Montalvão, N. M. M. Maia, and A. M. R. Ribeiro. A summary review of vibration–based damage identification methods. *Shock and Vibration Digest*, 38(4):295–324, 2006.

D. C. Montgomery. *Introduction to Statistical Quality Control.* John Wiley & Sons Inc., 2nd edition, 1991.

A.A. Mosavi, D. Dickey, R. Seracino, and S. Rizkalla. Identifying damage locations under ambient vibrations utilizing vector autoregressive models and mahalanobis distances. *Mechanical Systems and Signal Processing*, 26:254–267, 2012.

K. K. Nair and A. S. Kiremidjian. Time series based structural damage detection algorithm using gaussian mixtures modeling. *Journal of Dynamic Systems, Measurement, and Control*, 129:285–293, 2007.

K. K. Nair, A. S. Kiremidjian, and K. H. Law. Time series–based damage detection and localization algorithm with application to the asce benchmark structure. *Journal of Sound and Vibration*, 291:349–368, 2006.

H. G. Natke and C. Cempel. *Model-Aided Diagnosis of Mechanical Systems: Fundamentals, Detection, Localization, Assessment.* Springer-Verlag, 1997.

H. T. Nguyen and G. S. Rogers. *Fundamentals of Mathematical Statistics: Vols. I and II.* Springer–Verlag, New York, 1989.

C. K. Oh and H. Sohn. Damage diagnosis under environmental and operational variations using unsupervised support vector machine. *Journal of Sound and Vibration*, 328:224–239, 2009.

Z. K. Peng and F. L. Chu. Application of the wavelet transform in machine condition monitoring and fault diagnostics. *Mechanical Systems and Signal Processing*, 18:199–221, 2004.

A. G. Poulimenos and S. D. Fassois. Vibration-based on-line fault detection in non-stationary structural systems via a statistical model based method. In *Proceedings of the 2nd European Workshop on Structural Health Monitoring (EWSHM)*, pages 687–694, Munich, Germany, 2004.

P. S. Rao and C. Ratnam. Health monitoring of welded structures using statistical process control. *Mechanical Systems and Signal Processing*, 27:683–695, 2012.

D. D. Rizos, S. D. Fassois, Z. P. Marioli-Riga, and A. N. Karanika. Vibration–based skin damage statistical detection and restoration assessment in a stiffened aircraft panel. *Mechanical Systems and Signal Processing*, 22:315–337, 2008.

A. Rytter. *Vibration based inspection of civil engineering structures.* PhD thesis, Department of Building Technology and Structural Engineering, Aalborg University, Denmark, 1993.

J. S. Sakellariou and S. D. Fassois. Stochastic output error vibration–based damage detection and assessment in structures under earthquake excitation. *Journal of Sound and Vibration*, 297:1048–1067, 2006.

J. S. Sakellariou and S. D. Fassois. A functional pooling framework for the identification of systems under multiple operating conditions. In *Proceedings of 15th Mediterranean Conference on Control and Automation*, Athens, Greece, 2007.

J. S. Sakellariou and S. D. Fassois. Vibration based fault detection and identification in an aircraft skeleton structure via a stochastic functional model based method. *Mechanical Systems and Signal Processing*, 22: 557–573, 2008.

J. S. Sakellariou, K. A. Petsounis, and S. D. Fassois. Vibration analysis based on–board fault detection in railway vehicle suspensions: a feasibility study. In *Proceedings of First National Conference on Recent Advances in Mechanical Engineering*, Patras, Greece, 2001.

J. S. Sakellariou, K. A. Petsounis, and S. D. Fassois. On–board fault detection and identification in railway vehicle suspenions via a funcitonal model based method. In *Proceedings of the ISMA 2010 International Conference on Noise and Vibration Engineering*, Leuven, Belgium, 2002.

O. S. Salawu. Detection of structural damage through changes in frequency: a review. *Engineering Structures*, 19(9):718–72, 1997.

T. Söderström and P. Stoica. *System Identification*. Prentice–Hall, 1989.

H. Sohn. Effects of environmental and operational variability on structural health monitoring. *The Royal Society – Philosophical Transactions: Mathematical, Physical and Engineering Sciences*, 365:539–560, 2007.

H. Sohn and C. R. Farrar. Statistical process control and projection techniques for structural health monitoring. In *Proceedings of the European COST F3 Conference on System Identification and Structural Health Monitoring*, Madrid, Spain, 2000.

H. Sohn and C. R. Farrar. Damage diagnosis using time series analysis of vibration signals. *Smart Materials and Structures*, 10:446–451, 2001.

H. Sohn, C.R. Farrar, N.F. Hunter, and K. Worden. Structural health monitoring using statistical pattern recognition techniques. *Journal of Dynamic Systems, Measurement, and Control*, 123(4):706–711, 2001.

H. Sohn, D. W. Allen, K. Worden, and C. R. Farrar. Statistical damage classification using sequential probability ratio tests. *Structural Health Monitoring*, 2(1):57–74, 2003a.

H. Sohn, C. R. Farrar, F. M. Hemez, D. D. Shunk, D. W. Stinmates, and B. R. Nadler. A review of structural health monitoring literature: 1996–2001. Technical Report LA-13976-MS, Los Alamos National Laboratory, 2003b.

H. Sohn, D. W. Allen, K. Worden, and C. R. Farrar. Structural damage classification using extreme value statistics. *Journal of Dynamic Systems, Measurement, and Control*, 127:125–132, 2005.

M. D. Spiridonakos and S. D. Fassois. Vibration based fault detection in a time-varying link structure via non-stationary FS-VTAR models. In *Proceedings of the International Operational Modal Analysis Conference (IOMAC)*, Ancona, Italy, 2009.

W. J. Staszewski. Advanced data pre-processing for damage identification based on pattern recognition. *International Journal of Systems Science*, 31(11):1381–1396, 2000.

W. J. Staszewski. Structural and mechanical damage detection using wavelets. *Shock and Vibration Digest*, 30:457–472, 1998.

W. J. Staszewski and A. N. Robertson. Time-frequency and time-scale analyses for structural health monitoring. *The Royal Society – Philosophical Transactions: Mathematical, Physical and Engineering Sciences*, 365: 449–477, 2007.

W. J. Staszewski, C. Boller, and G. R. Tomlinson, editors. *Health Monitoring of Aerospace Structures: Smart Sensor Technologies and Signal Processing*. John Wiley & Sons, Chichester, U.K., 2004.

A. Stuart and J. K. Ord. *Kendalls Advanced Theory of Statistics: Vol 1. Distribution Theory*. Oxford University Press, New York, 5th edition, 1987.

T. Uhl and K. Mendrok. Overview of modal model based damage detection methods. In *Proceedings of the ISMA 2004 International Conference on Noise and Vibration Engineering*, Leuven, Belgium, 2004.

A. Wald. *Sequential Analysis*. Wiley, New York, 1947.

Z. Wei, L. H. Yam, and L. Cheng. NARMAX model representation and its application to damage detection for multi-layer composites. *Composite Structures*, 68:109–117, 2005.

K. Worden. Structural fault detection using a novelty measure. *Journal of Sound and Vibration*, 201(1):85–101, 1997.

K. Worden and G. Manson. Experimental validation of a structural health monitoring methodology: part I. novelty detection on a laboratory structure. *Journal of Sound and Vibration*, 259(2):323–343, 2003.

K. Worden, G. Manson, and N. R. J. Fieller. Damage detection using outlier analysis. *Journal of Sound and Vibration*, 229(3):647–667, 2000.

A.M. Yan, P. de Boe, and J. C. Golinval. Structural damage diagnosis by kalman model based on stochastic subspace identification. *Structural Health Monitoring*, 3(2):103–119, 2004.

Q. W. Zhang. Statistical damage identification for bridges using ambient vibration data. *Computers and Structures*, 85:476–485, 2007.

H. Zheng and A. Mita. Two–stage damage diagnosis based on the distance between arma models and pre–whitening filters. *Smart Materials and Structures*, 16:1829–1836, 2007.

Y. Zou, L. Tong, and G. P. Steven. Vibration-based model-dependent damage (delamination) identification and health monitoring for composite structures. *Journal of Sound and Vibration*, 230:357–378, 2000.

A Appendix: Central limit theorem and statistical distributions associated with the normal

A.1 The Central Limit Theorem (CLT)

Theorem A.1 (The Central Limit Theorem (Stuart and Ord, 1987; Nguyen and Rogers, 1989; Montgomery, 1991)). *Let $Z_1, Z_2, \dots Z_n$ designate mutually independent random variables each with mean μ_k and (finite) variance σ_k^2. Then, for $n \longrightarrow \infty$ the distribution of the random variable $X = \sum_{k=1}^n Z_k$ approaches the Gaussian distribution with mean $E\{X\} = \sum_{k=1}^n \mu_k$ and variance $var[X] = \sum_{k=1}^n \sigma_k^2$.* □

A.2 The χ^2 distribution

Let $Z_1, Z_2, \dots Z_n$ designate mutually independent, normally distributed, random variables, each with mean μ_k and standard deviation σ_k. Then the sum:

$$X = \sum_{k=1}^{n} \left(\frac{Z_k - \mu_k}{\sigma_k} \right)^2 \tag{A.1}$$

is said to follow a (central) chi-square distribution with n degrees of freedom ($X \sim \chi^2(n)$). Its mean and variance are $E\{X\} = n$ and $\text{var}[X] = 2n$, respectively. Notice that imposing p equality constraints among the random variables $Z_1, Z_2, \dots Z_n$ reduces the set's effective dimensionality, and thus the number of degrees of freedom, by p (Stuart and Ord, 1987, pp. 506–507).

For $n \longrightarrow \infty$ the $\chi^2(n)$ distribution tends to normality (Stuart and Ord, 1987, p. 523).

The sum $X = \sum_{k=1}^n Z_k^2/\sigma_k^2$ is said to follow non-central chi-square distribution with n degrees of freedom and non-centrality parameter $\lambda = \mu_k^2/\sigma_k^2$. This distribution is designated as $\chi^2(n; \lambda)$ (Nguyen and Rogers, 1989, Vol. II p. 33).

Let $\boldsymbol{x} \in \mathbb{R}^n$ follow n-variate normal distribution with zero mean and covariance $\boldsymbol{\Sigma}$ ($\boldsymbol{x} \sim \mathcal{N}(\boldsymbol{0}, \boldsymbol{\Sigma})$). Then the quantity $\boldsymbol{x}^T \boldsymbol{\Sigma}^{-1} \boldsymbol{x}$ follows (central) chi-square distribution with n degrees of freedom (Söderström and Stoica 1989, p. 557, Stuart and Ord 1987, pp. 486–487, Gertler 1998, p. 120).

A.3 The Student's t distribution

Let Z be the standard (zero mean and unit variance) normal variable. Let X follow a (central) chi-square distribution with n degrees of freedom and be independent of Z. Then the ratio:

$$T = \frac{Z}{\sqrt{X/n}} \tag{A.2}$$

is said to follow a Student or t (central) distribution with n degrees of freedom (central because it is based on a central chi-square distribution; Nguyen and Rogers 1989, Vol. II p. 34). Its mean and variance are $E\{T\} = 0$ $(n > 1)$ and $\text{var}[T] = \frac{n}{n-2}$ $(n > 2)$, respectively (Stuart and Ord, 1987, p. 513).

The (central) t distribution approaches the standard normal distribution $\mathcal{N}(0,1)$ as $n \longrightarrow \infty$ (Stuart and Ord, 1987, p.523).

A.4 The Fisher's F distribution

Let X_1,X_2 be mutually independent random variables following (central) chi-square distributions with n_1, n_2 degrees of freedom, respectively. Then the ratio:

$$F = \frac{X_1/n_1}{X_2/n_2} \tag{A.3}$$

is said to follow a (central) F distribution with n_1, n_2 degrees of freedom ($F \sim \mathcal{F}(n_1, n_2)$; central because it is based on central chi-square distributions; Nguyen and Rogers 1989, Vol. II p. 34). Its mean and variance are $E\{F\} = \frac{n_2}{n_2-2}$ $(n_2 > 2)$ and $\text{var}[F] = \frac{2n_2^2(n_1+n_2-2)}{n_1(n_2-2)^2(n_2-4)}$ $(n_2 > 4)$, respectively (Stuart and Ord, 1987, p. 518).

Note that for the distribution's $1-\alpha$ critical point $f_{1-\alpha}(n_1, n_2) = 1/f_\alpha(n_2, n_1)$.

The (central) F distribution approaches normality as $n_1, n_2 \longrightarrow \infty$. For $n_2 \longrightarrow \infty$ $n_1 F$ approaches a (central) chi-square distribution with n_1 degrees of freedom (Stuart and Ord, 1987, p. 523).

FIBER OPTICS SENSORS

Alfredo Güemes[*] and Julián Sierra-Pérez[†]
[*] Department of Aerospace Materials and Processes, Polytechnic University of Madrid, Madrid, Spain
[†] Aerospace Engineering Research Group, Pontificia Bolivariana University, Medellín, Colombia

1 Introduction

Fiber optic technology started in 1970's, for long distance telecommunications, and it has experienced an exponential growth during the last four decades. Sensing applications are a small spin-off from this technology, taking profit of developments done for optoelectronic components and concepts.

Optical fibers are cylindrical dielectric waveguides for the propagation of the light, made out from high purity, low loss optical materials, usually silica (Optical fibers made from polymers and other transparent materials are also commercially available). The refractive index of the core (about 1.46 for silica) is slightly higher than surrounding material or cladding, due to the presence of dopants. Light is confined at the core, because optical rays arriving to the interface core/cladding with an angle higher than the total reflectance angle, as defined by the Snell law, will follow total reflections and will remain confined at the core. Only when the fiber is locally bent with a sharp radius, the light may escape. The OF is externally protected from mechanical scratches with a polymeric coating (the 'buffer'), and frequently several optical fibers are bundle together, and assembled with high strength mechanical fibers, such as Kevlar, to make a robust product known as optical cable, that can withstand rough industrial manipulation.

Glass optical fibers have an external diameter of 125 μm (human hair diameter range from 50 to 100 μm), and classify in two main groups: monomode and multimode optical fibers.

Monomode OF, of small core (about 10 μm , depending on the intended optical wavelength). The electromagnetic waves traveling by the core must satisfy the Maxwell equations; the cylindrical contour conditions only allow for a discrete number of solutions, V, dependent on the core diameter, and the wavelength. When V is less than 2.4, only one mode (2 orthogonal

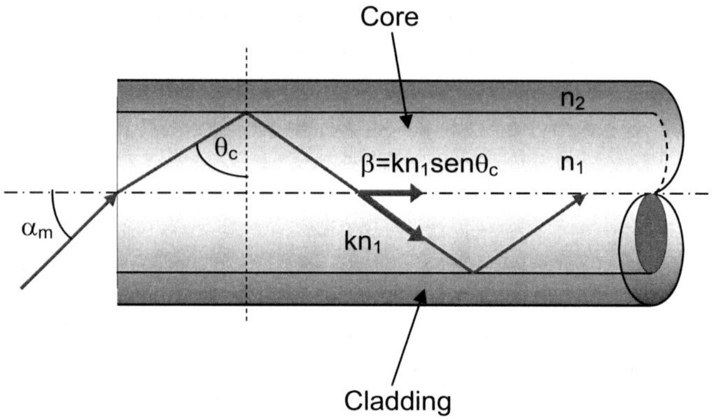

Figure 1. Schematic representation of an optical fiber.

polarizations) propagates through the optical fiber, it is called monomode. Monomode OF offer some advantages for optical communication, such as:

- Optical attenuation is smaller for monomode OF, due to the lower difference of refractive index between core and cladding, requiring a lower concentration of dopants.
- Multimode OF will suffer group dispersion due to different modes traveling at different speed.

The multimode optical fibers have as only advantage a larger core diameter (30 to 100 μm). The larger core makes simpler the alignment with the optical sources and connections. They are preferred when used only as guides for light, as in many medical applications, or in LAN. For sensing purposes they can only be used for intensity based methods.

Optical power loss is extremely small with current OF, in the order of 0.03 dB per kilometer, meaning that 100 km will only divide optical power by two; this is an important requirement for telecommunications. Optical power loss is minimized near the wavelengths of 1300 nm or 1550 nm, where minimum Rayleigh dispersion and minimum infrared absorption is found. From a sensing point of view, optical losses at the fiber are irrelevant, but still 1550 nm is the preferred windows, because optoelectronic components are more easily available.

Our discussion will be mainly restricted to strain and temperature measurements, but fiber optic sensor may be found for many other applications. This would be a first classification for optical sensors, according the physical parameter sensed: chemical, mechanical, electrical, etc. The optical fiber itself is a pipe for light, transmitting the information, but it may also be sensitive to local changes in the external environment surrounding the fiber, like temperature. With special equipments it is possible to measure the light reflected back along the fiber length and consequently the temperature distribution. The fiber would then act simultaneously as a distributed sensor and as the cable for data transmission, all along its length. This seems to be the ideal approach; it is called 'distributed sensing'; but accuracy and spatial resolution are not good enough for many applications. The approach based on local sensors is more common, either by including into the optical path an external device (extrinsic sensor), or by modifying the fiber during a short length (about 1 cm.), to make it sensitive to the strain or any other parameter. These are the so-called intrinsic sensors, engraved into the fiber, which would go undetectable to the naked eye.

Common advantages of every kind of optical fiber sensors come from its small size and weight; its non-electrical nature, making them immune to EM interference and to electrical noise, also allowing them to work into explosive environments; they usually have a very high sensitivity and a wide operating temperature range. For Smart Structures, local intrinsic sensors are the most appreciated, because of simplicity, minimal perturbation to the host material and accuracy of reading. Multiplexing capability, or the possibility to 'write' several sensors on the same optical fiber, is another highly desired feature. A single optical fiber, 0.25 mm diameter, may afford information of strain readings from 10 to 20 different points at the structure. When compared to the 2 leads required by each electrical strain gauge, a clear advantage is found.

2 Classification of Fiber Optic Sensors

As a wide definition, a sensor is an artifact able to transform a certain physical or chemical magnitude into readable information. Besides the previous topological classification (local/distributed, intrinsic/extrinsic), a more basic classification may be done by the optical parameter. Light is an electromagnetic wave, defined by its frequency, intensity, phase and polarization. The external action to be monitored must change some of the magnitudes that define the optical wave. Based on the kind of these changes, different kind of optical sensors have been developed.

As illustrated at Figure 2, the sketch of a fiber optic sensing system is always very simple: A solid state light source, either monochromatic (laser) or LED (light emitting diode), containing white light as a mixture of frequencies. Light goes though splices and optical connectors, used for assembly/disassembly the optical fibers, and also through couplers and circulators; a 3 dB coupler is a simple artifact where two optical fibers have been fused together, the light arriving by one end is leaving by the other opposite two ends; the couplers work identically in both directions. Light detectors use to be again solid-state, pig tailed optoelectronic components (pig tailed means that an optical fiber is leaving the solid state casing, doing very easy to connect it to the optical circuit).

2.1 Intensity-based sensors

These are the simplest devices among the FOS, and consequently they were among the first attempts, and still are in use as proximity sensors, for damage detection, cure monitoring and hydrogen detection. A stable light source, white or monochromatic, an optical fiber, preferably multimode for higher light power transmission and a sensitive photodetector may compose the whole system.

Commercially available proximity sensors are used for non-contact monitoring of the displacement of rotating shafts with accuracy better than 3 μm. The distance between the cleaved fiber optic end and the moving surface is correlated to the amount of light reflected by this surface, and captured again by the optical fiber (Figure 2).

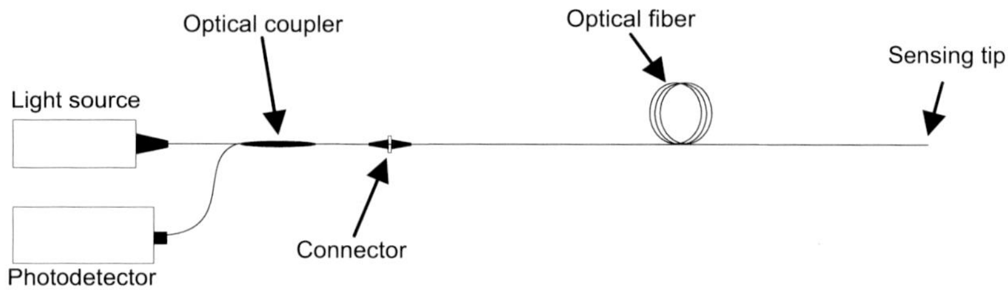

Figure 2. Fiber optic proximity sensors.

Microbending sensors were popular intrinsic sensors for pressure monitoring; they were based on intensity measurements. It is known that light is guided inside the optical fiber because its incidence angle is higher the critical reflection angle at the core-cladding interface, that is always true except for very sharp bends. If the OF is placed between two rough surfaces, a higher pressure promotes higher bending, and consequently higher optical losses. If adequately calibrated, they would afford a measurement of the transversal load on the fiber, by using a local plate, or even the natural roughness of the composite fabrics. Early works were done with this concept, which is currently abandoned; fluctuations of optical power at the light source, connectors, temperature, etc., make the system rather unreliable. Still the microbending losses must be keep in mind when embedding optical fiber into composite fabrics, because they can fade out the optical signal.

The amount of light reflected at the tip of the optical fiber, according the Fresnel law, is proportional to the refractive index of the external surrounding media (Figure 3); this phenomena has been used for cure monitoring, (the refractive index of the resin is known to change with the cure advancement state), and for hydrogen leakage detection, if the tip was coated with and hydrogen sensitive material, as palladium.

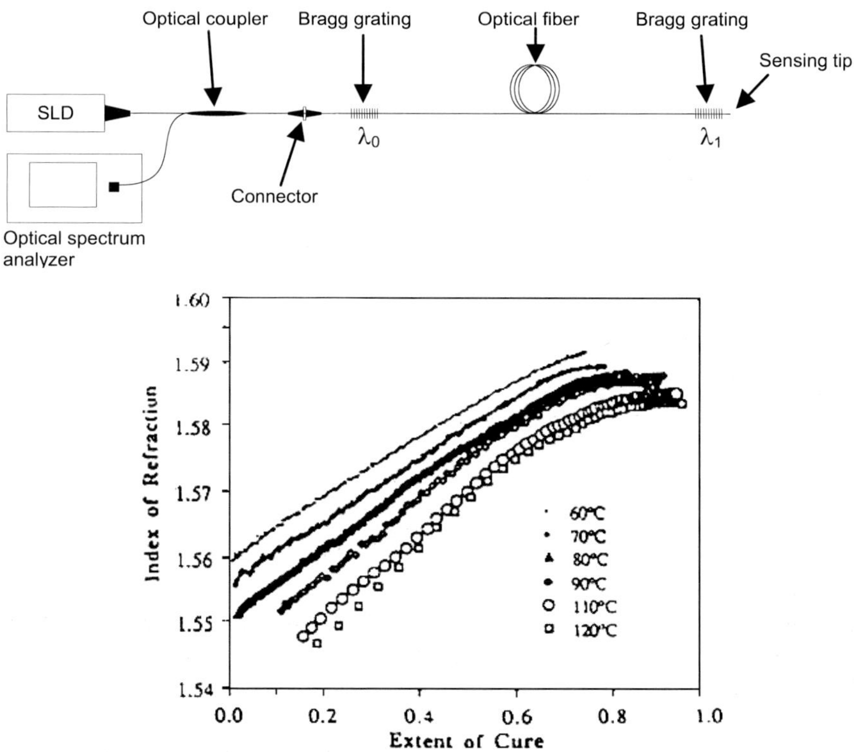

Figure 3. Sketch of cure monitoring systems by intensity based FOS.

2.2 Phase modulated optical fiber sensors, or Interferometers.

Interferometry is the most accurate laboratory technique for precise distance measurements. It has to be bear in mind that direct phase measurements, or the display of the electrical field vs. time cannot be done for the optical wave, as it is done with oscilloscopes for low frequency electrical signals; only light intensity, averaging the power of the electromagnetic field, can be measured. Instead of it, what is done is to split and recombine again the electromagnetic wave. If a continuous monochromatic light wave is divided in two beams (by a partial mirror when dealing with conventional optics, or by a coupler in the case of optical fibers), and these waves travels through different paths before being recombined, any slight difference of path-lengths will cause a delay from one wave to the other, and consequently their electromagnetic fields would not summed up, even they may counteract, and the final result is that output intensity may even cancel.

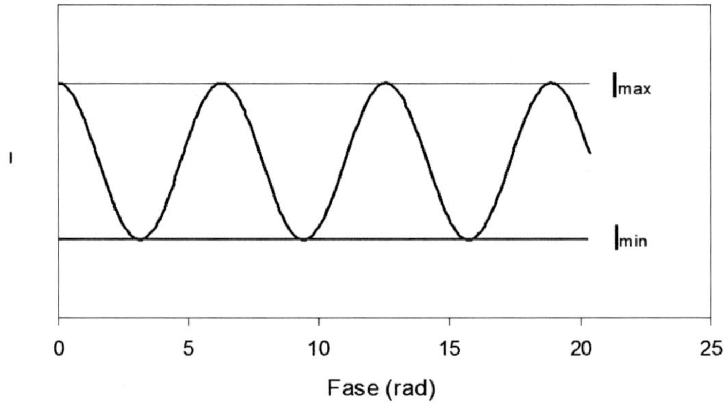

Figure 4. Power output from interferometric measurements.

Interferometers output will move from the power input level to zero level, as sketched in Figure 4, each time the length of any of the two paths increment or decrement in half wavelength (380 nanometers, if the red light of a He-Ne laser was used). This extreme sensitivity to any perturbation in the optical path of any of the two arms makes measurements difficult outside the laboratory environment, because just a local temperature change may cause several maxima and minima drifting. The sensor is the whole length of the optical path of one arm; it is an 'averaging distributed sensor'. Another drawback is that measurements are related to the length when the equipment is put ON, any interruption will lose the reference; also, there is no direct indication if increment or decrement is happening, and a quadrature technique is needed to identify it.

The Figure 5 sketches the most common optical fiber interferometric architectures, known as Mach-Zender, Michelson, Fabry-Perot, from the names of classical optical interferometry. At the Michelson and Fabry-Perot systems, the light is reflected at the end of the optical fibers, traveling back and recombining again. It has to be emphasized that interferometry requires monochromatic light (laser). It is very simple to arrange these optical sketches with optical fibers, all that is needed are the optoelectronic components, and one or two couplers. In spite of this simplicity, they are very seldom used. A detailed discussion of each of these systems is found at the literature. Fabry-Perot is special in the sense that light travels most of the time by a single fiber, interference is caused among the waves reflected at a partial mirror, and the waves that pass this mirror and are reflected back later. For Smart Structures, a single optical fiber was preferred, because only one ingress point into the structure is required. A lot of work was done for developing a rugged sensor head. Soldering optical fibers after partial coating of the fiber ends did intrinsic sensors. With the development of the Bragg grating sensors, this line of work was abandoned. Extrinsic Fabry-Perot heads, known as 'microcavities' are still commercially available; a precision machined wafer is bonded at the end of the optical fiber, leaving an small air gap, making miniature pressure and temperature sensors.

The Michelson arrangement is employed in the 'white light interferometry', a widely used technique for strain monitoring in civil engineering, which is commercially available by a Swiss company, under the name SOFO. Instead of using a continuous wave monochromatic light, a short pulse is launched through the two arms, the arrival times will be different if their lengths are different, and the photo detector would receive two pulses. If the reference arm can be mechanically stretched to get both pulses coincident, the increase of length in the sensing arm would be known, and from it, its averaging strain is obtained.

2.3 Wavelength based sensors, or Bragg gratings.

These are a kind of local, intrinsic, absolute, multiplexable, interruption immune fiber optic strain sensors, which have concentrate most of the attention since its discovery, at the beginning of the 1990's.

Basic idea is to engrave, at the core of the optical fiber and for a short length (standard FBG length goes from 1 to 10 mm), a periodic modulation of its refractive index. It will behave as a set of weak partial reflecting mirrors, which by an accumulative phenomenon called diffraction, will reflect

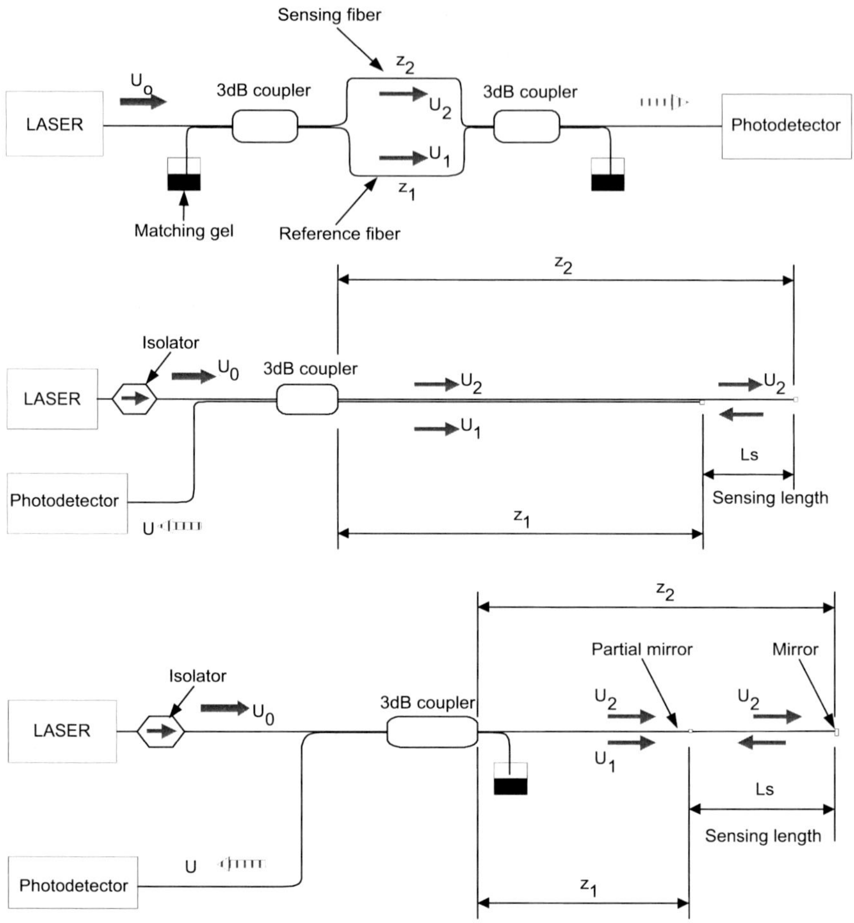

Figure 5. Typical interferometric arrangements (Mach-Zender, Michelson, Fabry-Perot).

back the optical wavelength that is exactly proportional to their spacing. The diffraction law, first established by Bragg and widely used for crystal structure analysis, simplifies under normal incidence to the simple expression:

$$\lambda_b = 2\overline{n_e}\Lambda_0 \tag{1}$$

Where λ_b is the Bragg wavelength, Λ_0 is the pitch or period of the modulation and $\overline{n_e}$ is the average refractive index. The Bragg grating behaves as a very narrow optical filter, as shown at the Figure 6. When a broadband light is traveling through the fiber, the grating promotes that a very narrow wavelength band is reflected back. If the grating is subject to a uniform axial strain, or a uniform thermal increment is applied, the central wavelength of the spectrum reflected by the grating will shift due to changes in the pitch and the refractive index.

Any Optical Spectrum Analyzer (OSA) will be able to detect these changes, and transform it into readable information (Figure 7).

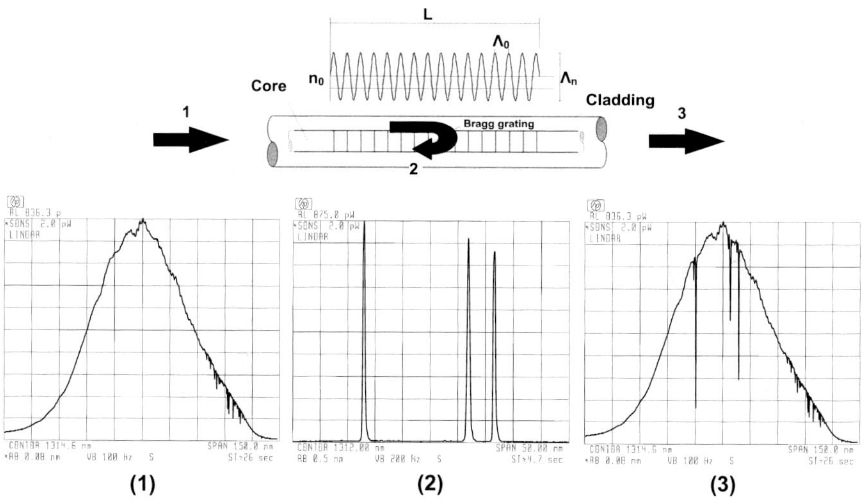

Figure 6. Fiber Bragg grating and operating principle schemes. (1) Intensity spectrum of a broadband source launched into the fiber. (2) Spectra reflected back by three fiber Bragg gratings. (3) Transmissive spectrum after passing the three Bragg gratings.

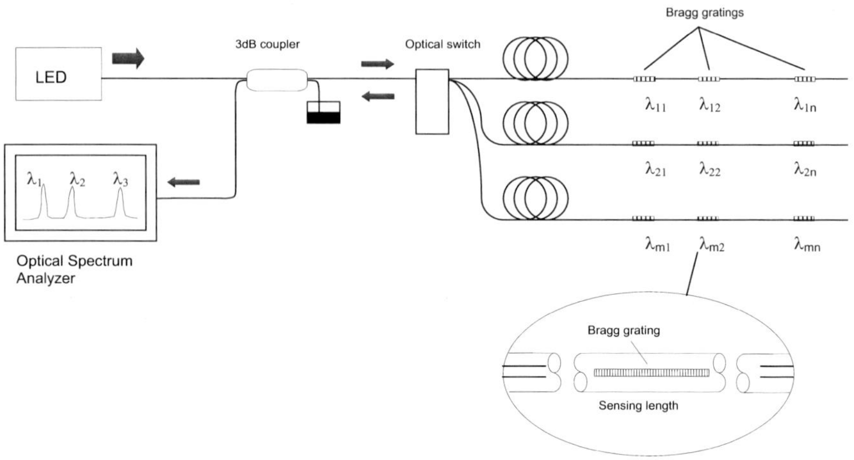

Figure 7. A multiplexed interrogation system for Bragg gratings.

Furthermore, commercially available white light sources have an spectral width around 40 to 60 nm. As the maximum excursion in wavelength is 10 nm for 10.000 $\mu\varepsilon$, several Bragg gratings, centered at different wavelengths, can be written at the same optical fiber, and interrogated at the same time. Multiplexing is easily implemented. The fact that the information is wavelength encoded makes the sensor very stable to aging, allowing absolute measurements of strain after long terms without recalibrating, a common nightmare with standard strain gauges.

3 Fiber Bragg Gratings as Strain and Temperature Sensors

FBG may be bought from several commercial sources, tuned at user defined central wavelength. The only ability required to the user would be to make the optical connections, normally by fusion splices, which require minor investments and training. This is the common approach for mechanical engineers, mainly interested into the usage of this technology for strain measurements. Procedures for manufacturing FBG require high investments and long trained people, and it is useful only for those people interested into the sensor development.

3.1 Response of the FBG to uniaxial uniform strain fields.

When a free FBG is submitted to a uniform uniaxial tension, the photoelastic equations afford a linear relationship among the wavelength drifting and the axial strain, the proportionality coefficient is called sensitivity factor to the axial strain: S_ε

$$S_\varepsilon = \frac{\delta \lambda_b}{\varepsilon_1} = \lambda_b \left[1 - \frac{\overline{n_e}^2}{2} \left[p_{12} - v_f \left(p_{11} + p_{12} \right) \right] \right] \tag{2}$$

Being λ_b the Bragg wavelength, $\overline{n_e}$ its refractive index, v_f the Poisson coefficient and p_{11}, p_{12} its photoelastic coefficients. As an example, for a Bragg grating with the following parameters: Λ_0 of 456.68 nm, $\overline{n_e} = 1.4496$, $v_f = 0.17$, $p_{11} = 0.113$ and $p_{12} = 0.252$ it is found $S_\varepsilon = 1.0598$ x 10^{-3} nm/$\mu\varepsilon$ (943.6 $\mu\varepsilon$/nm) or about 1 pm of wavelength shifting for an strain of 1 $\mu\varepsilon$ (when the central wavelength is about 1300 nm. For 1550 nm, the sensitivity is approximately 1.2 pm)

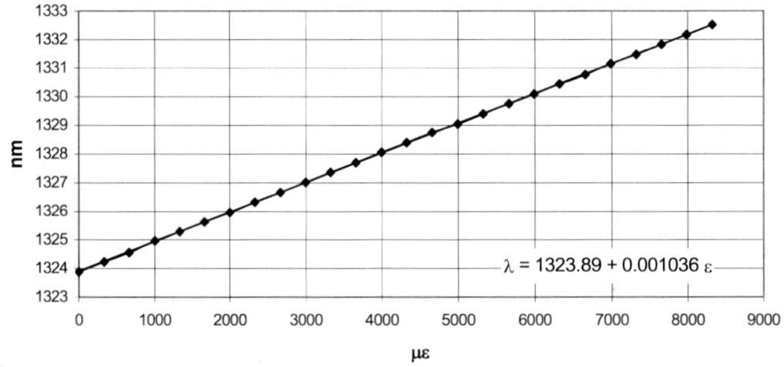

Figure 8. Wavelength shift vs. Axial strain. FBG submitted to uniaxial strain.

Validity of this model is verified with the following experiment, with a FBG bonded on the surface of a unidirectional specimen made from graphite/epoxy AS4/8552. Results are shown at Figure 8, showing an excellent linearity and low dispersion of individual data (standard deviation: 0.01767 nm or about 17.05 $\mu\varepsilon$). Experimentally obtained sensitivity was 1.0363 x 10^{-3} nm/$\mu\varepsilon$, quite close to those previously calculated.

3.2 Sensitivity of the FBG to temperature.

Temperature influences the behavior of the FBG by a linear thermal expansion of the grating, and also by changes in its refractive index. Both effects may be summarized in the following expression:

$$\frac{\delta\lambda_b}{\lambda_{b,0}} = (\alpha + \xi)\,\delta T \tag{3}$$

Typical values for these coefficients are 5 to 5.5 x 10^{-7} for the thermal expansion of the silica, and one order of magnitude higher (8 to 10 x 10^{-6}) for the thermo-optic coefficient. Changes in the refractive index are the main factor, and both coefficients are rather constants for a wide temperature range (-50 °C to 500 °C). These values mean a drifting of about 1 nm per 100 °C, for a grating with an unperturbed wavelength of 1300 nm. For accurate strain measurements, temperature effects need to be corrected, similarly to common strain gauges.

First exposure of the grating to high temperatures (over 100 °C) causes a partial loss of reflectivity of the grating, and also a permanent drifting in central wavelength. To get a repetitive response at high temperatures, gratings need to be annealed. Once this is done, FBG work reliably at high temperatures, up to 800 °C.

At cryogenic conditions, for temperatures down -223 °C, the coefficients are no longer constants, and the linearity is lost. At liquid helium (-268.95 °C) both coefficients are almost zero and the wavelength variation with temperature is negligible under these conditions.

3.3 Response of the FBG to non-uniform uniaxial strain field.

The reflected spectra by an unloaded standard Bragg grating is a very narrow (0.1 nm) symmetric Gaussian peak. This spectra may be acquired by an OSA (Optical Spectrum Analyzer), representing the collected power vs. wavelength. This peak would simply moves back or forwards when the FBG is uniformly loaded, and its pitch, or internal periodic modulation, is changing accordingly. If the FBG would be submitted to two different strain levels along its length, two different peaks of lower height must be expected. In general terms, an strain gradient along its length will distort the reflected peak. The direct problem, or how to calculate the shape of the reflected peak corresponding to a given strain field can be easily done by the Transfer Matrix Formalism (TMF). The TMF is a numerical method to solve the coupled mode equations for aperiodic grating structures. Figure 9 shows the calculated spectra for a linear gradient, and the experimentally obtained values with a simply supported beam reproducing this strain gradient are shown in the Figure 10. The agreement is very acceptable.

The inverse problem, or to obtain the strain profile from the acquired spectrum, is not as straightforward, information of the intensity and phase

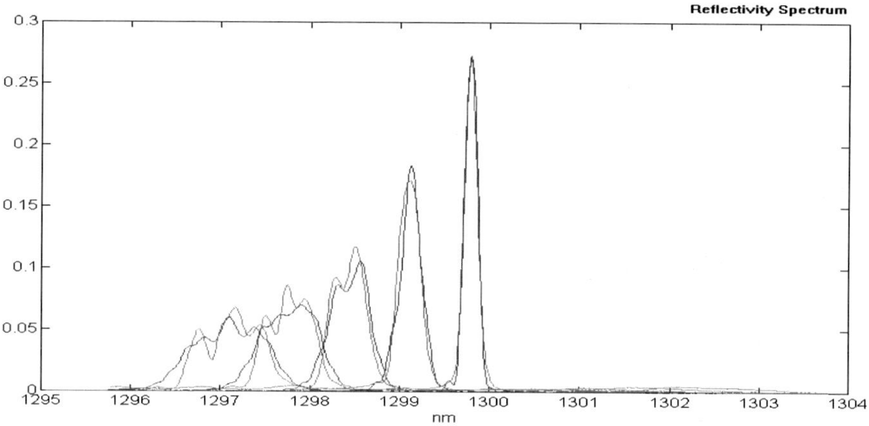

Figure 9. Real and simulated spectra of the grating attached to the beam in several load conditions.

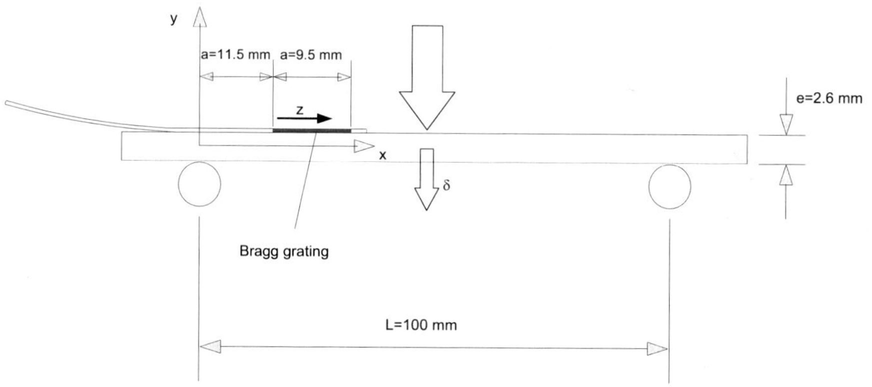

Figure 10. Experimental arrangement for the calibration of the sensor.

would be required. Even some algorithms have been proposed, a simpler approach to obtain the strain gradient is to engrave several short length FBG (1 mm each) as closely located as requested by the experiment.

3.4 Response of the FBG to transverse stresses.

The situation that complicates the suitability of fiber Bragg gratings is their response to non-unidirectional stress. The first sign of this effect appears in fiber Bragg gratings embedded in a thick graphite/epoxy laminate cured at high temperature. Several gratings used to monitor strain and temperature during the cure process split in twin spectra during the cooling of the laminate. Bending of the fiber shows an exchange of optical power between the two sides of the twin spectra, pointing out a birefringence of the fiber promoted by the presence of strong residual stresses inside the laminate.

A transverse stress field applied to the grating promotes a birefringence that modifies the isotropic optical structure of the fiber, generating two principal directions of polarization, with two different refractive indexes. The fiber Bragg grating verifies two different Bragg conditions, and the optical spectrum of the grating splits in two. By using a fiber Bragg grating written in the core of a standard low-birefringence optical fiber, immersed in a general stress field, it would not be possible to separate the components of the stress field using the spectral information of the grating. These phenomena have been experienced by many authors and several methods have been proposed to avoid, or minimize them: use of special fiber, encapsulate the gratings, etc.

Assuming an FBG submitted only to transversal stresses, a FEM analysis can be done to obtain the strains at the core in radial directions, the two refractive indexes and the corresponding wavelength drifting. The most immediate result that can be induced is that distance between the two peaks of the splitted peak is only affected by the transverse stress applied to the fiber, and is proportional to it (Figure 15), where $k \cong 70$ x 5.8514 GPa is the coefficient of proportionality, λ_0 is the Bragg wavelength of the non-perturbed grating, $\delta\lambda_2$ and $\delta\lambda_3$ are the wavelength shifts of the splitted peaks.

It is also possible to obtain an expression for the longitudinal strain of the grating as a function of the displacement of the wavelength of one of the peaks and the distance between the two peaks, $\delta\lambda = \delta\lambda_2 - \delta\lambda_3$, that is:

$$\varepsilon_1 = 1.25\frac{\delta\lambda_2}{\lambda_0} + 0.46\frac{\delta\lambda}{\lambda_0} \tag{4}$$

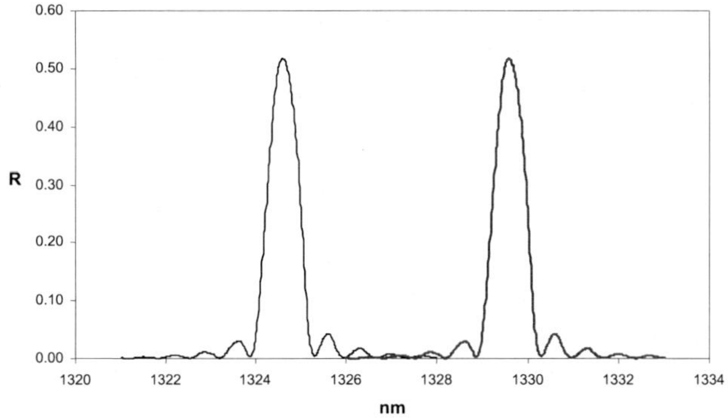

Figure 11. Spectra of a fiber Bragg grating bonded in an aluminum beam: unloaded (narrow line) and submitted to a uniform longitudinal strain of 5000 mε (thick line). The 'ideal' behavior of a grating.

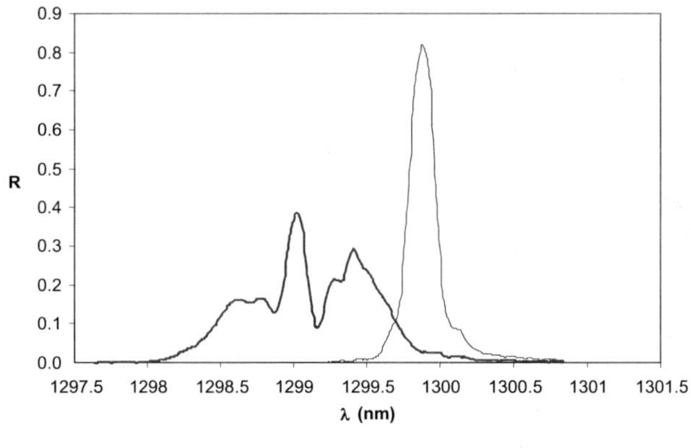

Figure 12. Spectra of a fiber Bragg grating embedded in a unidirectional composite laminate near a bonded joint: unloaded (narrow) and submitted to longitudinal load (thick). The distortion is due to strain gradients.

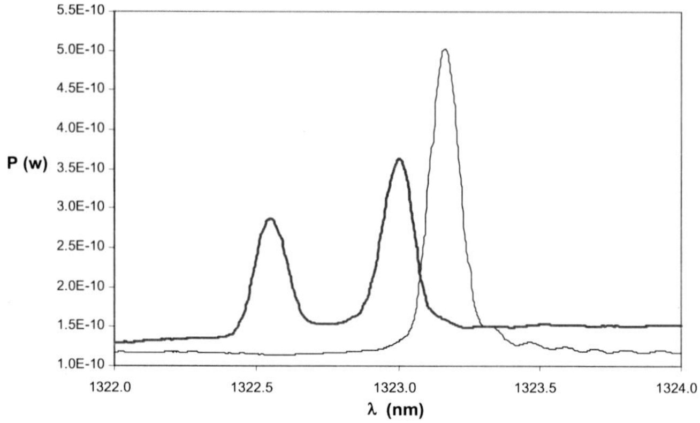

Figure 13. Spectra of a fiber Bragg grating before (narrow) and after being embedded in a thick quasi-isotropic laminate cured in autoclave (thick); the splitted red spectrum shows the effect of strong transverse stresses over the grating.

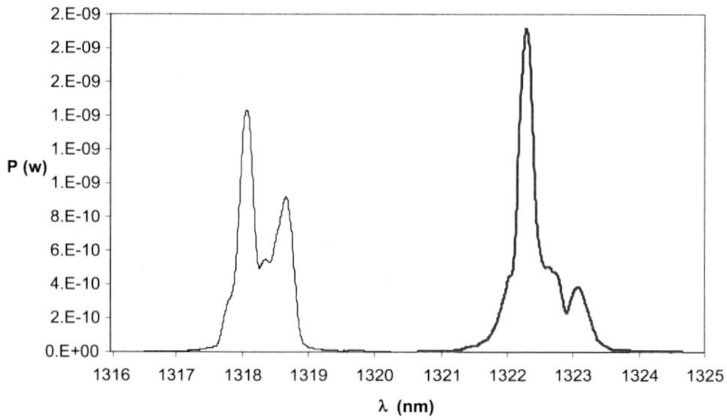

Figure 14. Spectra of a FBG embedded in a composite beam made with fabric and cured in autoclave: unloaded (narrow), shows the typical splitting of a biaxial stress state; loaded (thick), shows the effect of a complex non-uniform biaxial stress field.

$$\sigma_2 \cong k\left(\frac{\delta\lambda_2}{\lambda_o} - \frac{\delta\lambda_3}{\lambda_o}\right)$$

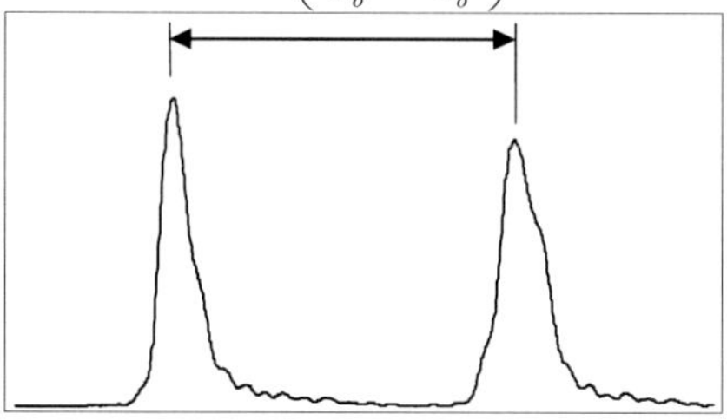

Figure 15. The distance between the two peaks of the splitted peak is only affected by the transverse stress applied to the fibre, and is proportional to it.

In order to prove the validity of these results, two different experimental tests have been implemented. The first one, very simple, uses the well-known thermal expansion of the aluminum to prove the importance of the effect of transverse stresses over fiber Bragg gratings in apparently very simple applications. The second test, with four gratings embedded in different positions in a graphite/epoxy laminate, will allow evaluating the effect of the residual stresses generated in a laminate during the cure processing.

The use of fiber Bragg gratings to monitor strain distributions in bonded joints in aluminum elements using high temperature adhesive film was one of the first applications in which the importance of transverse stress in their spectral response appears clearly. The following test has been configured as an illustrative example of this behavior, and shows the necessary precautions that have to be taken when using this technology in applications in which temperature or external loads can promote transverse strain fields over the gratings.

Figure 16 shows a bare fiber Bragg grating 10 mm long written in photosensitive fiber and submitted to thermal stabilization at 300 °C during 2 hours has been bonded in a beam of AA5083 aluminum using an epoxy adhesive film. The adhesive has been cured using a standard cycle shown in the Figure 17(1).

Figure 17(2) shows the evolution of the Bragg wavelength of the grating with the temperature during the curing cycle. The wavelength of the grating increases during the first part of the heating stage, according the curve wavelength/temperature of a free grating, as a single peak.

At 170 °C, there is a discontinuity in the slope promoted by the fast

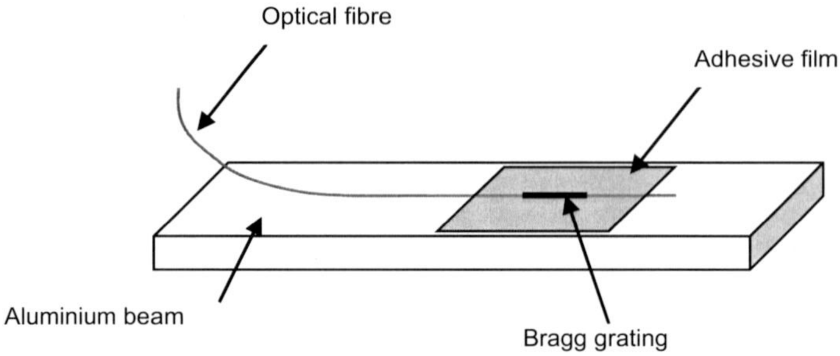

Figure 16. Schematic of aluminum beam with a fiber Bragg grating bonded using adhesive film.

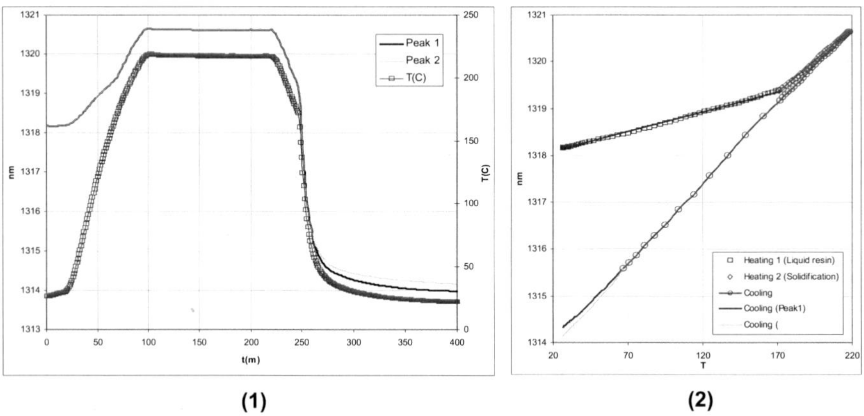

Figure 17. (1) Evolution of temperature and the wavelength of the Bragg wavelength of the grating during the curing cycle. The spectrum of the grating split in two during the cooling of the specimen; (2) Evolution of the Bragg wavelength with temperature during the curing cycle.

gelification of the resin, which forces the grating to follow the thermal expansion of the specimen. During the cooling, the wavelength decreases with this last slope, but the grating split in two at 70 °C (the discrimination of two different peaks depends on the resolution of the spectral analyzer used) due to the thermal transverse strain of the beam.

The wavelength shift of a grating bonded to an aluminum specimen with a linear expansion coefficient α_{AL} is given by:

$$\delta\lambda_b = \delta\lambda_{bT} + \lambda_{b0}\left(\alpha_{AL} - \alpha_f\right)\delta T \tag{5}$$

Where $\delta\lambda_{bT}$ is the wavelength shift of the free Bragg grating with the temperature and α_f is the linear expansion coefficient of the optical fiber. The values of linear expansion coefficient for the AA5083 aluminum and the optical fiber are known (α_{AL} = 23.5 x 10^{-6} K^{-1} and α_f =5.5 x 10^{-7} K^{-1}), then the value calculated for the final Bragg wavelength of the grating after the cooling is 1314.20 nm, whereas the empirical result obtained was 1314.19 nm for a cooling from 180 °C to 20 °C, demonstrating the accuracy of the model.

In this case, it is difficult to find an immediate quantitative relationship between the transverse shrinkage of the aluminum specimen and the shift of the spectrum of the grating due to the complex host/adhesive/sensor interaction.

A graphite/epoxy laminate specimen made of 24 plies of unidirectional T300/F155 prepreg tape with the following configuration: [[90]8[0]4]s, 150 x 125 x 8 mm, cured in a hot plates press at 180 °C in an quasi-isothermal cycle (the specimen was introduced in the press with the plates at the curing temperature. Four bare fiber Bragg gratings were embedded in the laminate parallel to the reinforcing fibers, two in the middle plane (gratings 1 and 2) and two in the outer laminae (gratings 3 and 4). Spectra of the four gratings at different stages of the process are shown in Figure 18.

With these data, and using the expressions obtained from the photoelastic model, the values reported in Table 1 of strains and stresses are calculated:

In the Table 1 a 'f' has been added as a subscript to remark than these results are the strains and stresses at the core of the grating. Strains in the laminate and the optical fiber should be roughly the same, but the agreement in this case is only qualitative for the transverse direction, and how to use FBG to obtain internal residual stresses in composite laminates is still an interesting research topic. The most common approach, to avoid these complexities, is to cover the FBG with some coating that dampen the transverse stresses, reducing or eliminating the peak splitting.

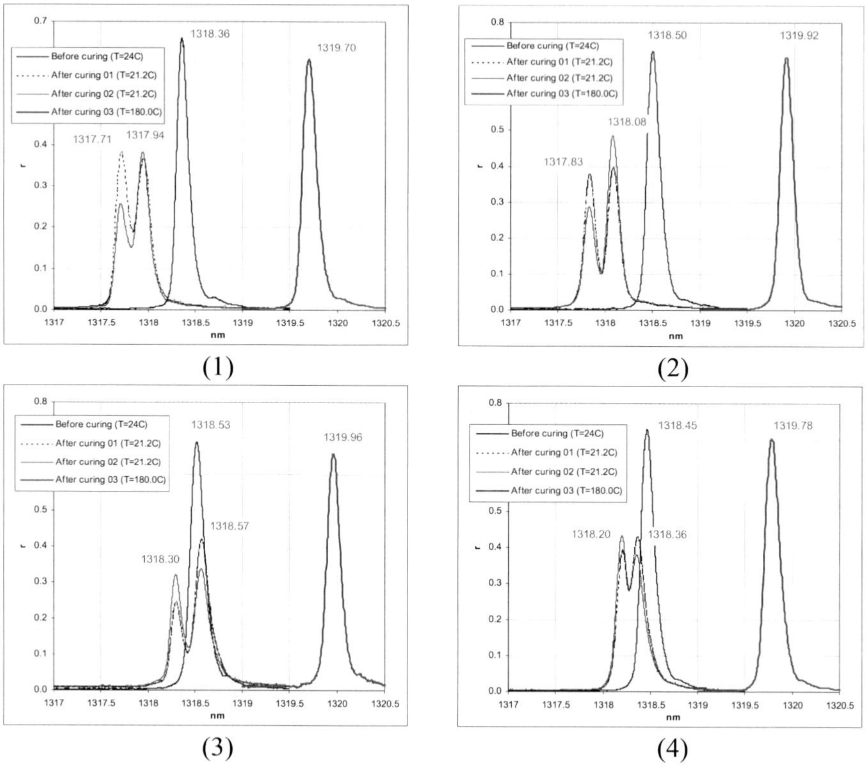

Figure 18. Spectra of the four gratings in different moments of the process: grating at 24 °C before curing, grating at 180 °C after curing (still single peak), and grating at 21.2 °C after curing (two peaks).

Table 1. Residual strains and stresses calculated using the photo elastic model.

Grating Nº	σ_{1f} (MPa)	σ_{2f} (MPa)	ε_{1f} ($\mu\varepsilon$)	ε_{2f} ($\mu\varepsilon$)
1	-15	71	-385	1057
2	-19	78	-453	1155
3	13	84	-20	1167
4	24	50	-87	705

3.5 Commercial FBG interrogation systems.

The interrogation scheme given at Figure 7, using an OSA (Optical Spectrum Analyzer), is too slow and bulky for practical applications; besides it, getting the detailed spectrum is not usually required, information of the peaks wavelength is enough for most of the practical cases.

FBG interrogation methods can be classified by the measurement frequency. Conventional systems use wavelength sweeping by mechanical moving parts by piezoelectric actuators, such as a tunable laser source or a tunable Fabry-Perot filter; they are used for the wavelength interrogation with the measurement frequency under 1 kHz, which is enough for most of the common mechanical applications.

For the measurement frequency over 1 kHz, which is required for strain measurements under impact conditions, or for the analysis of elastic waves generated into the structure, Bragg wavelength shifts should be converted into optical power through a certain type of optical filter without introducing any mechanical moving parts. A very simple and economic solution is obtained by a diffractive grating associated to a CCD array, even the resolution is not as good as with the former techniques.

The higher speed is obtained by using wavelength dependent filters, as a LPG or a AWG, being able to reach 1 MHz, as the only limit is the speed of the photo detector. In these cases, the sensing FBG must have a wavelength centered at the window of the filter, and the span is limited to the window width; these conditions limits severely the multiplexing capability, usually based on different central wavelengths.

4 Structures With Embedded Fiber Bragg Gratings

The possibilities of structural integration do not only depend on the sensor characteristics. The structural material, from now onwards denominate 'host', also have to fulfill a set of requirements to allow the integration of fiber Bragg gratings, without disturbing the manufacturing process. Host has to be immune to the presence of the optical fiber, so that its mechanical properties or its structural integrity is not modified.

Although the fiber optic, and therefore the sensors, can be bonded externally to any material on any structure, the concept of Smart Structure carries inherently the concept of integration. Some of the advantages associated to embedding fiber optic sensors are:

- Increased survival possibilities of the optical fiber and integrated sensors, intrinsically brittle, that would be protected by the external structure.

- The existence of an intimate sensor/host union.
- The absence of external wiring, important not only from the aesthetic point of view, but also allowing its use for applications that require clean surfaces, as it is the case of aeronautical structures.
- The possibility for internal monitoring of the structure, in places which cannot be reached by conventional sensors, or would be too intrusive by their size and by the high number of wires that usually are necessary to route until them.

A full integration of sensors into the host material is limited by the material nature and by the manufacturing processes. For example, the maximum temperature for silica fiber (without external shell) is about 900°C, limiting its possibility for embedding in metals and metal matrix composites. Concrete structures, and advanced composite materials with polymeric matrix are more adequate for integration purposes.

Manufacturing process of structural elements also limits the use of embedded fiber optics, because it is necessary to avoid potentially aggressive operations, such as high pressures application, shear stresses and very small chamfers. Another critical factor is the survival of fiber terminals to connect structural elements each other, for the output monitoring systems.

For using fiber Bragg grating as strain sensors it is necessary to have the maximum information possible about its surroundings in order to establish basic criteria to guarantee the accuracy, repeatability and stability of the measurements.

The main mission of the coating is the mechanical and environmental protection. Commonly used materials for coating optical fiber are polymers, acrylate or polyimide The coating characteristics are also very important because the stress and strain transmission from the structural element to the fiber is made through it, and its characteristics will impose the response of the sensor/structural element. Standard optical fiber has an acrylate coating, with low elastic moduli, low bare strength, oxidizing at temperatures used for curing composite materials (around 180°C). Polyimides have greater rigidity and bare strength, and do not degrade to these temperatures, reason why it is preferred for embedding in composite materials.

Some factors influencing the response are:

- The coating thickness: it has to be sufficient to isolate from cross-sectional stresses, but not excessive, since it could be intrusive and could delay the transmission of longitudinal deformations.
- Its tolerance to the adhesive cure process or the matrix in which the fiber is embedded, decreasing its degradation to high pressures and temperatures.

– Its chemical compatibility with host material, and degradation resistance in humidity conditions.
– Its rigidity, as nearest as possible to the resin matrix.

4.1 Orientation of the optical fiber optic respective to the reinforcement fibers.

The main factor in the behavior of fiber/host is, for the case of composite materials with long fiber reinforcement, the orientation of the optical fiber respective to the reinforcement. The monomode fiber used in Bragg grating sensors is 125 μm diameter, approximately the same thickness as a composite material lamina. In the surroundings of the optical fiber, a resin-rich zone appears, named resin pocket (Figure 19). In weave laminates the determination of resin pocket geometry is very difficult, by the complexity of the phenomena implied in its formation. In the case of tape laminates, resin pocket size is reduced when the optical fiber/reinforcement fiber angle is reduced, and disappears when it is embedded longitudinally to them.

The resin pocket geometry and size depends on the reinforcement fibers orientation of adjacent sheets to optical fiber. The best configuration is optical fiber embedded parallel to the reinforcing fibers. In this way the resin pocket size is null, and the optical fiber interference with the laminate is practically null. It will be necessary to set reduced points in which the angle optical fiber/reinforcement is not null and identify them as possible sensitive zones.

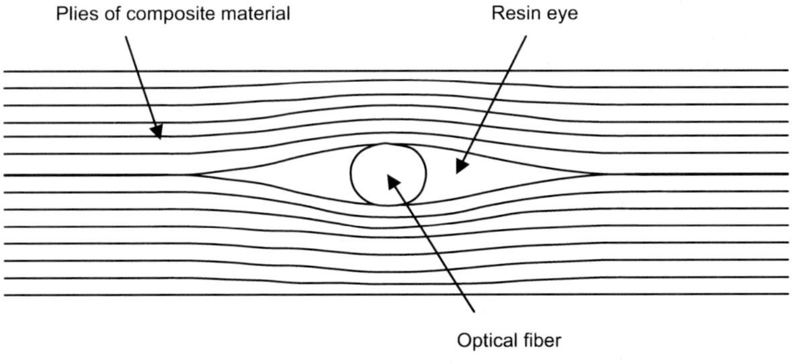

Figure 19. Sketch of the laminate cross section, showing the 'resin eye'.

It is evident that the resin pocket, besides to create a laminate zone without reinforcement material, causes a distortion of the stresses field which may initiate a delamination process.

In laminates made with fabrics, the fiber weft, added to the applied pressure during the curing process and the differential contraction that appears during cooling may cause a fiber breakage or, if the fiber has survived the application of pressure, the build-up of a complex stresses fields and residual strain around the Bragg grating, making very difficult to analyze its response. After a lot of tests on integration of Bragg gratings in materials we concluded that best way to use Bragg gratings in this type of laminates is surface bonding or embedded between unidirectional tapes that protect the optical fiber.

4.2 Ingress/egress from the laminate.

The input/output ends of the fiber are considered critical points for mechanical interference of the optical fiber with the structural host element. Standard procedures for input/output of the optical fiber from the composite materials still do not exist. It is due to the difficulty to edge machining the composite materials parts with optical fiber terminals or connectors leaving the laminates edges. The developments carried out during the last decade aim at two possibilities:

Use of composite materials manufacturing techniques that avoid the need of later edge trimming (RTM, SMC, RIM, etc.) and would allow semi-embedded connectors on any surface of a laminate. Their effect on the mechanical behavior of the part, and therefore the precautions to take from a point of view of the fiber survival, will be similar to conventional rigid inserts.

Output by means of pig-tails through the laminate flat surfaces, which obviously do not require any later manipulation. The effect on the laminate characteristics will depend on several factors: laminate configuration in the surroundings of the output, size and rigidity of the cable, etc.. The existence of a low module inclusion in a contour piece zone is not an optimal solution for high structural requirements applications, mainly in aeronautical applications, reason why it would have to be restricted to light loaded elements or laboratory applications. In the case of outputs by rigid inserts, the precautions to be taken will be the adapted to these kinds of elements.

5 Examples of Applications in Aeronautics and Civil Structures

Although not still incorporated to series production, neither accepted for Certification procedures, a large number of demonstrations can be found in the open literature. The practical details for the implementation of two structural elements instrumented with fiber Bragg gratings are detailed next, illustrating the special precautions to be taken, and the results obtained.

5.1 Stiffened panels with embedded fiber Bragg gratings.

Three stiffened panels with co-cured stiffener webs were manufactured by means of manual lay-up and autoclave curing. This structure could simulate a portion of the skin panel of a tail plane or a wing of a small aircraft. Two of the panels (panels number 1 and 2) were instrumented with optical fiber Bragg grating sensors bonded on the stiffeners and back-to-back on the skin surface. The other panel (panel number 3) was provided with two embedded optical fiber sensors in each stiffener. Additionally FBGs and strain gauges were bonded on the stiffener webs and on the skin back to back.

The panels have a dimension of 560 mm x 280 mm with 3 blade stiffeners of 20 x 20 mm cross-section situated in the middle line and at 105 mm on the left and on the right from the middle line. Skin thickness is 1 mm and stiffener thickness 1.6 mm. Fiber lay-up of the panels has been optimized, being in panel 3 skin: $(\pm 45^{\circ}_{(f)}, \pm 45^{\circ}_{(f)}, 0^{\circ}, \pm 45^{\circ}_{(f)}, \pm 45^{\circ}_{(f)})$ and in the stiffener: $(\pm 45^{\circ}_{(f)}, 0^{\circ}, 0/90^{\circ}_{(f)}, 0^{\circ})_S$ using a combination of unidirectional and fabric (f) carbon fibers.

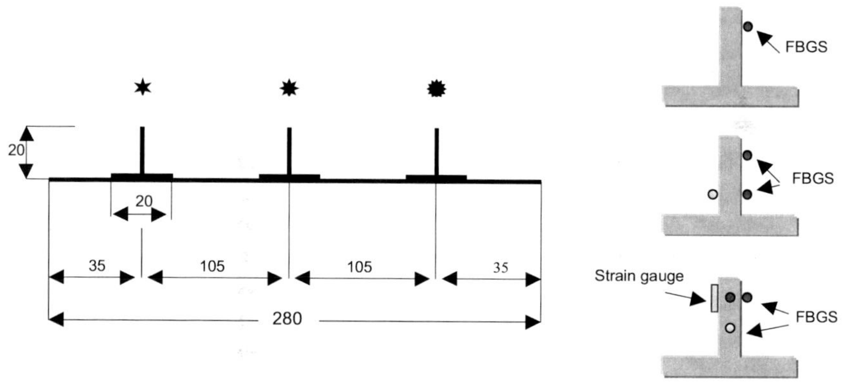

Figure 20. Dimensions of the panels with details of the sensors position in panel 1, 2 and 3.

For manufacturing the stiffened panels a toughened epoxy prepreg material Hexcel A-193-P/8552S carbon fabric and AS4/8552 unidirectional tape with AS4 high-strength carbon fiber were employed.

The panels were manually laid-up over a reinforced elastomeric mold made of Airpad polyacrylic rubber (Airtech). The elastomeric mold turned out to be very suitable for embedding the optical fibers in the laminate. The elastic rubber surface of the mold enables the optical fibers covered by their protection tubes to be guided between the laminate and the mold and protects them from high-pressure concentrations. The applied curing cycle was the standard one recommended by the prepreg manufacturer (180 °C/ 6 bar, 2h). For panel 3 a lower pressure of 4 bars was applied. Panel edges were machined with a diamond disc saw. After edge machining, panel edges were potted into an epoxy/gypsum mixture inside a normalized U-transverse section steel profile (80 mm width, 45 mm high) to ensure uniform loading (Figure 21).

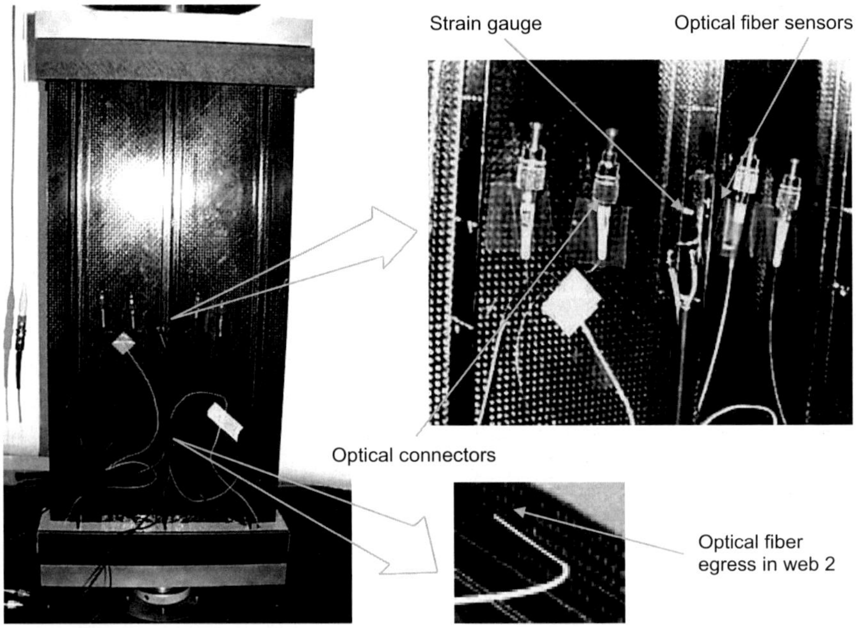

Figure 21. Integration of optical fibers in the stiffener webs

The optical fiber sensor used is a 230 mm long portion of photosensitive optical fiber with a 10 mm long Bragg grating sensor zone located at 65 mm from the end of the photosensitive fiber. For panels 1 and 2 sensors were bonded by means of a cyanate adhesive to the slightly grounded and degreased surface of the panel. The photosensitive fiber was spliced with a conventional single-mode fiber provided with an optical connector.

In panel 3 a 165 mm long photosensitive fiber was spliced with an 800 mm long single-mode conventional optical fiber. Two FBGs were embedded in the web of each one of the three stiffeners of the panel. The optical fibers were placed between the two central 0° layers and parallel to the reinforcement fibers. Their sensing zone was located at mid-span of the stiffener. Embedding the optical fibers between 0° prepreg layers ensures that they will not be destroyed during the consolidation process and that they will provide a reliable sensor signal that will not be spoiled by the lobed curve surface of the fabric layers. A tiny hole was made passing through the laminate to enable the optical fiber to get out of it. A 0.8 mm diameter PTFE tube was used to protect and guide the two optical fibers in each web out of the laminate. This tube was sealed with silicone to prevent resin contamination. A lead of about 25 mm of this PTFE tube was left inside the laminate to guaranty a good protection of the fiber at the point where it comes out from the laminate. The rest of this PTFE tube was guided between the laminate and the mold coming out of the mold at its front side. A long portion of the protected optical fiber was left outside the laminate and a connector was fixed to it after curing.

Additional FBGs and conventional strain gauges were surface bonded on panel 3 to verify the strain measurements of the embedded FBGs. Conventional strain gauges were also bonded on the skin to determine the skin buckling. The strain gauges were bonded in all the webs on the left web site at the same location than the upper embedded FBGs; FBGs were bonded on the right web side. In the middle web also additional gauges were bonded in front of the lower FBGs on both sides of the web. All the sensors survived the manufacturing of the part, showing clearly defined spectral peaks with no significant distortion due to complex local strain fields caused by residual stresses.

To verify the exact geometrical position of the embedded optical fibers inside the stiffener webs several X-ray images have been taken from panel 3. The images showing 300 mm of the stiffener length including the fiber egress. A micro focus X-ray system has been used operating at a low voltage level to obtain sufficient high resolution that enables the detection of the small optical fiber inside the composite. Figure 22 shows the X-ray image of stiffener web 3. The two optical fibers can be clearly identified within the composite. On the left side the fiber egress with the protective Teflon tube can be seen. The resolution is high enough to detect not only the fiber path but also possible optical fiber cracking within the composite.

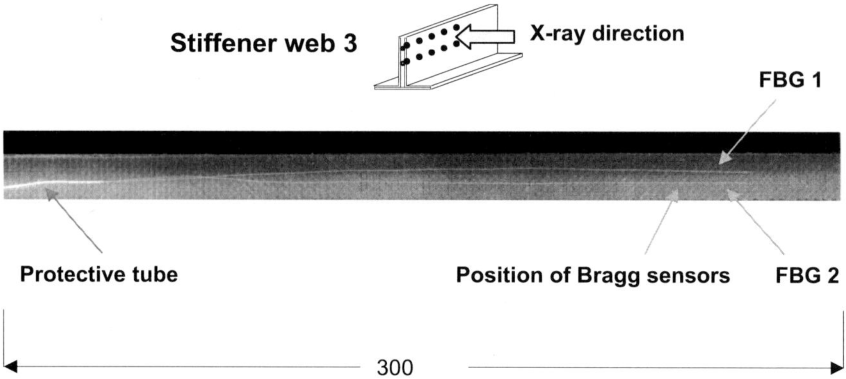

Figure 22. X-ray image of stiffener web 3 in panel 3 indicating the position of the optical fibers and the protective tube at fiber ingress.

Compression tests at room temperature were conducted for the three panels For panels 1 and 2 the sensors worked correctly up to the panel failure. In the explosive rupture the splice of the photosensitive and the conventional single-mode fiber broke at almost every sensor. Fixing an optical connector directly to the photosensitive fiber would leave the sensors operative even after rupture of the panel.

In the case of panel 3 the six FBGs were embedded in the centerline of the stiffener web so that they were able to measure tensile/compression strains but not transversal bending or twisting of the web. At the beginning of the compression test all the sensors detect the same compressive strain up to the moment when buckling occurs at a strain level of 800 $\mu\varepsilon$. A large bending strain in the skin can be noticed by the bifurcation of the sensor signals of the bonded strain gauges. Skin buckling increases the load that has to be carried by the stiffeners and induces also a twisting and transverse bending of the stiffeners. With increasing load, longitudinal bending of the web occurs. Figure 23 shows the strain measured with the upper embedded FBGs, the bonded FBGs and the bonded strain gauge of the stiffener web 1, 2 and 3. The measurement of the skin bonded strain gauges is also included in the graph to indicate the skin buckling.

5.2 Concrete beam repair

Advanced composites offer some advantages over traditional procedures for repairing concrete structures, due to their optimal corrosion properties,

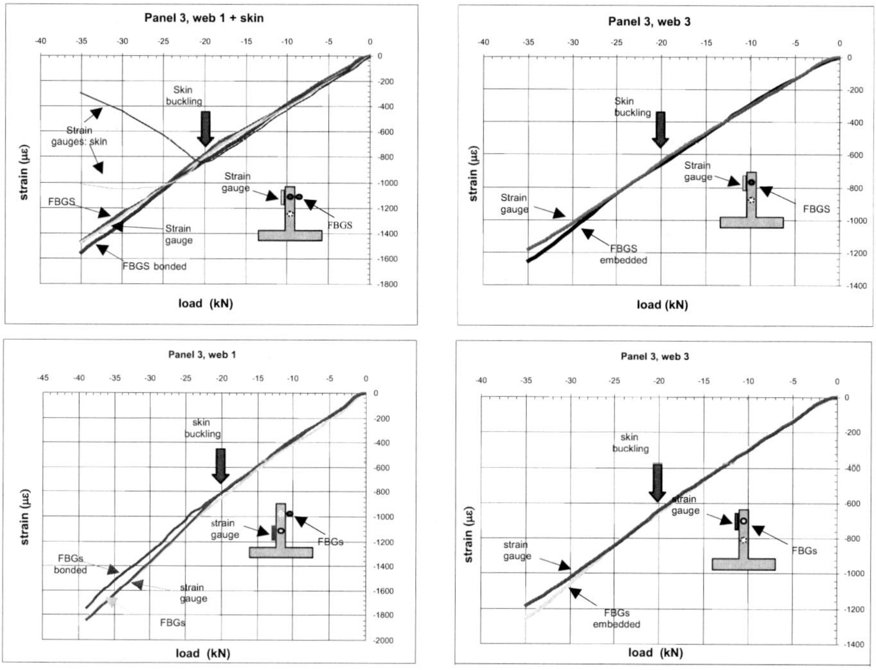

Figure 23. Strain vs. load values from the FBGs and strain gauges of stiffener web 1 and 3.

low weight and decreasing costs. Thin cured laminates may be externally bonded or dry fabrics can be applied in situ.

On the other hand, Bragg gratings show the aforementioned advantages when compared to conventional strain gauges: absolute measurements, spectrally encoded output, no EMI, no drift (long-term stability), low size, multiplexing capability and their ability to be embedded into laminates without degradation or detriment effects.

The combination of both techniques is easy and offers important advantages. The long- term mechanical behaviour of the repair may be checked and information on environmental degradation could be obtained. On the short term, information on the stress transfer from the concrete to the laminate is obtained, and the validity of models is verified.

Tests description. A reinforced concrete continuous beam with two spans (7.2 m + 7.2 m), representative of current in-service civil engineering structures, was loaded up to failure and repaired. After removing loose concrete near the cracks, new concrete was poured. Composite patches, made of two layers of carbon fabric, in-situ impregnated with room temperature epoxy resin, were extended over the crack locations, to transmit the load from the discontinued steel bars. Bragg gratings were placed with the fabrics, both in the internal and external surfaces. External pre-stressing was added to increase the failure load. Arrangement is shown in Figure 24

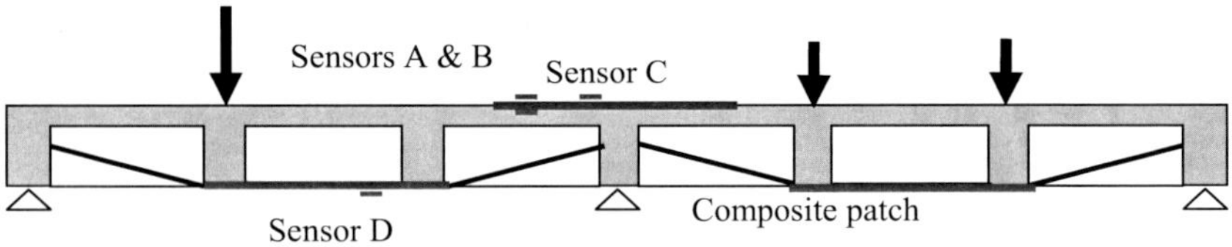

Figure 24. Test arrangement and Bragg grating locations.

Results. Strain vs. Load is plotted in Figure 25, obtained with strain gauges and FOS. Coincidence is very satisfactory, even at the non-linear region of the concrete beam. Damage starts near 80 kN, still with considerable residual strength. Worth to mention that ultimate load of the repaired beam (156.6 kN) was higher than undamaged beam (100 kN).

6 Distributing Sensing

Fiber Bragg grating sensors have been used for the last 20 years, and they have built up a confidence in its performances. FBGs can measure the strain with a similar accuracy to the standard strain gauges and extensometers, and they are also comparable in many aspects from a user's point of view. Indeed, their measurements are local and directional, they require compensation for temperature, they are commonly used by bonding them onto the structure surface and it is also possible to embed the optical fiber in the laminate in the case of composite structures. The main advantages of the FBG over the electrical strain gauges are its reliability for long term measurements; because it is frequency coded, without drifting by aging, and its ability for multiplexing; since several FBGs can be engraved on the same fiber at different positions, resulting in the simultaneous measurement points. The most common procedure for multiplexing is to use a different central frequency for each grating, allowing up to ten FBGs in a single optical fiber, for conventional applications. Another multiplexing

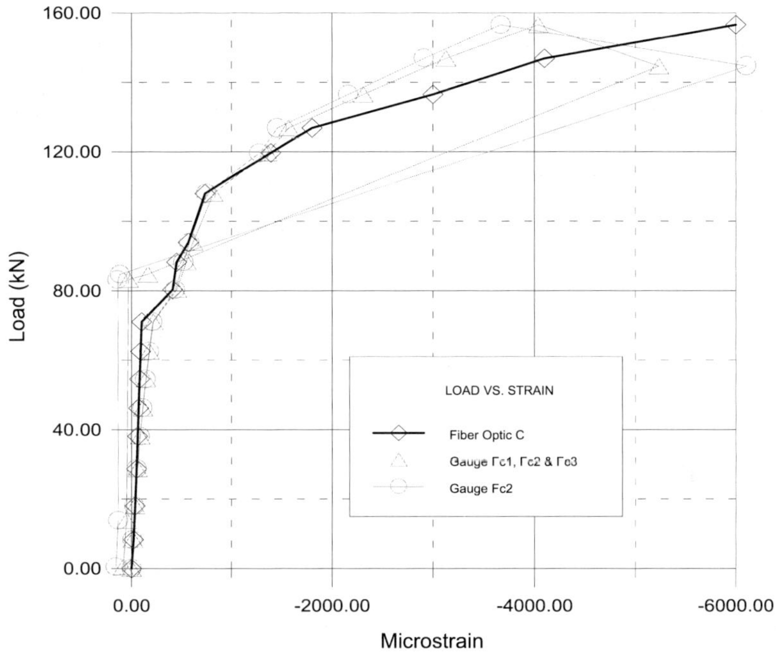

Figure 25. Strain measurements (FOS and strain gauges)

procedure is to have all the gratings with the same central frequency, with a low reflectivity; if gratings are adequately spaced, the return signal will be time-multiplexed, allowing a much larger number of sensors per fiber. The drawback of this procedure is that its response time is smaller, restricting the system to quasi-static applications.

Getting the strains all along the optical fiber, with adequate spatial resolution and strain accuracy, opens new possibilities for structural tests and for structural health monitoring. This is what is understood as 'distributed sensing', with the main difference that the fiber does not need to have local engraved sensors.

The starting point of fiber optic distributed sensing may be identified at the beginning of the 1980s, with the Optical Time-Domain Reflectometer (OTDR) technology, a technique widely used for testing optical cables in the telecommunications industry. The concept is to send a narrow pulse of light through the optical fiber and keep listening for the 'echo' of the backscattered radiation. The detected signal provides a detailed picture of the local loss distribution or reflections along the fiber caused by any of the attenuation mechanisms or some other non-homogeneities on the fiber. The location of the defect may be calculated by the time of flight. The resolution is in the range of meters, but the operating range was several kilometers, so the technique has been found to be very useful for locating fiber breaks. Figure 26 illustrates the typical graph obtained with conventional equipment. Note the X scale is distance, in km.

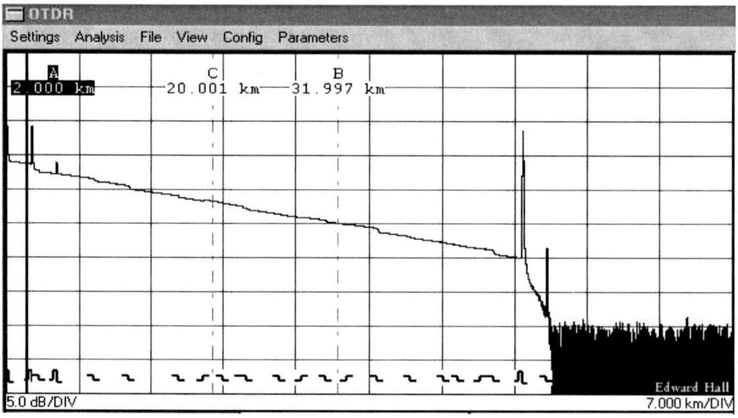

Figure 26. Typical graph obtained with an OTDR, distance vs. Optical losses.

The backscattered radiation. The backscattered light consists of different spectral components due to different interaction mechanisms between the propagating light pulse and the optical fiber. These backscattered spectral components include Rayleigh, Brillouin, and Raman peaks or bands. The Rayleigh backscattering component is the strongest one; it is due to density and composition fluctuations and has the same wavelength as the primary laser pulse. The Rayleigh component controls the main slope of the decaying intensity curve and may be used to identify the breaks and heterogeneities along the fiber.

A small percentage of photons inter-exchange energy with the atoms and new photons with lower (Stokes components) or higher energy are produced. This is known as Raman scattering, the spectral shift is related to the quantum energy absorption/desorption by the atomic bonds of the silica. The intensity of the anti-Stokes component of the Raman radiation increases with temperature, whilst the Stokes component remains stable; the power ratio between these two peaks, together with the time of the arrival provides information about the local temperature and position respectively. Many commercial systems are currently available to obtain temperature maps over long distances, with important hydrological and environmental applications, including the detection of leakages in buried pipes. These systems basically use a pulsed laser for sending a high power short pulse along the fiber (spatial resolution is related to the width of the pulse, and can hardly be better than 1 meter); at the returning path there is a wavelength separation module, filtering every radiation distinct from the two Raman peaks, and fast photodetectors for each peak. Computing the time for the distance to the light source, and the relative height, resolution in temperature in the range of 0.1 °C, for distances up to 50 km can be achieved.

Raman peaks height was insensitive to strain; in the case of the Brillouin scatter, the wavelength drifting is related to the local temperature and the strain in the fiber. Brillouin-based sensing techniques rely on the measurement of a frequency as opposed to Raman-based techniques which are intensity based. Brillouin based techniques are consequently inherently more accurate and more stable on the long term, since intensity-based techniques suffer from a higher sensitivity to drifts.

There are two types of Brillouin fiber optic sensors. Brillouin Optical Time Domain Reflectometers (BOTDR) resolves the strain or temperature based on Brillouin scattering of a single pulse. Brillouin Optical Time Domain Analysis (BOTDA) uses a more complicated phenomenon known as Stimulated Brillouin Scatter (SBS).

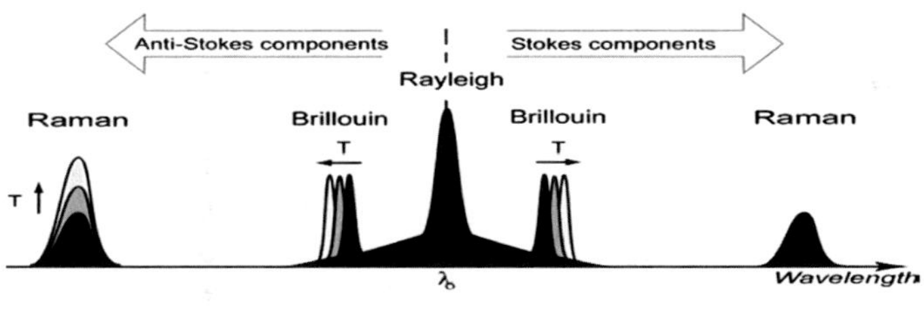

Figure 27. Wavelengths of the backscattered radiation.

Brillouin scattering occurs as a result of an interaction between the propagating optical pulse and the acoustic waves present in the silica fiber, generating frequency shifted components. The diffracted light experiences a Doppler shift since the atoms vibrate at the acoustic velocity in the fiber. The acoustic velocity is directly related to the medium density and depends on both temperature and strain.

The classical BOTDR is essentially an OTDR with a strong filter to avoid the Rayleigh radiation, and with an additional device to discriminate the wavelength of the Brillouin peak. Consequently, it will have the same advantages and limitations of OTDRs: very long interrogation distances, low spatial resolution.

Brillouin Optical Time Domain Analysis (BOTDA) uses a more complicated phenomenon known as Stimulated Brillouin Scatter (SBS). The typical sensor configuration requires two lasers that are directed in opposite directions through the same loop of fiber (one laser operating continuously, the other pulsed). When the frequency difference between the two lasers is equal to the 'Brillouin frequency' of the fiber, there is a strong interaction between the 2 laser beams inside the optical fibers and the enhanced acoustic waves (phonons) generated in the fiber.

This interaction causes a strong amplification of the Brillouin signal which can be easily detected and localized using an OTDR-type sampling apparatus. To make a strain or temperature measurement along the fiber, it is necessary to map out the Brillouin spectrum by scanning the frequency difference (or 'beat' frequency) of the two laser sources and fitting the peak of the Brillouin spectrum to get the temperature and strain information. Current existing BOTDA systems have a spatial resolution in the range of 10 cm, expecting to reach 1 cm in the next equipments generation.

The OBR (Optical Backscatter Reflectometry). The OBR uses swept wavelength interferometry (SWI) to measure the Rayleigh backscatter as a function of length in the optical fiber with high spatial resolution. An external stimulus (like a strain or temperature change) causes temporal and spectral shifts in the local Rayleigh backscatter pattern. These temporal and spectral shifts can be measured and scaled to give a distributed temperature or strain measurement. The SWI approach enables robust and practical distributed temperature and strain measurements in standard fiber with millimeter-scale spatial resolution over tens to hundreds of meters of fiber with strain and temperature resolution as fine as 1 $\mu\varepsilon$ and 0.1 °C.

Figure 28 shows the high spatial resolution achieved. Imagine we have a fiber under test, with four partial reflectors. Each reflector forms an interferometer with the light that goes directly through the other arm of the first coupler. When the system is interrogated by a continuous tunable laser, the reflected signal from each partial reflector is modulated by a unique frequency, longer distances implies faster beatings. A Bragg grating sensor is a partial reflector, with a spectral width much wider that the beating of interferometry, as depicted. Using the Fourier transform, the position of each grating is obtained. With SWI, hundreds of FBG could be multiplexed in a single optical fiber.

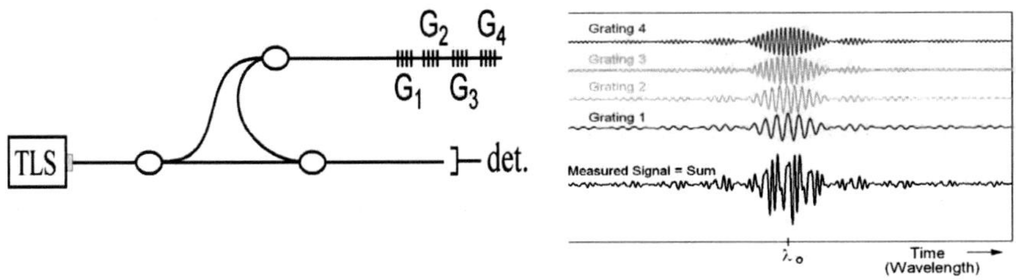

Figure 28. Sketch of the SWI, and the obtained signals.

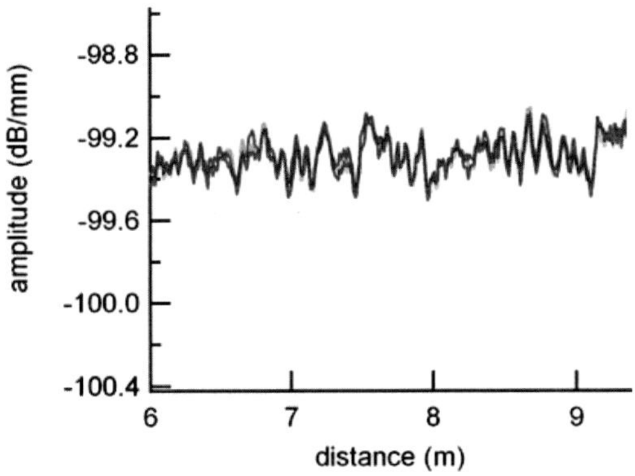

Figure 29. Backscatter Light profile represents a fingerprint that is imprinted in the fiber when it is drawn. Backscatter light presents a high repeatability in absent of external effects.

But there are still more possibilities with this technique. Rayleigh backscatter in optical fiber is caused by random fluctuations in the index profile along the length of the tested fiber. These perturbations will behave as very weak arbitrary FBGs engraved in the fiber all along its length. Shifts in the fiber index of refraction n or in the average perturbation period caused by an external stimulus (like strain or temperature) because shifts in the local spectral frequency of the Rayleigh backscatter. Accumulated changes in the refractive index along the optical path also manifest as a time shift of the Rayleigh backscatter return loss amplitude pattern. Performing a cross correlation on the backscatter amplitude time domain or frequency domain data accurately measures these spectral and temporal shifts, which are scaled to form distributed temperature or strain measurements.

Consequently, a plain optical fiber behaves as a continuous strain/temperature sensor, with a high spatial accuracy, because it was obtained by interferometric procedures, and with the accuracy in the strain/temperature measurements similar to FBGs. The low power of the reflected signals implies a poor SNR (signal to noise ratio), which can be partially improved by averaging, or repeating the measurements several times, limiting the applications of the technology to almost static tests.

Comparison of technologies. This comparison must be consider very rough, firstly because these technologies are quite new, less than ten years old, and the equipment's performances and prices are evolving very quickly. Secondly, because the definition of performance specifications for distributed sensors is more difficult than for traditional point sensors, since the performance depends on a combination of related measurement parameters. For example, accuracy depends on the spatial resolution, acquisition time, distance range or cumulated loss prior to measurement location.

Two clear conclusions may be drawn from the Table 2:

- Current distributed sensing systems are useful only for static tests. For dynamic tests a multiplexed FBG system is still the only choice.
- The cost of distributed systems is still deterrent if compared to classical extensometry.

7 Applications of Distributed Sensing

Structural tests. The distributed sensing technique may significantly improve the instrumentation of structural tests; instead of bonding a large number of sensors, it will be substituted by bonding one (or several) plain optical fibers in the regions of interest. The main advantage of this technology is that it provides a full coverage of the instrumented area. This issue is of special interest in composite structures, commonly designed with changes of thickness and different stacking sequences. As a consequence, the strain field is not uniform and it is cumbersome to monitor all the critical areas with point strain gauges, since this information is directly obtained with distributed sensing. This technology is also very suitable for detecting local buckling, a problem which presents a great level of difficulty when predicting its character.

In comparison with traditional extensometer sensors, as strain gauges, distributed optical sensors provide a lower price/performance ratio, as it

Table 2. Comparison of technologies.

	FBG	BOTDR	BOTDA	OBR
Strain accuracy	$\pm$ 1 $\mu\varepsilon$	$\pm$ 30 $\mu\varepsilon$	$\pm$ 2 $\mu\varepsilon$	$\pm$ 1 $\mu\varepsilon$
Spatial resolution	Related to Grating length	0.1 m	0.1 m	0.001 m
Length range	Point sensor	100 km	100 km	100 m
Acquisition time	3 kHz typical sampling rate	0-20 minutes	As low as 1 second	Related to accuracy and length. 30 sec typical
Temperature accuracy	$\pm$ 0.1 °C	N.A.	$\pm$ 0.1 °C	$\pm$ 0.1 °C
Sellers	MicronOptics, Fibersensing, Insensys	Yokogawa, Sensornet, NTT	OZ Optics, Omnisens, Neubrex	LUNA
System cost	3000-6000 Euro	Over 100.000 Euro	Over 200.000 Euro	150.000 Euro

is easy to install, either bonded or embedded, with significant savings in time and money. As no sensor needs to be engraved, standard optical fiber can be used, and it can be obtained with a very competitive price in the market. In spite of these enormous advantages, this technology has not replaced traditional extensometry, mainly due to two reasons; in first place, the purely static acquisition rate, which limits the technique to static tests. It is possible to increase the acquisition rate by using a special optical fiber much more expensive, but it is still unable to be used for aeronautical testing. The second reason is the directional nature of the optical fiber, which makes it extremely difficult to measure the strain field in more than one direction, whilst it is easy to make a rosette with strain gauges. Currently, the state of art of this technology allows it to be used in structural tests, but always as a support for traditional extensometry.

This technique is already used for structural tests of wind turbine blades. Wind turbine blades are the perfect candidates due to their enormous size, because a high number of sensors are required to cover the full structure and the material, with continuous changes of thickness and non-uniform properties.

The results of the structural test of a 43 meters long wind turbine blade are presented. Up to six standard optical fibers where bonded at the external surface of a wind turbine blade at the following locations: one in the leading edge, two in the pressure side, one in the suction side, and two each side of the trailing edge (see Figure 31). Only one interrogation system was used, the signal was shifted from one fiber to the next one by and optical switch. All these fibers were oriented in the direction of the spar, and run from the root to the edge. Temperature corrections were not needed because the temperature was kept uniform over the structure and constant since the start of load application. In case the temperature would have changed during the tests, a compensation procedure should be done on the data, as with any other strain gauges.

Figure 30. Structural test of a full wind turbine blade.

The concordance amongst the point data obtained with strain gauges and the continuous line obtained with the OBR is excellent; the crooked line is not due to noise, but to the abrupt changes due to ply drops and changes in thickness in the skin of the blade. OBR measurements help the designers detect all the stress concentration areas, even if they were not initially considered of interest. Also, the initiation of the elastic buckling can be clearly detected and located.

Another interesting application under study for wind turbine blades,

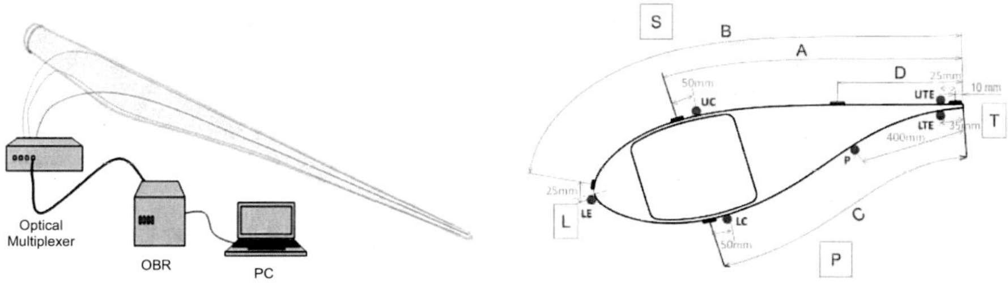

Figure 31. Scheme of fibers mounted in the full blade (left) and in one transversal section (right).

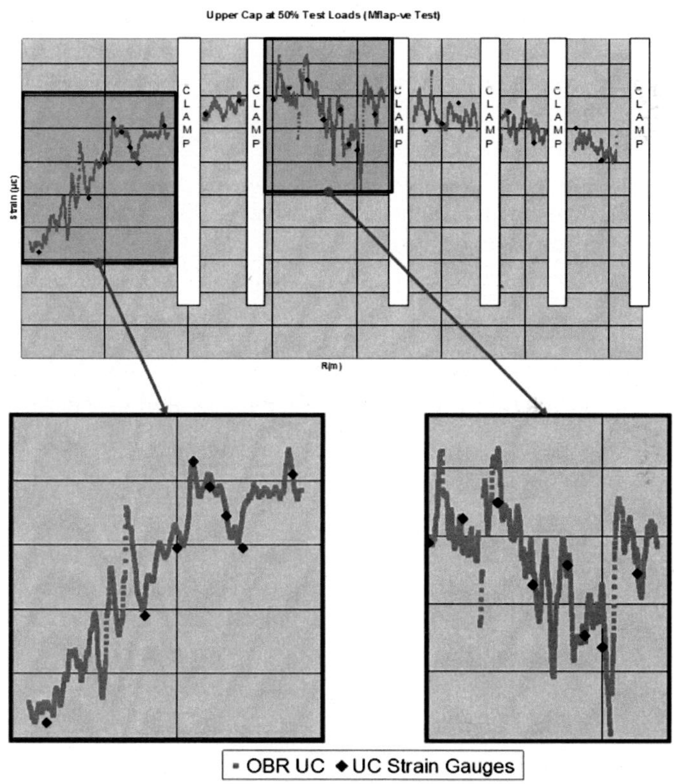

Figure 32. Strain measurements along the length of the blade (above) Details of some areas (below).

not related with structural test, is to monitor temperature during the curing along the full structure. One optical fiber can substitute several thermocouples with a better integration. The first results obtained in the laboratory are very promising.

Damage detection in civil structures. Another field in which distributed sensing can deliver important benefits is civil engineering, since it can be used for monitoring civil structures, such as big buildings, bridges or pipelines. These structures need to be monitored in order to prove their structural safety, and to prevent them from deterioration or damage. One of the key points when monitoring this kind of structures is their large size, bigger than the usual instrumented structures. Unless damages produce a significant change in the stresses path, local sensors, as strain gauges, can only detect damages in their nearest area, and a large amount of sensors is required to cover the full structure.

Distributed optical sensors provide a high density sensor network to cover the full structure at a low price, as standard single-mode fiber is used as sensor. However, the capability to easily manage a huge number of sensors using distributed sensing is not new. Up to now distributed sensing techniques have not found a widespread usage in civil structural applications due to their insufficient resolution and their limitation to be used in static examples to measure average strain, due to the poor resolution of previous distributed techniques. One of the applications that have attracted most interest is pipeline crack monitoring. Small cracks can be detected due to the large increase of the average strain at the region of the fiber where the crack was produced. Several papers have used this approach; the OBR improves the spatial resolution of the mentioned technologies.

A similar experimental test was carried out on a reinforced concrete slab in the Structural Technology Laboratory of UPC (Polytechnic University of Catalunya, Spain) . One fiber was bonded on the surface of a reinforced concrete slab of 5.60 x 1.60 m. The fiber was bonded along the longitudinal direction, with two lines in the upper side and another two lines bonded on the lower side.

Strain sensors are not commonly used in the surface of concrete slabs since material heterogeneity due to the presence of aggregates of several sizes promotes a non uniform strain field in the surface. In Figure 34 the strain field in the upper side at different load levels can be observed. Strain grows uniformly, with irregularities near the load application region.

At the bottom side, tensile stresses promote crack appearance, and the strain chart is quite different. Data measured with optical fibers detects strain concentration due to cracks, even at low load levels. It is possible to

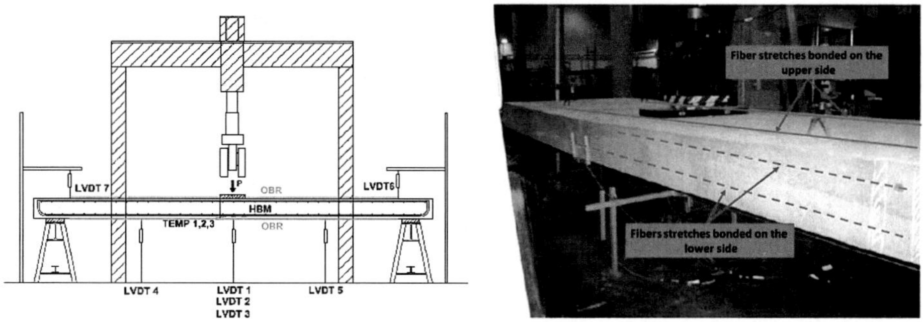

Figure 33. Test scheme (left). Location of optical fiber on concrete slab right).

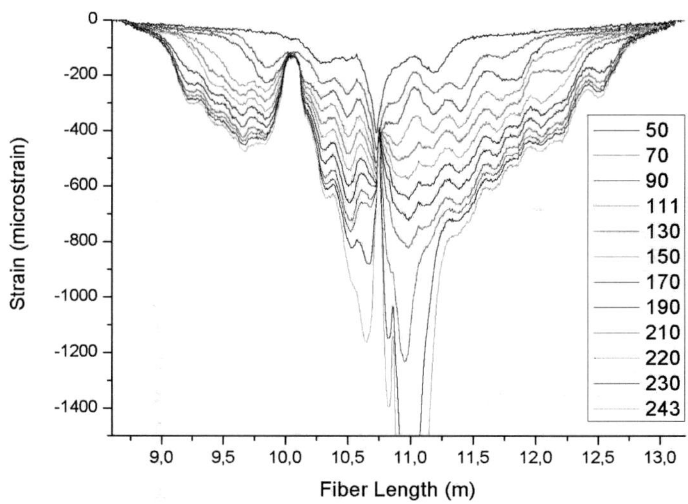

Figure 34. Strain field in the upper side of the slab at different load levels.

track crack appearance with increasing load levels, and detect crack location even in severe cracking conditions (Figure 35 and Figure 36).

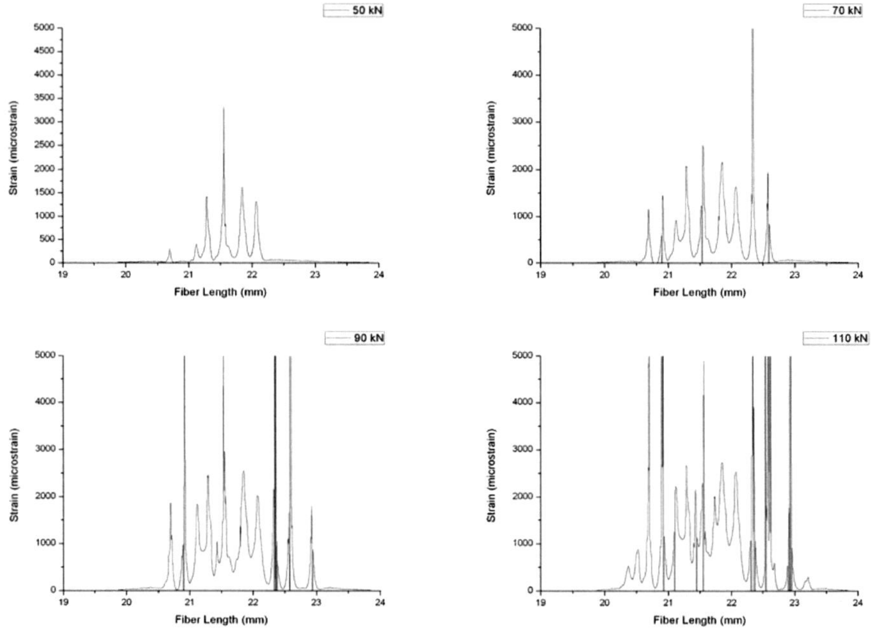

Figure 35. Strain measured with OBR along the lower side for different load levels: 50 kN (up, left), 70 kN (up, right), 90 kN (down, left) and 110 kN (down, right).

Damage detection in composites laminates. High accuracy in strain resolution and a spatial resolution better than 1 mm allows an ultra-high density sensor network with a single optical fiber, with such good performances that it may even detect the strain changes induced by a delamination.

Strain changes induced by a delamination are related to the residual strains built in the laminate during the curing and have only a local influence, limited to the delamination area. For these reasons a high-density strain sensor network is required, and formerly SHM techniques for delamination detection were not based on strain measurements.

To study the changes in the strain field promoted by impact damages, plates 150 mm long and 100 mm wide were manufactured with embedded

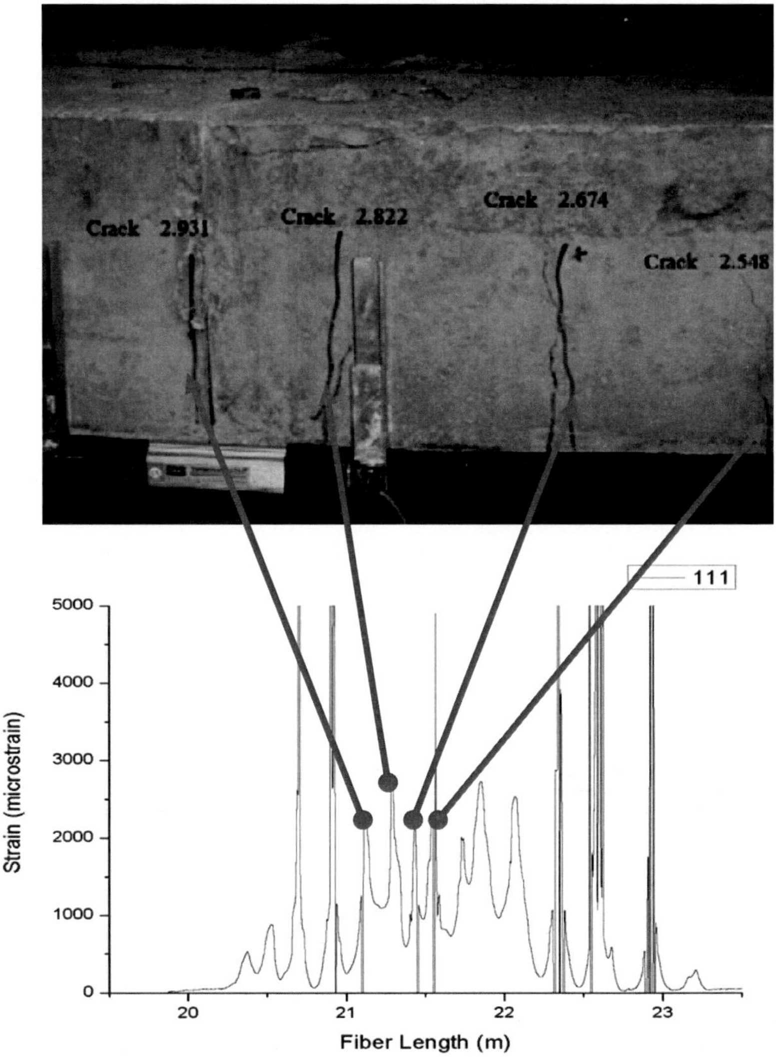

Figure 36. Correspondence between strain concentration peaks (down) and cracks in the concrete slab (up).

optical fiber. Plates were built with carbon/epoxy tape AS4/8552 with $[0_2,90_2,0_2]$ cross ply lay-up. Optical fibers were embedded along the longitudinal central line between plies with the same orientation. The embedded optical fiber has polyimide coating with a final diameter of $140 \pm 2\ \mu\varepsilon$. This type of coating is recommended for measuring the strain gradients due to their small thickness.

The first plate was subject to a 2.4 J impact at the middle of the plate. With this energy level, fiber C-1 (embedded in the opposite side to the impact) was broken. The strain readings at the other two fibers show a peak in damage location with similar strain level. In Figure 37 the C-Scan inspection after the damage can be observed. The strain field is mainly affected in the delamination area, with a maximum of nearly 500 $\mu\varepsilon$. One of the fibers, the one embedded in the middle of the plate, between the 90° plies, presents a strain field variation limited to the delamination area. The other fiber, embedded between two 0° plies, is mainly affected in the delamination area, but strain field changes affect also the close delamination region. With an impact of 4.4 J, all the fibers were broken. Test performed with fibers with acrylate coating of 250 $\mu\varepsilon$ did not show an improvement of resistance to impacts.

To avoid fiber breakage, impacts were performed 10 mm away from the fiber line (see Figure 38 Left). In this case, a growing delamination was monitored. In spite of the delamination size growing, the fiber length inside the delamination remained nearly constant. Figure 39 shows the strain along the different fibers: strain fields are similar in all of them, and present a maximum strain level in the delamination area. In these tests, all of the fibers show also an influence in the close strain field. Additionally, impact energy has no effect in the strain field level. Impacts were done for increasing energy levels (2.4, 4.4, 6.3 and 12 J).

Strain results can be explained by the influence of the cracks induced in the 90° plies, as the residual strain of these plies in the direction of the fiber is mainly compression. After the cracks appear, residual stresses are relaxed and affect the 0° plies too.

Strain measurements obtained from the full surface can be used to create a strain map of the structure. In case that damage promotes a change in the strain field, it is possible to obtain an image that represents the state of the structure. Figure 41 shows the effect of a delamination in a cross ply laminate of 200 x 200 mm. Strain data was obtained from a fiber in a zig-zag configuration that covered one of the sides.

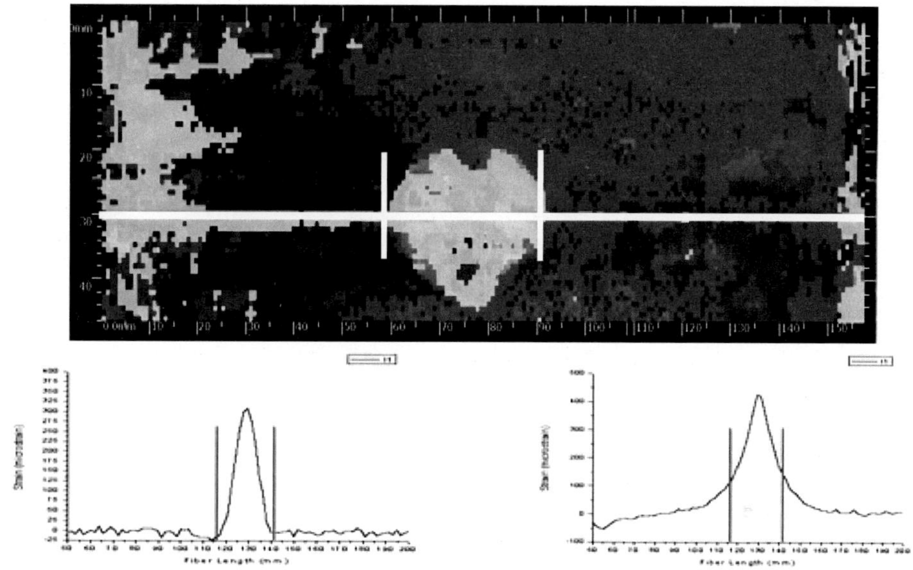

Figure 37. Ultrasonic inspection C-Scan after 2.4 J impact (Up). Strain measures of fibers C-2 (down, left) and C-3 (down, right).

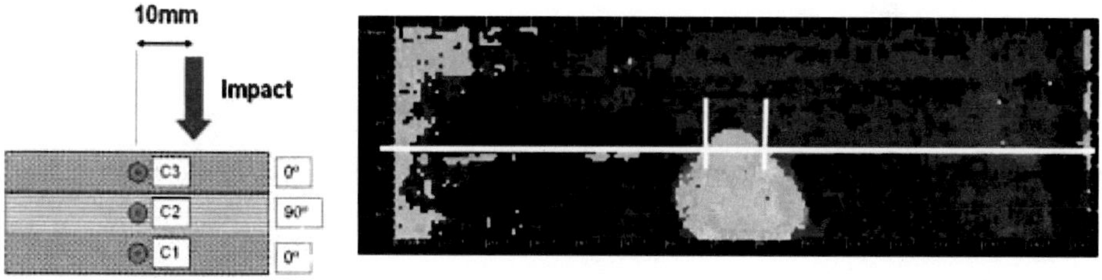

Figure 38. Scheme of test (left). Ultrasonic inspection C-Scan after 4.4 J impact (Right).

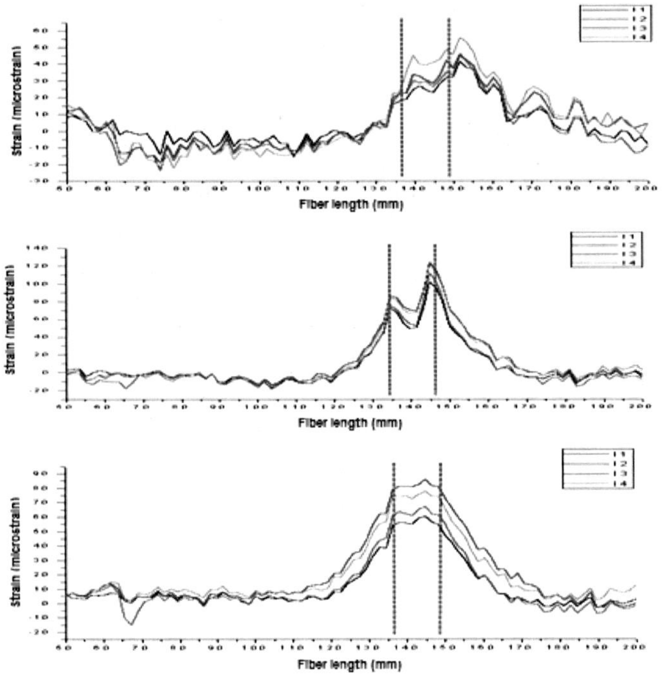

Figure 39. Strain distribution along the fiber after different impact levels for fiber C-1 (Up), C-2 (Middle) and C-3 (down).

8 Final Comments

The technology of fiber optic sensors, and particularly of the fiber Bragg gratings, is well matured for strain monitoring at a few points, and can be used for load monitoring in conventional and advanced structures. It offers several important advantages over conventional strain sensing, namely:

- Low size and weight, embeddable capability, single ended cabling.
- Long term stability, it can be used for load monitoring during the whole structural life.
- Inherent multiplexability, typical 10 sensors/fiber without decreasing reading speed.
- Immune to electromagnetic noise, able to work in harsh or explosive environment

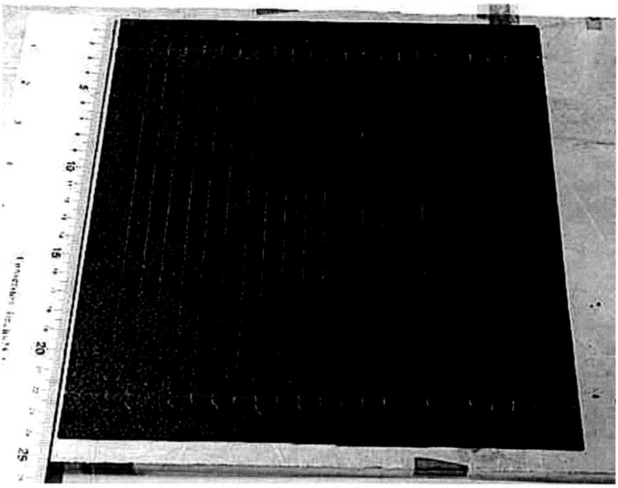

Figure 40. Fiber in a uniform zig-zag distribution in a composite laminate.

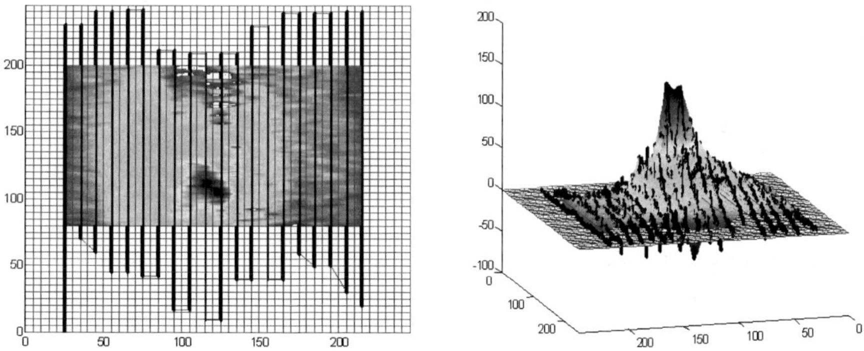

Figure 41. Strain map due to a delamination in 2D (left). And 3D (right).

Current efforts address to development and demonstration of large sensor arrays, affording a detailed map of the strain field. in a complex structure, such as a ship, an aircraft or a bridge. This would require new optoelectronics and signal processing systems, able to handling at high speed the information coming from several hundreds of sensing points, compressing this information to the significant events. This is needed if damage detection is intended, because in continuous structures a local damage will produce a very small change in the global strain field. Only considering the dynamic response or the transients strains, some information can be derived about damage occurrence.

While FBG does local strain measurements, similarly to strain gauges, new possibilities are opened by the distributed sensing, giving the information all along the optical fiber. Structural tests may be instrumented with greater details, and crack occurrence will be easily detected, as long as the crack crosses the optical fiber path.

Only the strain and temperature response of FBG was discussed in this chapter. Additionally, fiber optic sensors may afford very valuable information on chemical processes, such as corrosion in metals, degradation phenomena in concrete structures, or for the resin curing in composite materials. This is a very active research area, and combined to process modeling is paving the way to the intelligent materials processing.

Bibliography

M. Amano, Y. Okabe, N. Takeda, and T. Ozaki. Structural health monitoring of an advanced grid structure with embedded fiber bragg grating sensors. *Struct. Health Monit*, 6:309–316, 2007.

D.C. Betz, G. Thursby, B. Culshaw, and W.J. Staszewski. Acousto-ultrasonics sensing using fiberbragg grating. *Smart Mater. Struct*, pages 122–128, 2003.

D.D. Betz, W.J. Staszewski, G. Thursby, and B. Culshaw. Multi-functional fibre bragg grating sensors for fatigue crack detection in metallic structures. *J. Aerospace Eng*, 12:453–461, 2006.

K. Chandler, S. Ferguson, T. Graver, A. Csipkes, and A. Mendez. On-line structural health and fire monitoring of a composite personal aircraft using an fbg sensing system. In *Proc. SPIE*, 2008.

B.A. Childers, M.E. Froggatt, S.G. Allison, T.C. Moore, D.A. Hare, C.F. Batten, and D.C. Jegley. Use of 3000 bragg grating strain sensors distributed on four eight-meter optical fibers during static load testing of a composite structure. 2001.

C. Connolly. Fibre-optic-based sensors bring new capabilities to structural monitoring. *Sensors Rev*, 26:236–243, 2006.

G. Coppola. Analysis of feasibility on the use of fiber bragg grating sensors as ultrasound detectors. In *Proc. SPIE*, pages 224–232, 2001.

B. Culshaw and A. Kersey. Fiber-optic sensing: A historical perspective. *IEEE/OSA J. Lightwave Tech*, 26:1064–1078, 2008.

M. Frövel, I. Fernandez, J. M. Pintado, J. A. Güemes, and J. M. Menéndez. Monitoring the buckling and postbuckling behaviour of stiffened cfrp panels with fibre optic sensors. In *European COST F3 Conference on System Identification and Structural Health Monitoring, Proceedings of the Conference*, pages 539–548, 2000.

N. Fuerstenau, D. D. Janzen, and W. Schmidt. Flight tests of a fiber-optic interferometric strain gauge. In *Fiber optic smart structures and skins V; Proceedings of the Meeting, SPIE-1798*, pages 295–301, 1992.

B. Glisic and D. Inaudi. *Fibre Optic Methods for Structural Health Monitoring*. John Wiley and Sons, Ltd, 2007.

J.A. Güemes and J. M. Menéndez. Fiber-optics sensors. In D. Balageas, C.P. Fritzen, and J.A. Güemes, editors, *Structural Health Monitoring*. Wiley, 2006.

J.A. Güemes, S. Díaz-Carrillo, and J. M. Menéndez. Measurement of strain distribution in bonded joints by fiber bragg gratings. In *Smart Structures and Materials; Proceedings of the Meeting, SPIE-3330*, pages 264–271, 1998a.

J.A. Güemes, S. Díaz-Carrillo, J. M. Menéndez, C. Pardo de Vera, P. Vionis, R. Scherer, D. Bercebal, and A. Cuerva. Strain and damage monitoring of wind turbine blades by piezoelectrics and fiber optic sensors. In *Proceedings of the ECCM-8 European Conference on Composite Materials*, pages 357–364, 1998b.

J.A. Güemes, J. M. Menéndez, M. Frovel, I.Fernandez, and J.M Pintado. Experimental analysis of buckling in aircraft skin panels by fibre optic sensors. *Smart Materials and Structures.*, 10:490–496, 2001.

J.A. Güemes, A. Fernandez, and B. Soller. Optical fiber distributed sensing. physical principles and applications. *J. Structural Health Monitoring*, 9: 233–245, 2010.

H. Guo, Y. DaiÂ¡, G. Xiao, N. Mrad, and J. Yao. Interrogation of a long-period grating using a mechanically scannable arrayed waveguide grating and a sampled chirped fiber bragg grating. *Opt. Lett.*, 33:1635–1637, 2008.

K.O. Hill and G. Meltz. Fiber bragg grating technology fundamentals and overview. *IEEE/OSA J. Lightwave Tech*, 15:1263–1276, 1997.

K.O. Hill, Y. Fujii, D. D. Johnson, and B. S. Kawasaki. Photosensitivity in optical fiber waveguides: application to reflection filter fabrication. *Applied Physics Letters*, 32:647–649, 1978.

A. Hongo, S. Kojima, and S. Komatsuzaki. Applications of fiber bragg grating sensors and high-speed interrogation techniques. *Struct. Control Health Monit.*, 12:269–282, 2005.

R. L. Idriss, M. B. Kodindouma, A. D. Kersey, and M. A. Davis. Multiplexed bragg grating optical fiber sensors for damage evaluation in highway bridges. *Smart Materials and Structures*, 7:209–216, 1998.

R. Isago and K. Nakamura. A high reading rate fiber bragg grating sensor system using a high-speed swept light source based on fiber vibrations. *Meas. Sci. Tech.*, 20:1–5, 2009.

M. Leblanc, S. Y. Huang, M. M. Ohn, A. Güemes, and A. Othonos. Distributed strain measurement based on a fiber bragg grating and its reflection spectrum analysis. *Optical Letters*, 21:1405–1407, 1996.

B. Lee. Review of the present status of optical fibre sensor. *Opt. F iber Tech*, 9:57–79, 2003.

Z.Q. Luo, C.C. Ye, Z.P. Cai, X.Z. Dai, Y. Kang, and H.Y. Xu. Numerical analysis and optimization of optical spectral characteriscs of fiber bragg gratings modulated by a transverse acoustic wave. *Appl. Opt*, 46:6959–6965, 2007.

M. Majumder, T.K. Gangopadhyay, A.K. Chakraborty, K. Dasgupta, and D.K. Bhattacharya. Fibre bragg gratings in structural health monitoring: Present status and applications. *Sens. Actuat. A*, 147:150–164, 2008.

R. M. Measures. Smart structures with nerves of glass. *Progress on Aerospace Science*, 26:289–351, 1989.

R.M. Measures. *Structural Monitoring with Fiber Optic Technology*. Academic Press, 2004.

G. Meltz, W. Morey, and W. H. Glenn. Formation of bragg gratings in optical fibers by a transverse holographic method. *Optics Letters*, 14: 823–825, 1989.

J. M. Menéndez and J. A. Güemes. Bragg-grating-based multiaxial strain sensing: its application to residual strain measurement in composite laminates. In *Smart Structures and Materials 2000; Proceedings of the Meeting, SPIE-3986*, pages 271–281, 2000.

J. M. Menéndez and J.A. Güemes. Strain measurements inside thick cfrp laminates at the vicinity of bolted joints. In *Smart Structures and Materials; Proceedings of the Meeting, SPIE-3670*, pages 184–194, 1999.

J. M. Menéndez and J.A. Güemes. Bragg grating-based multiaxial strain sensing: its application to residual strain measurement in composite laminates. 2008.

A. Minardo, A. Cusano, R. Bernini, L.G. Zeni, and M. Giordano. Response of fiber bragg gratings to longitudinal ultrasonic waves. *IEEE Trans. Ultrason. Ferroelect. F req. Control*, 52:304–312, 2011.

W.W. Moorey, G.A. Ball, and H. Singh. Applications of fibre grating sensors. *Proc. SPIE*, pages 2–7, 1996.

N. Mrad. Potential of bragg grating sensors for aircraft health monitoring. *Trans. CSME*, 31:1–17, 2007.

N. Nakamura, T. Ogisu, H. Yoneda, Y. Okabe, N. Takeda, and Sakurai. Impact monitoring of the aircraft composite structure using fbg sensor/pzt actuator hybrid sensor system. In *Proc. SPIE*, 2007.

I. Perez, H.L. Cui, and E. Udd. Acoustic emission detection using fiber bragg gratings. pages 209–215, 2001.

D. Philen, I. White, J. Kuhl, and S. Mettler. Single-mode fiber otdr: Experiment and theory. *IEEE J. Quant. Elect*, 18:1499–1508, 2003.

X.L. Qing, A. Kumar, C. Zhang, I.F. Gonzalez, G.P. Guo, and F.K. Chang. A hybrid piezoelectric/fiber optic diagnostic system for structural health monitoring. *Smart Mater. Struct.*, 14:98–103, 2005.

F. Rodríguez-Lence, P. Muñoz-Esquer, J. M. Menéndez, C. Pardo de Vera, S. Díaz, and J. A. Güemes. Smart sensors for resin flow and composite cure monitoring. In *Proceedings of the Ninth Conference on Artificial Intelligence, ICCM-12*, 1999.

B. E. A. Saleh and M. C. Teich. *Fundamental of Photonics*. Wiley Interscience, 1991.

P. Schindler, R. May, and R. O. Claus. Optical fiber sensors for damage analysis in aerospace materials. *NASA Technical Report NASA-CR-199981*, pages 1–64, 1995.

L.Y. Shao, X.Y. Dongand A.P. Zhang, H.Y Tam, and S.L. He. High resolution strain and temperature sensor based on distributed bragg reflector fiber laser. *IEEE Photon. Tech. Lett.*, 19:1598–1600, 2007.

H. Soejima, T. Ogisu, H. Yoneda, Y. Okabe, N. Takeda, and Y. Koshioka. Demonstration of detectability of shm system with fbg/pzt hybrid system in composite wing box structure. 2008.

D.M. Spirit and L.C. Blank. Raman-assisted long-distance optical time domain reflectometry. *Elect. Lett*, 25:1687–1689, 2002.

M.D. Todd, J.M. Nichols, S.T. Trickey, M. Seaver, C.J. Nichols, and L.N. Virgin. *Bragg grating-based fibre optic sensors in structural health monitoring*. Phil. Trans. Roy. Soc. A, 2007.

E. Udd. Embedded sensors make structures 'smart'. *Laser Focus*, pages 135–, 1988.

G. Wild and S. Hinckley. Acousto-ultrasonic optical fiber sensors: Overview and state-of-the-art. *IEEE Sens. J*, 8:1184–1193, 2008.

G. Xiao, P. Zhao, F. Sun, Z. Lu, Z. Zhang, and C. Grover. Interrogating fiber bragg grating sensors by thermally scanning an arrayed waveguide grating based demultiplexer. *Opt. Lett.*, 29:2222–2224, 2004.

G. Zhou and L.M. Sim. Damage detection and assessment in fibre-reinforced composite structures with embedded fibre optic sensors review. *Smart Materials and Structures.*, 11:925–939, 2002.

Damage localisation using elastic waves propagation method. Experimental techniques.

Wieslaw Ostachowicz [*‡] Pawel Malinowski [‡] and Tomasz Wandowski [‡]

[*] Faculty of Navigation, Gdynia Maritime University, Gdynia, Poland
[‡] Department of Mechanics of Intelligent Structures, Polish Academy of Sciences, Gdansk, Poland

Abstract In this chapter methods of Structural Health Monitoring using elastic wave propagation method are presented. Different techniques are presented that are used for elastic wave generation and sensing. In the case of elastic waves sensing presented methods are divided into contact and non contact (optical). Wide description of measuring methods that are based on laser vibrometry can be also find. This chapter also includes description of damage localisation algorithms. Also phenomenon of elastic wave propagation in the structural elements is widely described. In this chapter different types of elastic wave propagating in structural elements are presented however attention is focused on Lamb waves. Chapter includes lots of experimental results related to measurements and visualisation of elastic waves propagation as well as results of damage localisation.

1 Structural Health Monitoring based on elastic waves propagation

Structural Health Monitoring (SHM) is technique that allow to conduct continuous monitoring of engineering structures. The monitoring process is realized in real time during the normal usage of structures. Structural Health Monitoring systems are very often compared with human nervous system in which nerves provide information about pain. These information are next processed in the human brain. In SHM systems network of transducers can be compared with human nerves and signal processing unit with the human brain. SHM systems very often utilize elastic wave propagation method in order to assess the state of the whole structure or its single elements. Elastic wave propagation method is based on fact that any kind of

discontinuities in elements of structure cause changes in elastic wave propagation. These changes in the form of reflections or amplitude decrease can be sensed and used in damage detection or localisation algorithms. This method is very sensitive therefore damage in early stage of growth before it endangers the safety of structures can be detected. For comparison vibration based SHM systems are less sensitive to damage. Therefore elastic wave propagation method is so promising technique.
Elastic waves are very often excite and sensed using piezoelectric transducers that are thin, light and can be simply mounted on or embedded in the structural elements. Except piezoelectric transducers also other techniques are used but descriptions of these method are included in the next section. SHM system based on elastic wave propagation phenomenon can be divided into two types: passive and active. In the passive systems network of transducers (sensors) only listen the structure and sensing signals representing elastic waves that propagate in the structure (Gangadharan et al., 2009), (Ding et al., 2004). Passive systems for example are utilized for impact localisation. Examples of impact in aircraft structures are: bird strike or spanner dropped by the mechanic. In active systems diagnostic signals are excited and sensed using transducers network and such systems are used for assessment of structural health (Kirikera et al., 2006). In SHM five levels of diagnose can be distinguished (Rytter, 1993), (Worden and Dulieu-Barton, 2004):

- detection,
- localisation,
- identification,
- damage size extraction,
- prediction of remaining structure lifetime.

Many researchers all around the world are interested in application of phenomenon of elastic wave propagation to assess the state of real structures without the needs of exclude the structure from its normal exploitation. However numerous difficulties with real structures cause that research is very often connected only with parts of the structure or with simple real structures. Complexity of real structures cause lot of problems with damage localisation using elastic waves. Complex geometry of structure, stiffeners, bolts, rivets and welds are sources of wave reflections and amplitude decreases. Many problems are related to sensor integration with structure, reliability of sensors in spite of mechanical loads influence, material fatigue, corrosive environment, wide range of exploitation temperatures. In the case of aerial structures very important is also problem of very restrictive airworthiness directives that limit the possibilities of installation of SHM systems. However in the case of structural elements some algorithms for damage detection, localisation and identification of type or size of damage were developed with success. Problem of damage detection and localisation in beams was solved by (Giurgiutiu, 2008), (Lestari and Qiao, 2005), (Palacz et al., 2005), (Su et al., 2007). Damage localisation algorithms were also developed for the structures in the form of pipes or pipelines (Cau et al., 2006), (Qing et al., 2009), (Siqueira et al., 2004), (Thien, 2006). Elastic waves are also utilized in order to detect and localise damage in rails (Lee et al., 2009), (Zumpano and Meo, 2006). Moreover elastic waves are utilized for damage localisation in metallic plates and plates made out of composite materials: (Giurgiutiu, 2008), (Giurgiutiu et al., 2007), (Ihn and Chang, 2008), (Ihn and Chang, 2004), (Ostachowicz et al., 2009), (Ostachowicz and Kudela, 2007), (Salas and Cesnik, 2009), (Yu et al., 2008).

2 Elastic waves

As was mentioned in previous section elastic waves are very often utilized in SHM systems. Elastic waves propagating in the structure carry important information about damage presence, its localization, size or type. Therefore it is so important to understand the types of elastic waves that can propagate in the structural elements. Different types of elastic waves can be characterized by different sensitivity on the specific types or location of damage.

In unlimited three dimensional solids two types of wave can propagate: longitudinal waves and shear waves. The classification is connected with direction of particles vibration in relation to direction of elastic wave propagation. In the case of longitudinal waves particle displacement direction is compatible with direction of propagation see Figure 1a). For the shear waves direction of particles displacement is perpendicular to direction of waves propagation as it is shown in the Figure 1b).

Situation is more complicated if one surface of three dimensional solid is limited. In such a case due to internal reflection of elastic waves raise up elastic waves with complex displacement field. Two types of elastic waves can be distinguished: Rayleigh waves and Love waves. Rayleigh waves are the type of elastic waves that propagate on the surface of solids which thickness is many times larger than length of propagating waves. Vibrating particles due to wave motion move along the trajectories in the form of circles Figure

1c). Characteristic feature of Rayleigh waves is effect of amplitude decay with the growing depth. Largest values of Rayleigh waves amplitudes are observed on the surface. Love waves have more complex displacement field than in the case of Rayleigh waves. Displacement field of Rayleigh waves can be presented in 2D however for illustration of displacement related to Love waves three dimensional plot is needed Figure 1d). In the case of Love waves similarly like for Rayleigh waves amplitude decay with the growing depth of structure.
In the case of 3D solid limited by two surfaces (e.g. plates, shells) situation changes radically. In such structures longitudinal waves and shear waves propagate simultaneously. But in this case shear waves can be divided into shear horizontal waves (SH) and shear vertical waves (SV). For both types of waves direction of particles vibration is perpendicular to direction of elastic wave propagation. In the case of SH waves particles of structure vibrate in horizontal direction and in the case of SV wave particles vibrate in vertical direction. As result of internal reflections of longitudinal waves and shear vertical waves new forms of waves are created - Lamb waves. Its name is connected with the Horacy Lamb that discovered these waves and formulated theory of its propagation in year 1917. These waves exist as symmetrical (Figure 1e)) and antisymmetrical modes (Figure 1f)). Number of this modes is infinite but number of propagating modes depends on product of excitation frequency and thickness of limited structure. Both symmetrical and antisymmetrical modes of Lamb waves are dispersive what means that its velocity of propagation changes with frequency. Dispersive character of both modes can be described by so called Rayleigh-Lamb equations 1, 2:

$$\frac{tan(qd)}{q} + \frac{4k^2 ptan(pd)}{(q^2 - k^2)^2} = 0 \tag{1}$$

$$qtan(qd) + \frac{(q^2 - k^2)^2 tan(pd)}{4k^2 p} = 0 \tag{2}$$

where: d - half of plate thickness $[m]$, k - wave number $[1/m]$.
Wave number can be calculated using formula:

$$k = \frac{\omega}{c_p} \tag{3}$$

Coefficients: p and q can be calculated using following formulas: 4 and 5.

$$p^2 = (\frac{\omega}{c_L})^2 - k^2 \tag{4}$$

$$q^2 = (\frac{\omega}{c_T})^2 - k^2. \tag{5}$$

where: c_L and c_T propagation velocities of longitudinal and shear wave respectively calculated based on: 6 i 7.

$$c_L = \sqrt{\frac{\lambda + 2\mu}{\rho}}. \tag{6}$$

$$c_T = \sqrt{\frac{\mu}{\rho}}. \tag{7}$$

where: ρ - density $[kg/\mathrm{m}^3]$, λ and μ Lame constants calculated from equations:

$$\lambda = \frac{\nu E}{(1+\nu)(1-2\nu)}, \tag{8}$$

$$\mu = \frac{E}{2(1+\nu)}, \tag{9}$$

where: E - Young modulus $[N/m^2]$, ν - Poisson ratio [-].

Solution of Rayleigh-Lamb equations 1 and 2 allow to plot dispersion curves. These curves illustrate the changes of group and phase velocity Lamb wave modes in relation to plate thickness and excitation frequency. Computer code solving Rayleigh-Lamb equation has been developed in MATLAB environment and has been based on algorithm presented in (Rose, 2004).

Dispersion curves were calculated for aluminium alloy EN AW-5754 with following material properties: Poisson ratio $\nu = 0.33$, density $\rho = 2680$ kg/m^3, Young modulus $E = 70500$ N/m^2, thickness $d = 0.001$ m. For these parameters velocities of longitudinal and shear waves were calculated: $C_L =$ 6420 m/s i $C_T = 3040$ m/s. Dispersion curves for group and phase velocities were presented respectively in Figure 2 and Figure 3. After analysis it can be noticed that up to frequency 15 MHz in 1 mm thick plate six symmetrical (S_0-S_5) and anti-symmetrical (A_0-A_5) modes propagate. Up to almost 2 MHz in the 1 mm thick plate only two fundamental modes symmetrical and antisymmetrical denoted as S_0 and A_0 propagate. For the higher frequencies higher modes propagate in the plate.

When phase velocities of propagating Lamb wave modes are known in simple way displacement distributions through the thickness for chosen mode and frequency can be calculated. Displacement distributions can be

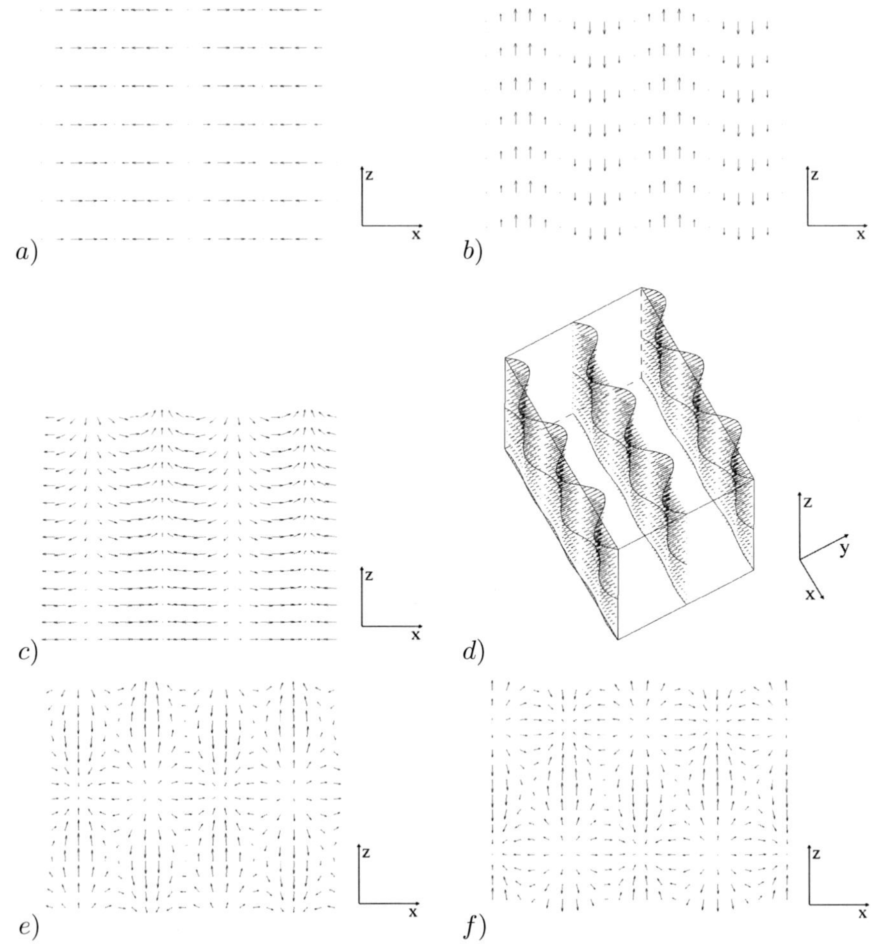

Figure 1. Displacement fields for: a) longitudinal waves, b) shear waves, c) Rayleigh waves, d) Love waves, e) symmetric Lamb waves mode, f) anti-symmetric Lamb waves mode

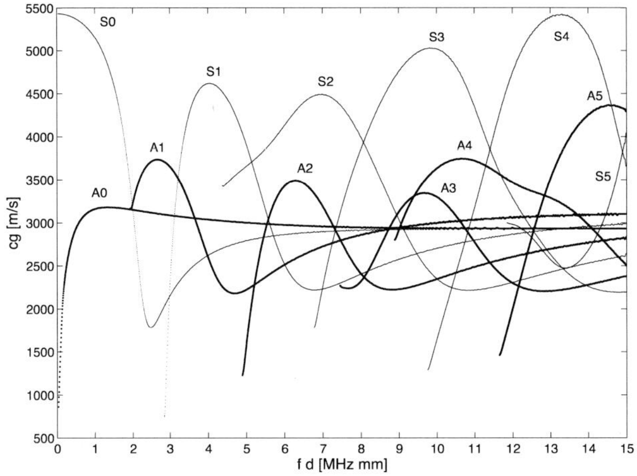

Figure 2. Dispersion curves for group velocity of Lamb waves ($C_L = 6420$ m/s i $C_T = 3040$ m/s); S - symmetric modes, A - antisymmetric modes

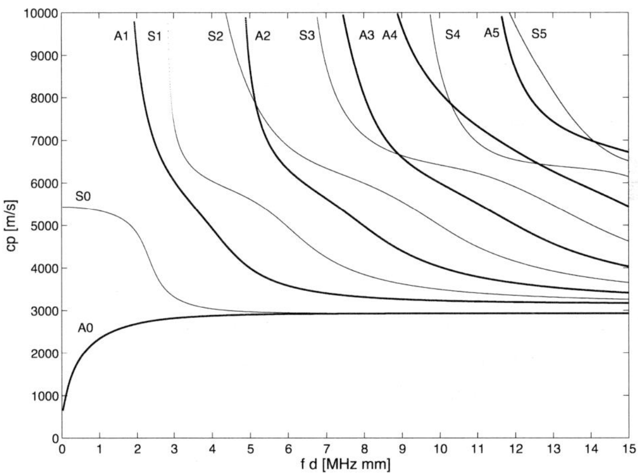

Figure 3. Dispersion curves for phase velocity of Lamb waves ($C_L = 6420$ m/s i $C_T = 3040$ m/s); S - symmetric modes, A - antisymmetric modes

simply obtained using formulas: 10, 11 for symmetric modes and formulas 12, 13 for antisymmetric modes:

$$u_S = -2k^2 q \cos(qh)\cos(pz) + q(k^2 - q^2)\cos(ph)\cos(qz) \tag{10}$$

$$w_S = -2ikpq \cos(qh)\sin(pz) - ik(k^2 - q^2)\cos(ph)\sin(qz) \tag{11}$$

$$u_A = -2k^2 q \sin(qh)\sin(pz) + q(k^2 - q^2)\sin(ph)\sin(qz) \tag{12}$$

$$w_A = 2ikpq \sin(qh)\cos(pz) + ik(k^2 - q^2)\sin(ph)\cos(qz) \tag{13}$$

where: d - half of plate thickness $[m]$, z - coordinate of point in which displacement is calculated $[m]$, i - imaginary unit.

Calculated displacement fields for both Lamb wave modes with different frequencies are presented in Figure 4.

Let's get back to the dispersion curves especially those presenting phase velocity. These dispersion curves are very useful in the case of elastic wave generation. Wave packet consists of some number of waves with frequencies from certain frequency band. Dispersion curves for phase velocities show velocities with which ingredient waves from wave packet propagate. In the case of different values of phase velocities related with ingredient waves the shape of wave packet to be changed with growing distance of waves propagation. This fact is caused by dispersion phenomenon. In order to reduce effect of dispersion excitation signal used during elastic wave generation must be characterized by narrow frequency band. To improve the result of dispersion effect reduction frequency band of excitation signal must be located in the flat part of phase velocity curve. In order to limit the span of frequencies in the excitation signal modulation windows are used. Additionally small number of cycles of sine function (usually 3 to 15) is used as excitation signal. In the Figure 5 three types of modulation window were used in order to modulation of excitation signal in the form of sine with five cycles (carrier frequency 300 kHz).

In the case of Hann window (Figure 5a,b) the narrowest frequency band will be obtained. A little bit wider frequency band will be obtained in the case of triangular window modulation (Figure 5c,d). Definitely the widest frequency band will be obtained in the case of rectangular window (Figure 5e,f). Many additional peaks around main frequency band can be observed (Figure 5f).

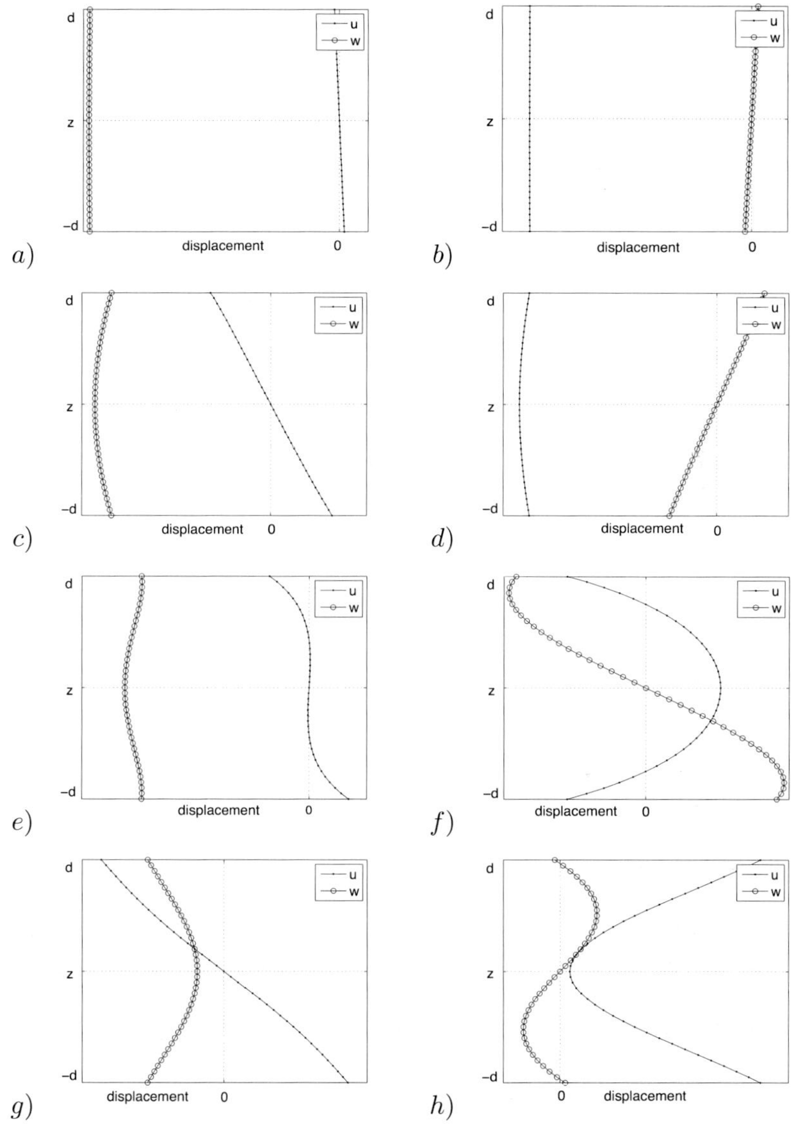

Figure 4. Distributions of in-plane (u) and out-of-plane (w) components of displacements of antisymmetric (left column) and symmetric (right column) Lamb wave modes for frequencies: a,b) 100 kHz, c,d) 250 kHz, e,f) 800 kHz, g,h) 3000 kHz

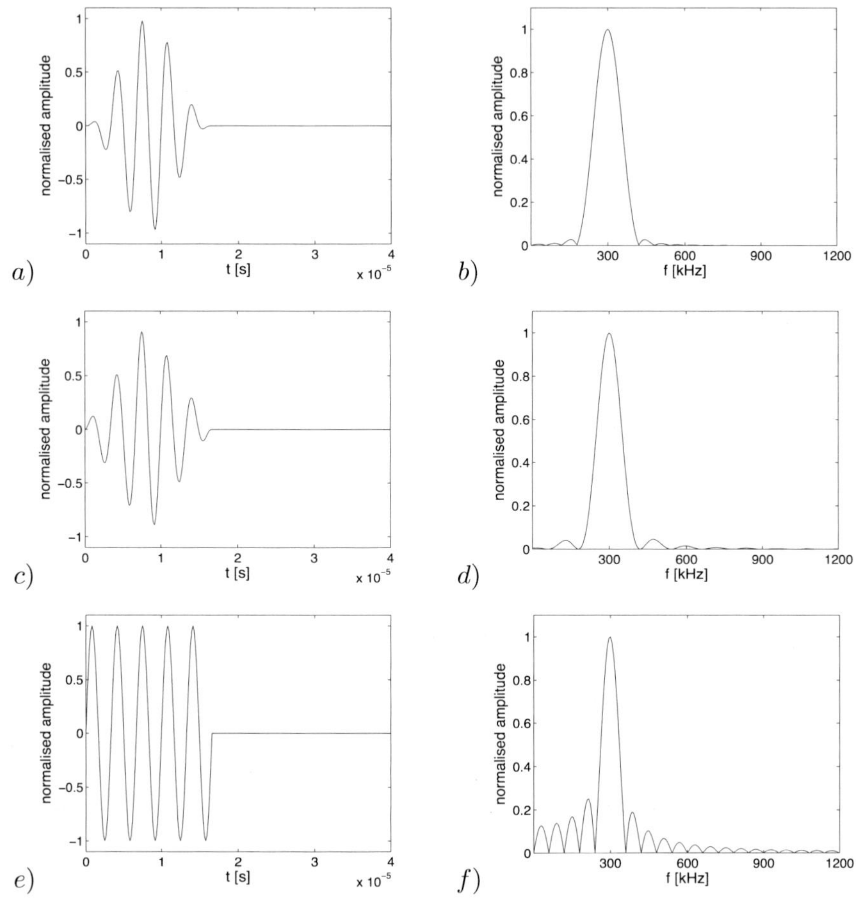

Figure 5. Different modulation types: a,b) Hann window, c,d) triangular window e,f) rectangular window. Left column - signal in time domain, right column - signal in frequency domain

3 Elastic wave generation and sensing

Many techniques can be utilized in order to excite or sense. However each of known method have some advantages and drawbacks that limit its applications. In this section a short review of various techniques for elastic wave generation and sensing will be presented.

3.1 Elastic waves generation techniques

Simplest well known method of elastic waves generation is method utilizing impulse of force. In practice impulsive character of force is realized by modal hammer or impact of steel ball (Kundu et al., 2008), (Jeong, 2001). These two methods are used during the scientific research but in the case of real structures impact can be caused for example by bird strike. In case of scientific research related with composite structures so called Hsu-Nielsen method is utilized that really on breaking the pencil lead (Berthelot et al., 1992). The common disadvantage of mentioned method is fact that impulsive character of excitation cause that wide range of frequencies are generated. This fact is very undesirable in the case of Lamb type elastic waves because wide frequency band of excitation bring about problem of dispersion phenomenon besides more than two fundamental modes of these waves will be excited. Because of mentioned restrictions new method that allow to control frequency band of excitation have been developed. Most of all ultrasonic technique is known which utilizes conventional plane bulk longitudinal wave transducers (Wilcox, 1998). These transducers are immersed in liquid (water for example) together with the structure in which elastic waves are to be generated - Figure 6. Transducer generates a longitudinal waves that propagate through the liquid.

When propagating waves reaches the structure due to internal wave reflections Lamb waves are generated in the structure. This method very often is precluded from application in many engineering structures because of need to use liquid in which transducer and structure must be immersed. Therefore instead of liquid air as a medium is used but only small part of longitudinal wave is converted into Lamb waves for this combination. This problem is caused by different acoustic impedance of air and the most of materials which is used for structural element manufacture. Problem of liquid or air usage can be eliminated by the using of ultrasonic transducers integrated with coupling material named Perspex (Plexiglas). These transducers are called wedge coupled angle-adjustable ultrasonic probes (Wilcox, 1998). Advantage of mentioned transducers it that by choosing the angle of ultrasonic transducer orientation it is possible to actively tune in order to excite chosen mode of elastic waves. Well known are also comb-type ultra-

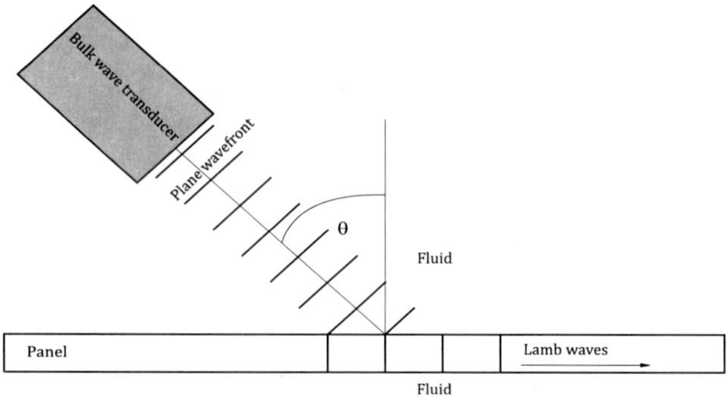

Figure 6. Idea of elastic wave generation in plate using a conventional ultrasonic transducer

sonic transducers (Rose, 2004) which scheme is presented in Figure 7. These transducers consist of some number of periodically spaced tips. Spacing of tips must correspond with length of generated wave. The main disadvantage of this transducer application is fact that tips spacing determine the specific wavelength and this is done in permanent fashion. Described methods are not used nowadays.

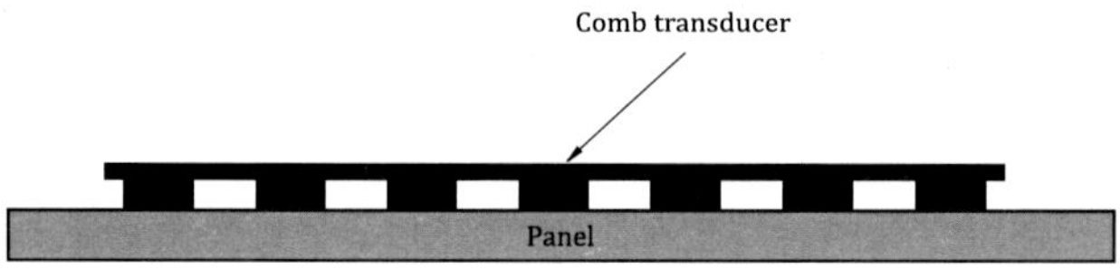

Figure 7. Scheme of comb-type ultrasonic transducer

Nowadays new methods are developed for elastic wave generation. First of such methods is based on electro-magnetic phenomenon. Transducers utilizing mentioned phenomenon are called Electro Magnetic Acoustic Transducers (EMATs) (Park et al., 2006), (Murayama and Mizutani, 2002), (Dixon and Palmer, 2004). These transducer can be used only in the case of metallic structures, moreover material of structure must be ferromagnetic. EMATs consist of fixed magnet with solenoid through which flows electrical current. As result eddy currents are induced in structure and these currents generate elastic waves. The most important disadvantage is that EMATs are very large and heavy.

The most popular transducers that are widely use for elastic wave generation are based on piezoelectric effect (Giurgiutiu, 2008), (Ihn and Chang, 2004), (Ostachowicz et al., 2009). Piezoelectric transducers are very thin, light and can be simply integrated with structure however these transducers are very fragile. Solid-state solution of two perovskities (lead titanate) and (lead zirconate) is well known as PZT. Different types of piezoelectric transducers utilized in authors laboratory are presented in Figure 8.

Figure 8. Different types of piezoelectric transducers in authors laboratory; from the left: MIDE QP22B, T216-A4NO-273X Piezo Systems, Inc., Noliac CMAP06 and Noliac CMAP11, Ceramtec Sonox P502 disc

During the elastic wave generation inverse piezoelectric effect is utilized. This effect allow to generation strain in piezoelectric materials due to external electrical field application. This effect can be described by constitutive equation 14.

$$\begin{Bmatrix} S_{11} \\ S_{22} \\ S_{33} \\ S_{23} \\ S_{31} \\ S_{12} \end{Bmatrix} = \begin{Bmatrix} s_{11} & s_{12} & s_{13} & 0 & 0 & 0 \\ s_{21} & s_{22} & s_{23} & 0 & 0 & 0 \\ s_{31} & s_{32} & s_{33} & 0 & 0 & 0 \\ 0 & 0 & 0 & s_{44} & 0 & 0 \\ 0 & 0 & 0 & 0 & s_{55} & 0 \\ 0 & 0 & 0 & 0 & 0 & s_{66} \end{Bmatrix} \begin{Bmatrix} T_{11} \\ T_{22} \\ T_{33} \\ T_{23} \\ T_{31} \\ T_{12} \end{Bmatrix} + \\ + \begin{Bmatrix} g_{11} & g_{21} & g_{31} \\ g_{12} & g_{22} & g_{32} \\ g_{13} & g_{23} & g_{33} \\ g_{14} & g_{24} & g_{34} \\ g_{15} & g_{25} & g_{35} \\ g_{16} & g_{26} & g_{36} \end{Bmatrix} \begin{Bmatrix} E_1 \\ E_2 \\ E_3 \end{Bmatrix} \tag{14}$$

where: S - strains [-], s - compliance coefficients $[m^2/N]$, T - stresses $[N/m^2]$, d - piezoelectric coupling coefficients for strain-charge form $[C/N]$,

E - electric field $[N/C]$.

In order to protect piezoelectric transducers from aggressive environment and humidity a special layers were developed. Such a layers were made out the polyamide foil in which transducers were embedded. Such a commercial solution is known as SMART Layer (Stanford Multi-Actuator-Receiver Transduction Layer) and was developed in Stanford University and produced by Acellent Technologies, Inc. (Lin and Chang, 2002), (Qing et al., 2009), (Qing et al., 2006). Very similar to them is solution named SAL (Smart Active Layer) developed in Korea Research Institute of Standards and Science (Lee et al., 2005).

Elastic wave are also generates using inter-digital transducers (IDT) (Mustapha et al., 2007) they are very similar to ultrasonic comb transducers (Figure 9, 10). IDT transducers very often are made of PVDF (polyvinylidene fluoride) which has very important advantage - is very flexible. Thanks to this it can be simply integrated with curved shape structures like pipes, tanks, shells, etc.

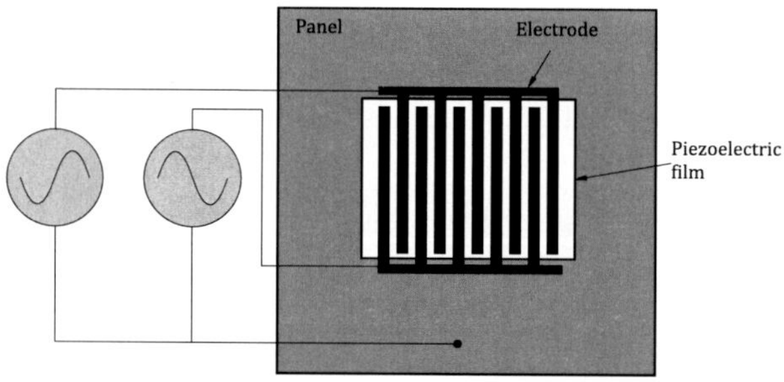

Figure 9. Scheme of inter-digital transducer IDT - top view

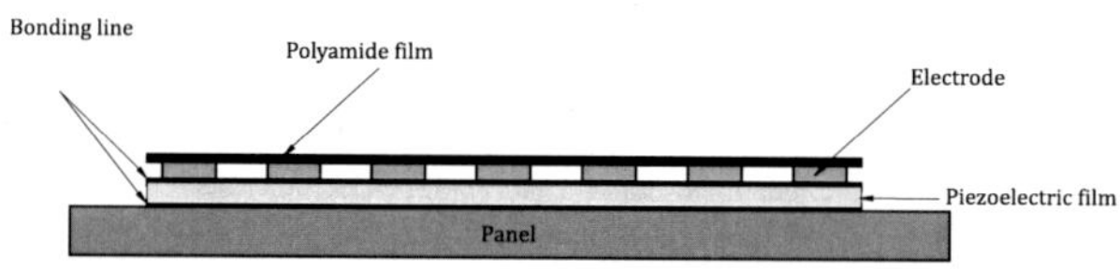

Figure 10. Scheme of inter-digital transducer IDT - side view

A little bit similar solution are MFC (Macro Fibre Composite) (Wagg et al., 2007) and APC (Active Fibre Composite) ((Melnykowycz et al., 2006) and (Melnykowycz et al., 2010)) transducers made of piezoelectric fibres embedded into the foil (Figure 11). The only one difference in these transducers are cross-section of fibres: square in the case of MFC and circular for AFC. Both types are also very flexible like in the case of IDT transducers.

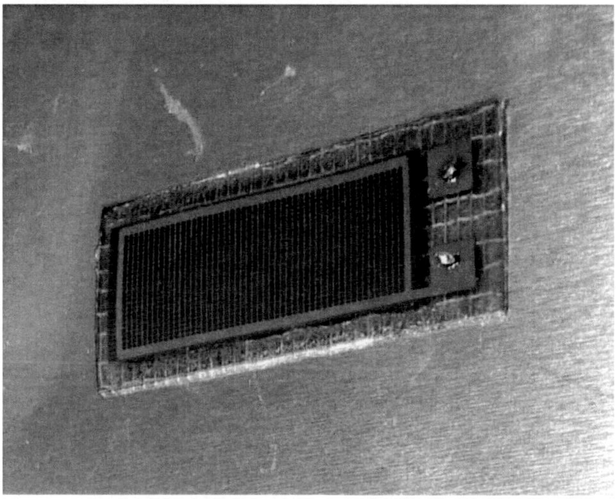

Figure 11. Macro Fibre Composite (MFC) transducer in authors labo-rathory

It also should be mentioned that magnestostrictive effect can be utilizes for elastic wave generation. Transducers employing this effect consist of fixed magnets, solenoids and nickel gratings. These transducers are able to excite high-power elastic wave. Besides this type of transducers can be used in the case of non-ferromagnetic materials. Elastic wave can be also excited using photo-thermal non contact methods. In this approach Nd:YAG laser sources is used for elastic wave excitation (Valle and Littles, 2002), (Hongjoon et al., 2006a), (Hongjoon et al., 2006b). Main advantage of this approach is the possibility of generation of waves with chosen shape of source and in chosen site of structures.

3.2 Elastic waves sensing techniques

Apart from elastic waves generation such a same important is the process of elastic wave sensing and visualization of its propagation in the structures. Similarly like in the case of elastic waves generation in the case of sensing many techniques can be distinguished. Some of them are the same for both cases but few methods can be only utilized in the purpose of wave sensing. In the past, very often for elastic wave sensing conventional plane bulk longitudinal wave transducers were used however wedge coupled angle-adjustable ultrasonic transducers, comb-type ultrasonic transducers were also commonly used. In order to sense elastic waves until now transducers leveraging magnetic phenomenon (EMATs) are used. This type of transducer was described in previous section. At present very interesting is method that is based on piezoelectric transducers. During the elastic wave sensing direct piezoelectric effect is utilized. As result of transducer mechanical deformations electrical charges appear on the electrodes of piezoelectric element. Direct piezoelectric effect can be described using constitutive equation 15.

$$\begin{Bmatrix} D_1 \\ D_2 \\ D_3 \end{Bmatrix} = \begin{Bmatrix} g_{11} & g_{12} & g_{13} & g_{14} & g_{15} & g_{16} \\ g_{21} & g_{22} & g_{23} & g_{24} & g_{25} & g_{26} \\ g_{31} & g_{32} & g_{33} & g_{34} & g_{35} & g_{36} \end{Bmatrix} \begin{Bmatrix} T_{11} \\ T_{22} \\ T_{33} \\ T_{23} \\ T_{31} \\ T_{12} \end{Bmatrix} + $$
$$+ \begin{Bmatrix} \epsilon_{11} & \epsilon_{12} & \epsilon_{13} \\ \epsilon_{21} & \epsilon_{22} & \epsilon_{23} \\ \epsilon_{31} & \epsilon_{32} & \epsilon_{33} \end{Bmatrix} \begin{Bmatrix} E_1 \\ E_2 \\ E_3 \end{Bmatrix} \tag{15}$$

where: D - charge-density displacement $[C/m^2]$, ϵ - electric permittivity $[F/m]$.

In the Figure 12 some real application of piezoelectric transducer network is presented. In this case piezoelectric transducers are used for excitation and sensing of elastic wave propagation during the fatigue test of the part of helicopter main rotor blade. During the fatigue tests crack was propagating from the previously introduced notch - Figure 13.

Likewise in the case of elastic wave excitation for sensing purpose elastic SMART Layers are used as well as pure PVDF material, inter-digital transducers IDT and transducers based on piezoelectric fibres laminated in polyamide film: Macro Fibre Composites MFC and Active Fibre Composites AFC. Very interesting is method based on paint with piezoelectric properties. This paint can be simply sprayed on the structural element. The main disadvantage of this method is the low sensitivity.

In the case of elastic waves registration also magnetostrictive transducers are used that were described in previous subsection. Some method allow only to register elastic waves without possibility of its excitation. These methods are based on optical phenomena. In 1864 August Toepler developed a method for visualisation of flows with variable density. This method was called schlieren photography. Optical system that is used in this method includes collimated light source illuminating a flowing fluid. Changes of

Figure 12. Elastic wave propagation measurements in the part of helicopter main rotor blade during the fatigue tests

refraction index resulting from fluid density gradient have influence on the collimated light beam. In results spatial changes of light intensity that in turn allow to visualise phenomena in the fluid flow. This technique allowed to determine velocity of ultrasonic wave propagation in solids. Such a experiments were conducted by Barnes and Burton in 1949, Chinnery, Humphrey and Beckett in 1997, and Neubauer in 1973 (Rose, 2004). Some theoretical work on visualisation of elastic wave propagation was also performed in 1922 by Brillouin. He investigated interactions between elastic waves in solids and electromagnetic waves. These interactions can be noticed as a changes of medium permittivity due to elastic wave propagation. Experimental research for this topic was done by Lucas and Biquard in France and

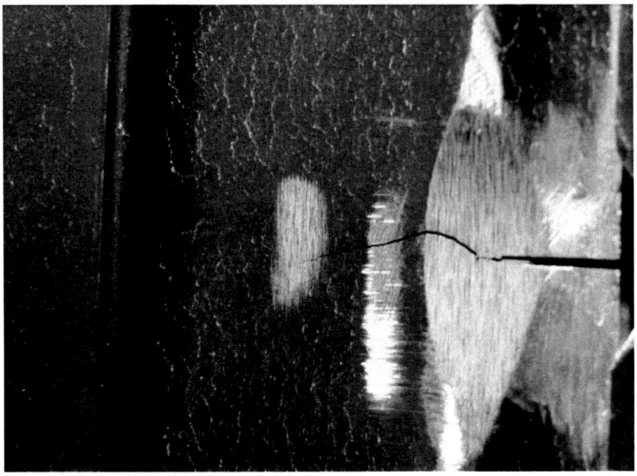

Figure 13. Growing crack in the part of helicopter main rotor blade during the fatigue tests - back side view

Debye and Sears in the US (Royer and Dieulesaint, 1999). Later the same topic was investigated by Raman and Nath in India (Royer and Dieulesaint, 1999).

Very interesting technique for visualisation of elastic wave propagation is photoelasticity method. Mentioned method can be used for visualisation of propagating elastic waves in glass. This method is based on fact that polarised light can be divided into the components that propagating with different velocities. As result colourful fringes associated with deformations of material can be observed. Photoelasticity method can be used in materials exhibiting birefringence resulting from material stress. As result of birefringence material to exhibits distinct refractive indexes. Value of refractive index at given point of material is strictly correlated with level of mechanical stress. Described method was used for ultrasonic waves visualisation by Zhang, Shen and Ying (1988) and by Li and Negishi (1994) (Rose, 2004).

In recent years new techniques are being developed for elastic wave registration. One of the newly developed method is based on polarimetric fibre optic sensors and fibre Bragg grating sensors (FBG) (Thursby et al., 2004). Polarimetric sensors require simple equipment for elastic wave sensing however registered signals are very difficult in interpretation and analysis. In the case of FBG sensors more complicated equipment must be used but measurements are easy to interpret. Polarimetric sensors allow to take measurements from larger area of structure than in the case where FBG sensors are used. The great advantage of suing FBG sensors is its multiplexing capability. One measurement set-up can be used for generating and registering optical signal for the grid of sensors. The most important thing in the field of elastic wave sensing is that the length of fibre optic sensor must be at least seven times shorter than the length of elastic wave being sensed (Takeda et al., 2005). Fibre optics sensor are very light, have very small diameter however they are very brittle like a piezoceramic materials. Due to new technological achievements in optics-based measurement technologies non-contact optical techniques came back. First widely used method for visualisation of elastic wave propagation is shearography method. This method is non-contact deformation measurement technique that is based on laser source and image shearing camera. This camera is used to create pair of laterally shifted images. Laser source is used for illumination of structure being measured. As result shifted images interfere creating such a called speckle image. In the purpose of elastic wave sensing a combination of spatial phase shifting technique and shearography is used. This can be achieved using Mach-Zehnder interferometer. Very often for elastic wave sensing Electronic Speckle Pattern Interferometry (ESPI) method is used (Lammering, 2010).

In recent years more and more significant is method based on non contact laser vibrometer. Operational principle of laser vibrometer is based on Doppler effect. Laser vibrometer register changes in frequency of light beam reflecting from vibrating surface of measured structure. Single point vibrometers are equipped with one measurement head (1D) and allow to measure velocity of chosen point of vibrating structures along the direction of laser beam. In order to measure vibrations in few point laser head must be manually positioned. Scanning vibrometer allow to scan the vibration of structures in chosen points. Laser beam is directed to chosen point of mesh using set of mirrors controlled by system. In some cases information on vibration in three dimensions are important therefore 3D scanning laser vibrometer has been developed. This equipment consists of three independent laser measurement heads (Figure 14). These measurement heads are oriented at different angles relative to structure being measured. Three laser beams must be focused in one points. Before the vibration measurement system measure the geometry of structure. Knowing the structure geometry and positions of measurement heads system simply can measure components of vibration velocities and as result compose full vector of vibration velocity in chosen point of structures. Due to scanning feature of vibrometer vibration velocities for whole structure surface can be simply extracted.

Figure 14. Measurement heads of Polytec PSV-400 3D scannig laser vibrometer

The main advantage of laser vibrometry techniques is non contact measurement approach. Non contact measurement eliminate problem related to additional mass of sensors attached to the structure and the problem of contact sensor with structure. This is especially very important problem during the measurements of elastic wave propagation. Any additional mass like a transducers cause a changes of elastic wave propagation (wave reflections) and disturb the measurements. Laser vibrometer is very useful during the experimental research in laboratory or during the outdoor measurements but it cannot be utilized in real time SHM systems.
Now some experimental result will be presented related to 3D laser vibrometry technique utilized for elastic waves sensing. First example is related with measurement of three components of displacement (two in-plane and one out-of-plane) connected with elastic wave propagation in plate made out of EN AW-5754 aluminium alloy. Carrier frequency of excitation was equal 200 kHz. These components of displacement were measured in one point and are presented in Figure 15.

Second example is connected with visualisation of elastic wave propagation also in plate made out aluminium alloy. Displacement filed were measured in the mesh of points covering whole surface of plate. Excitation carrier frequency was equal 100 kHz. Results of measurements are

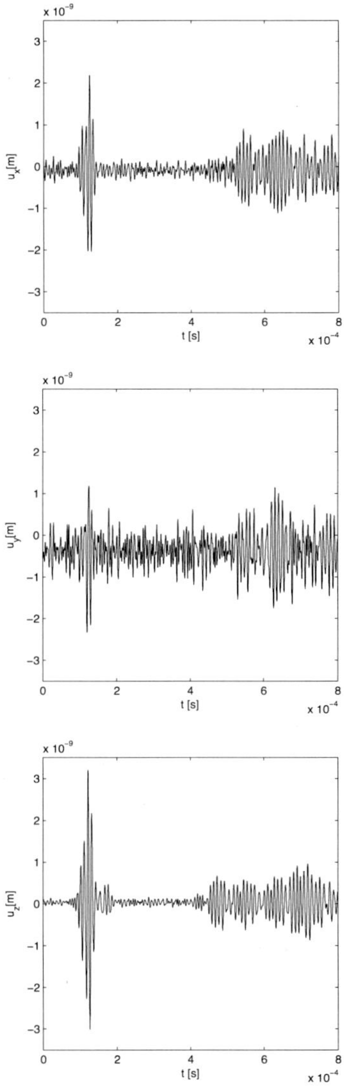

a)

b)

c)

Figure 15. Measured three components of displacements in one point using 3D laser scanning vibrometer: a) in-plane x component, b) in-plane y component, c) out-of-plane z component presented in Figure 16. Visualisation of in-plane and out-of-plane displacement components allow to understand better the phenomenon of Lamb wave propagation especially in structures with complicated geometry. Even in the case of so simple structure like plate 3D scanning measurements are very helpful because symmetric and anti-symmetric modes can be easy noticed. Both of these modes can be observed for the in-plane components Figure 16a,b) but symmetric mode (larger wavelength) has larger amplitudes. Anti-symmetric mode can be observed for the out-of-plane displacement component Figure 16c).

Last example is connected with visualisation of elastic wave propagation in the part of helicopter main rotor blade with crack. Measurement equipment and specimen are presented in Figure 17. In this case only out-of-plane displacements were measured. Results of elastic wave propagation are presented in the Figure 18.

4 Wave generation and sensing equipment

In this section examples of equipment for elastic wave generation and sensing are presented. It should be emphasized that all presented equipment is connected with piezoelectric transducers. The simplest way of elastic wave excitation and sensing can be realised using popular laboratory equipment like digital signal generator, oscilloscope and some equipment dedicated for piezoelectric transducers (Figure 19). Digital signal generator also called arbitrary waveform generator allow to define signal parameters (frequency, type of modulation, voltage) and generate signal that will drive piezoelectric transducers. Signal before to be driven piezoelectric tranducer have to be apmplified. For this purpose amplifier especially dedicated to piezoelectric transducer has to be used (capacitive electrical load). In the authors laboratory EPA-104 Piezo Linear amplifier produced by Piezo System, Inc. is used (Figure 19b). The last one is the digital oscilloscope used in order to register signal from piezoelectric transducers. It should be underlined that electrical signal from piezoelectric transducers can only be registered directly by oscilloscope if the transducer have large capacitance. In the case of very small capacitance of piezoelectric transducer additionally dedicated charge amplifiers should be used.

In practice better way is to use compact measurement equipment that allow to generate and sense signals. It is especially very important during the measurements conducted outside the laboratory. Therefore authors have designed compact equipment for this purpose, see Figure 20. This equipment allow to generate in one chosen channel from thirteen accessible channels and to register signals in the rest twelve. This equipment is con-

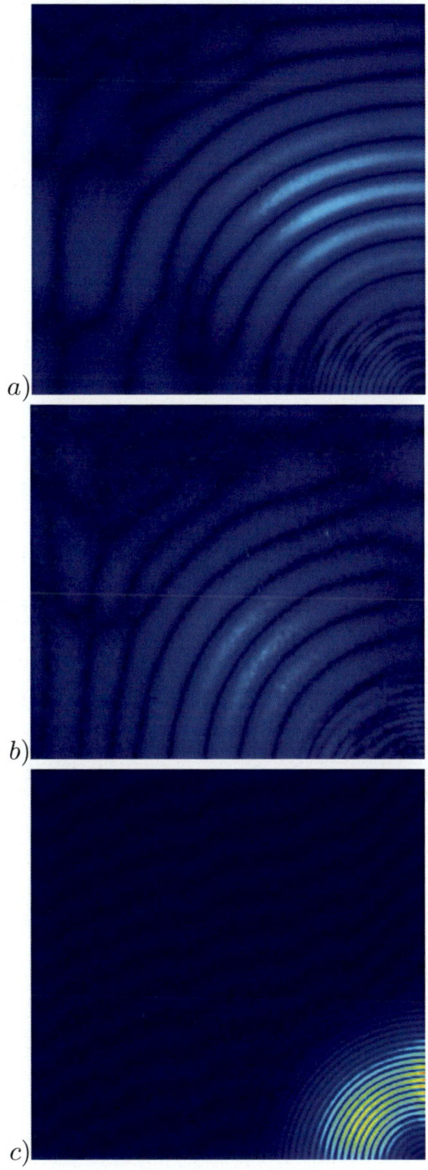

Figure 16. Visualisation of elastic wave propagation (displacements): a) in-plane x component, b) in-plane y component, c) out-of-plane z component

Figure 17. Measurements of elastic wave propagation in the part of helicopter main rotor blade using laser vibrometer

trolled by computer using MATLAB environment. Control software allow to set such a parameters as a channel for excitation, channels for registration, voltage of excitation, number of cycles of tone-burst excitation signal and modulation type.

Very interesting is also commercial compact system produced by Acellent Technologies, Inc. (Figure 21).

5 Transducer network configurations

In SHM systems very often piezoelectric transducers are utilized for elastic waves generation and sensing. Single piezoelectric transducers are arranged

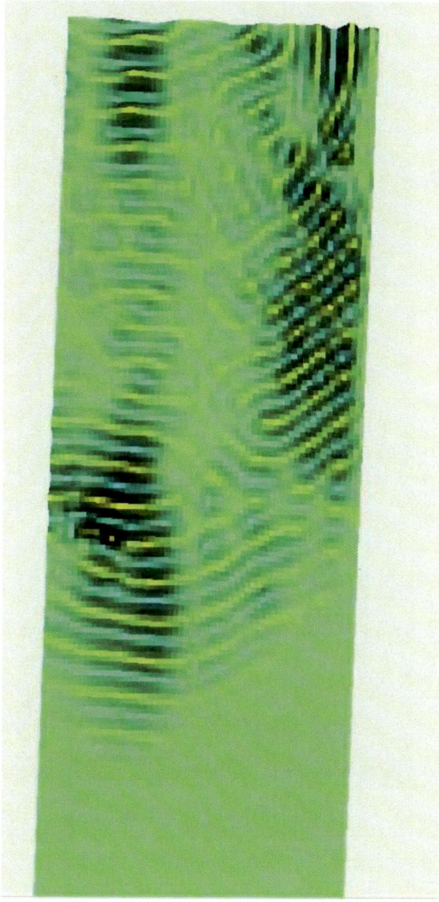

Figure 18. Visualisation of elastic wave propagation in part of helicopter main rotor blade with crack

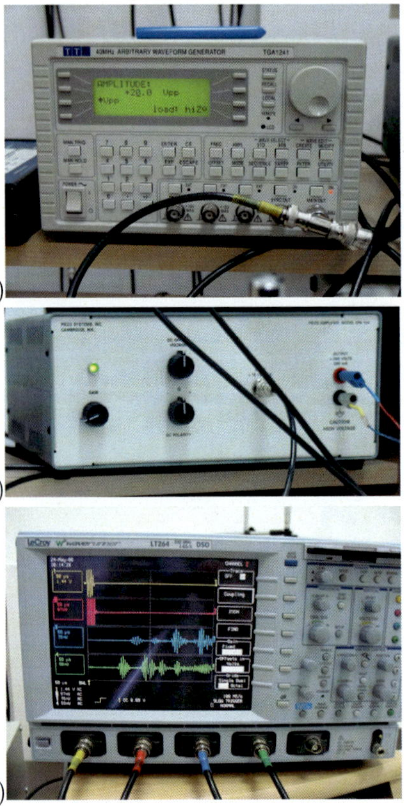

a)

b)

c)

Figure 19. Simplest laboratory measurement equipment: a) digital signal generator, b) piezo amplifier, c) oscilloscope

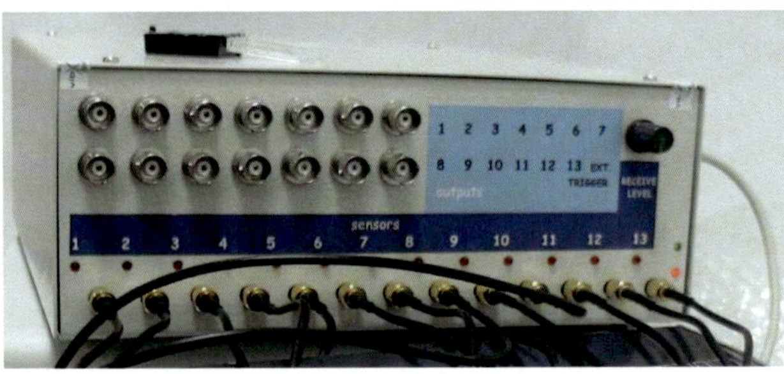

Figure 20. Propotype for elastic wave generation and sensing

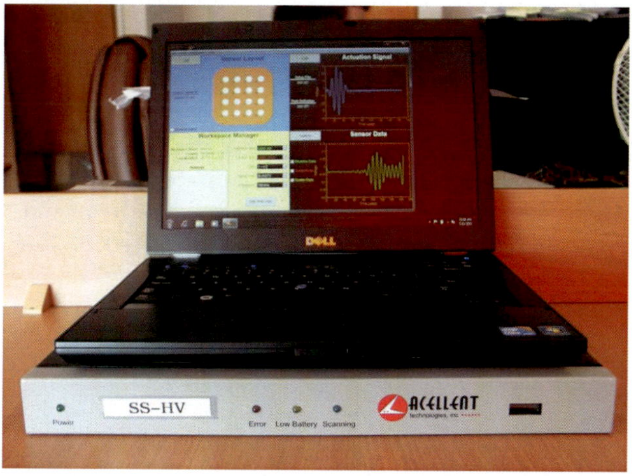

Figure 21. Elastic wave generation and sensing system available on the market

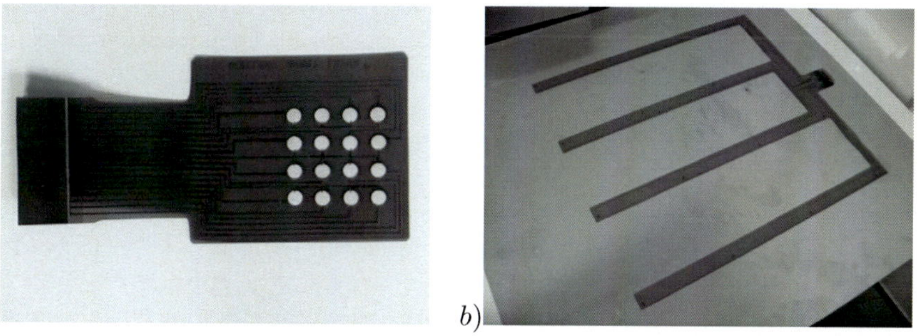
a) *b)*

Figure 22. Piezoelectric transducer configurations in authors laboratory: a) concentrated, b) distributed

in network with various configurations. Generally three types of network configurations can be distinguished: concentrated (Giurgiutiu, 2008), distributed (Ihn and Chang, 2004), mixed (Malinowski et al., 2011). In the first case transducers are placed close to each other in chosen site of structure (Figure 22a). In the case of distributed network transducers are placed uniformly (or not) on the surface of the structure (Figure 22b). In the case of mixed configuration network consists of uniformly (or not) placed group of transducers. In each part transducers are placed close to each other.

Type of transducer network configuration and manner of transducer

placement depends on method of elastic wave generation and sensing that is used. Three methods of elastic wave generation and sensing can be distinguished from the literature: pitch-catch, pulse-echo and phased array. In the pitch-catch method elastic waves are excited in one transducer and sensed in the second. If elastic wave propagates through the damaged area its amplitude decrease and propagation velocity of wave is changed. Such a method can be utilized with distributed or mixed transducer networks. In the pulse-echo method elastic waves are generated in one transducer. Elastic wave propagating in the structure reflects from damage and these reflections are sensed in second (or the same) transducer. This method can be used with concentrated or mixed transducer network. Phased array method is based on constructive elastic wave interference phenomenon. As result of firing transducers in concentrated network with specially adjusted time delay wave front is created that propagate in chosen direction. This method works like airborne radar. In this method wave reflections are amplified and thank to this damage placed far away in relation to transducers can be localised. Very important is assumption that distance between two neighbour transducers must be less or equal half of wavelength. It should be mentioned that this method can be also utilized with mixed transducers network. But in this case it will be combination of few individual phased array systems for each concentrated transducer subgroup from whole configuration.
The next two sections contain description of damage localisation algorithms that are used in Structural Health Monitoring application. The theoretical background is followed by examples based on experimental mesurements. The experimental investigation was conducted in author's laboratory. Both concentrated and distributed transducer arrays were considered. Provided examples cover the case of an aluminium alloy panel and a CFRP panel.

6 Energy summing algorithm

6.1 Basis for numerical algorithm

In this approach it is assumed that any discontinuity on the elastic wave propagation path causes wave reflection. In the general approach the sensing configuration consists of N transducers that excite and sense elastic waves. In order to monitor a surface with these transducers they are placed in special manner in points $T_i (i = 1, .., N)$. The transducer placement is not important at this point. For simplicity of explanation but without losing the generality a flat surface is assumed. In order to obtain information from a point P arbitrarily chosen from the surface the following distances

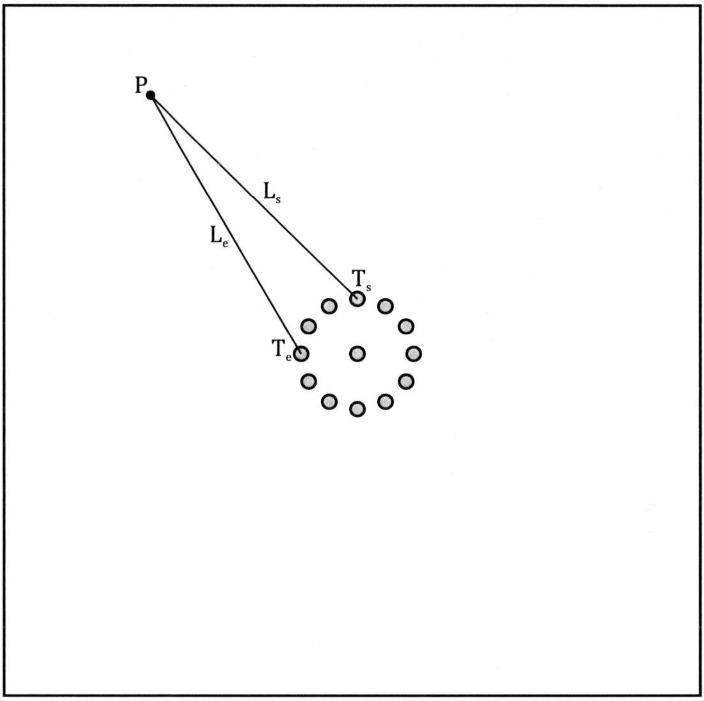

Figure 23. An example indicating transducer placement and distances needed for calculations.

are calculated (Figure 23)

$$\begin{aligned} L_e &= |T_e P|, \\ L_s &= |T_s P|. \end{aligned} \tag{16}$$

If the time of wave generation and the time of starting wave sensing is the same, the wave should be sensed in transducer T_s after the time:

$$t = \frac{L_e + L_s}{c} \tag{17}$$

where c is group velocity of wave propagation. Beginning at time instant t, segments of signals recorded in transducer T_s are extracted (Figure 24). Time width of this time segment is denoted by Δt. This segment is denoted by B_k, where k is an index ($k = 1, \ldots, K$). Damage indicator associated

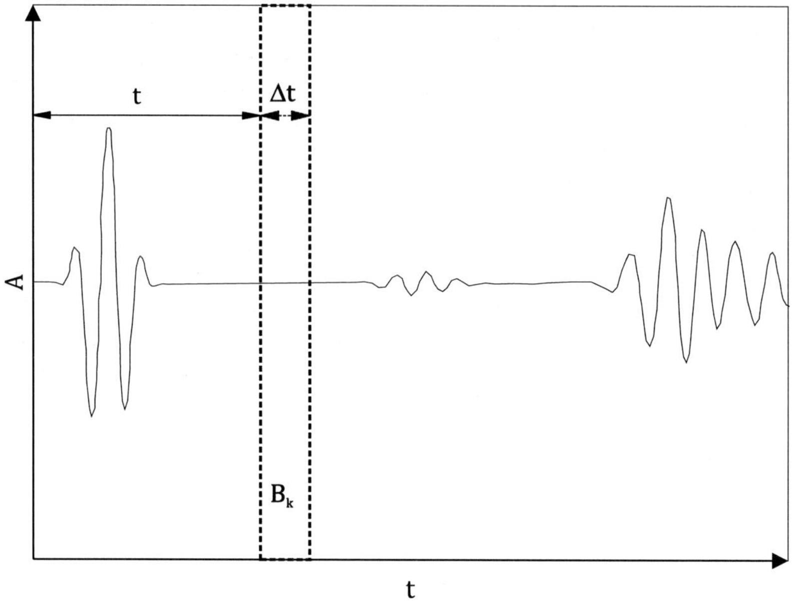

Figure 24. Illustration of signal segment extraction for creating a damage index.

with point P and based on two transducers (T_e, T_s) is defined as

$$DI(P) = \sum_{k=1}^{K} (D_k B_k)^2 \tag{18}$$

where D_k is a compensating coefficient taking into account amplitude decay with travelled distance.
The more general formula can be written if one consider all sensors. In this case two additional sums are needed. First for sensing and second for excitation transducers:

$$DI(P) = \sum_{i=1}^{N} \sum_{j=1}^{N} \sum_{k=1}^{K} (D_{ijk} B_{ijk})^2 \tag{19}$$

If the extracted segments contain reflected waves, value of the indicator DI for point P will be larger than for other points. This will clearly indicate a potentially damaged area. The procedure described above is repeated for other points of the monitored area.

Similar approach of mapping information contained in signals is based on adding either values of signal amplitudes or envelopes for a single time sample corresponding to a point within the monitored area. This means that for the DI only the time sample (or its envelope) corresponding to t is extracted (Michaels et al., 2008).
Another approach for damage localisation can use the information from the frequency domain. Damage localisation is preformed be exciting a structure with a known signal with frequency f_0. This signal is applied to the transducer electrodes. The frequency spectrum of such signal is known or can be calculated. Due to dispersive nature of elastic guided waves very often the excitation signal is chosen in the form of tone burst with energy concentrated in relatively narrow frequency band around f_0. And, this information can be used for creating the damage index. The procedure is following:

1. Perform the Fast Fourier Transformation on signal segments B_{ijk}.
2. Extract the amplitude value from FFT module for carrier frequency f_0 and denote it by B_{ijk}^f
3. Create new damage index

$$DI_f(P) = \sum_{i=1}^{N} \sum_{j=1}^{N} B_{ijk}^f \tag{20}$$

By analogy, the obtained frequency damage index DI_f is associated with point P. The procedure described above is repeated for other points of the monitored area. By its definition, the frequency damage index is a frequency filter. Extraction of the frequency response for f_0 guarantees that damage index is not influenced by measurement noise with other frequencies.
The results of the descirbed algorithm in the form of damage index DI are alligned along an ellipsis with foci in the exciting and recieving tranaducers. Of course, if the same transdcuers is used for excitation and sensing the ellipsis is a circle. Considering, for example, three transducers - one for excitation (T_e) and two for sensing (T_{s1} and T_{s2}) one can obtain two ellipses (Figure 25) related to higher DI. This will happen in case if there is damage in point P. For only one pair of transducers (T_e and T_{s1}) the result is ambigious and equal DI is obtained along the ellipsis. When an additinal signal for signal processing is used (from transducer T_{s2}) there are four points of ellipsis intersection at which DI increases. One is the correct point P and the remaining intersections are denoted by P' (Figure 25). This means that the result is more precisse - only four possible indications of damage. Adding more transducers results in intersecion of more ellipses and the result become unambigious.

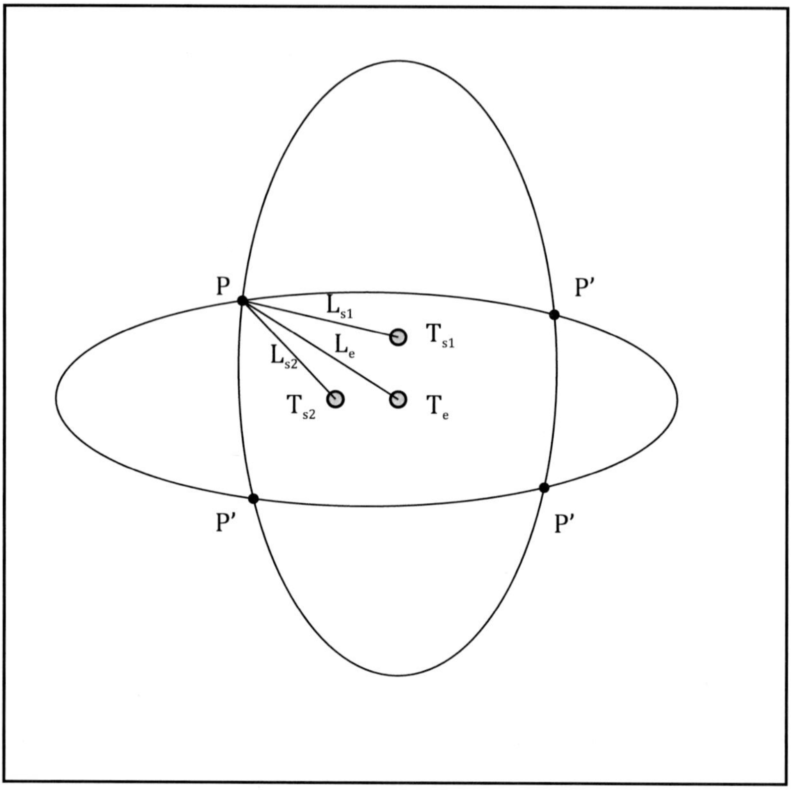

Figure 25. Illustration of the signal processing result.

6.2 Results

Concentrated configuration The described algorithm was applied to results of experimental investigation. Aluminium panel with dimensions of 1000 mm × 1000 mm × 1 mm was considered (Figure 26). Piezoelectric transducer with diameter of 10 mm and 0.5 mm thickness was used for exciting the elastic waves. It was attached at the centre of the panel surface. For adhesion a acceloremter wax was used. Additional mass was installed on panel surface 40 mm above and 250 mm to the right from the transducer (Figure 26). Excitation signal was in the form of tone burst with frequency 16.5 kHz and 5 periods. Hann window was used for modulation. Such excitation signal was chosen due to dispersive nature of Lamb waves. Thier velocity of propagation highly depends on frequency. The perfect situation

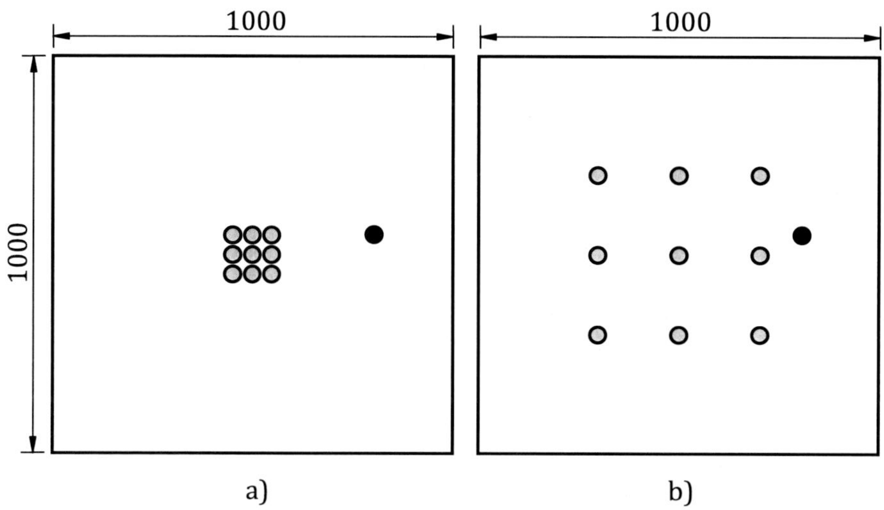

Figure 26. Aluminium panel with two types of investigated transducer configurations: a) concentrated, b) distributed. Black dot indicates the position of additional mass.

is an excitation of one freqency only, then it propagates with a defined velocity. However if a finite-length signal is conisdered the badnwith gets wider. This is the reason why signal modulation is used to limit the excitation bandwith and reduce wave dispersion.

Signals were measured by a contactless technique based on laser scanning vibrometer. The response was mesured on the other side of the specimen (not this one with attached transducer). Measurement points were defined in a way to form a shape of a transducer configuration. The central point of configuration was assumed to correspond to coordinates of piezoelectric transducer. Signals measured in these points represent changes of surface vibration velocity as function of time. Both concentrated and distributed square configurations were considered that comprise of 9 elements (Figure 26). Circular grid of angular spacing between points (P) of 2° and radial spacing of 1 mm was chosen for result visualisation. Uniform method of presenting results was chosen for all damage localization methods, by representing damage indicator using colour scale mapped on the surface of the investigated element. The maximum damage value colour is white. The minimum value was presented in black. 256 levels of grey were chosen for presenting intermediate values. Images presented according to these condi-

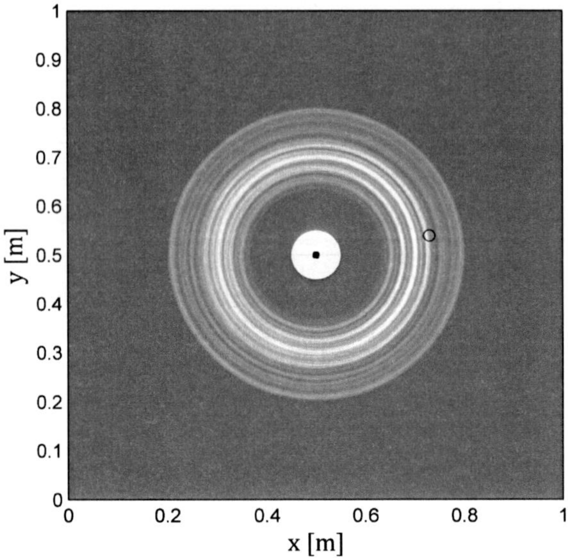

Figure 27. Localisation result for square 3 × 3 configuration with transducer spacing s=3 mm.

tions map damage indicator values onto specimen coordinates.

Concentrated square transducer cofiguration with 3 mm spacing was chosen as the first example. The value of damage indicator is visualised by damage influence map presented in Figure 27. Significant increases of damage indicator is visible in the form of white circle. Radius of this circle corresponds to distance from the centre of the configuration (piezoelectric transducer) to location of sought discontinuity. However, from the obtained result it is no clear where the wave reflection source lies excatly. One obtains only the radial position. Considering the diagnostic classification terminology the damage detection stage was obtained. Nevertheless, damage localization would be more useful and it was not achieved. In the course of further research the same concentrated layout (N =9) was tested for larger spacing of measurement points (30 mm). Result of damage localization can be seen in Figure 28. Comparing it to the result presented in Figure 27, one notices wider area with elevated damage indicator values. In the previous case these foci were located so close together that the indication was practically a circle. Increasing distance d results in more clear appearance of elliptic

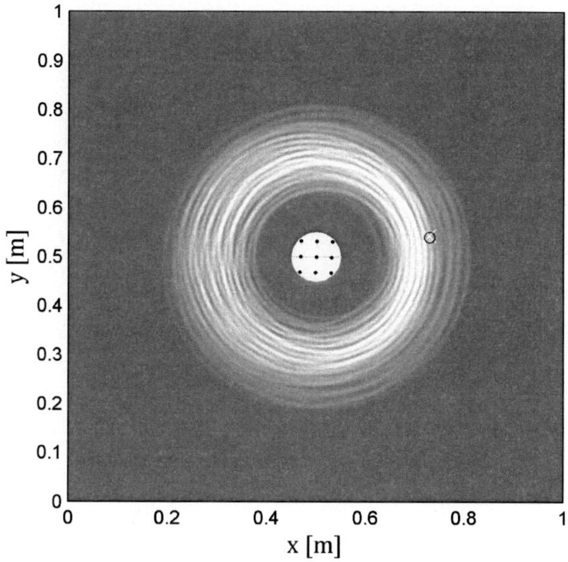

Figure 28. Localisation result for square 3 × 3 configuration with transducer spacing s=30 mm.

shapes.

In examples presented uptil now only nine signals were taken for signal processing because nine element arrays were considered with one source of wave excitation. It is interesting to invesigate the result of adding more information (signals) to the damage detection algorithm. In order to conduct such test the 3 × 3 configuration has to be extended by more sensors. To keep the symmetry in relation to the piezoelectric transdcuer, the new configuration was chosen as an 5 × 5 array. This means that 16 more signals were considered in comparison to the nine-element configuration. The first investigated example was with 3 mm-long spacing so the length of the configuration edge is 12 mm-long, that is two times longer than for 3 × 3 configuration. The result of the localisation is depicted in Figure 29. The DI increase due to damage reflections is more significant in this case than for the smaller configuration (Figure 27). In order to conisder a 25-element configuration of the same size as the one presented in Figure 28 the spacing needs to be set to 15 mm. The result for this case in presented in Figure 30. Characteristic elliptical shapes are visible. Sum of these ellipses around

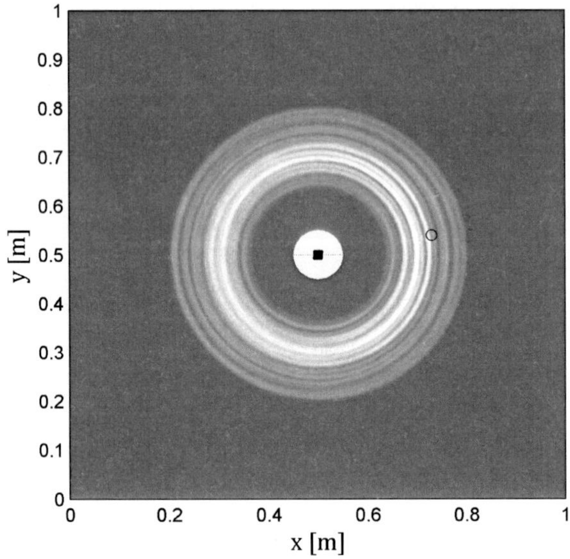

Figure 29. Localisation result for square 5 × 5 configuration with transducer spacing s =3 mm.

the sought discontinuity (additional mass position) results in large area of high DI value. After increasing the size of configuration even more to obtain 120 mm-long edge (30 mm-long spacing) one can observe more precise result (Figure 31). The ellipses intersect in the area near additional mass. Damage index at the opposite side of the sensor configuration did not give a focused response.

Distributed configuration A concurrent solution for damage localisation is based on sparse sensor arrays forming a distributed configuration. Just like in Figure 26b). In comparison to concentrated configuration the distributed one has a potential advantage. Assuming that damage is placed far away from the concentrated arrays the distance from the sensors to damage is comparable. In the case of small damage wave reflection can be weak and the returning wave, that is registered by sensors, can be burried in signal noise. As far as distributed configurations are concerned there is a greater probability that at least one of the sensors can register the damage-related response higher than the noise level.

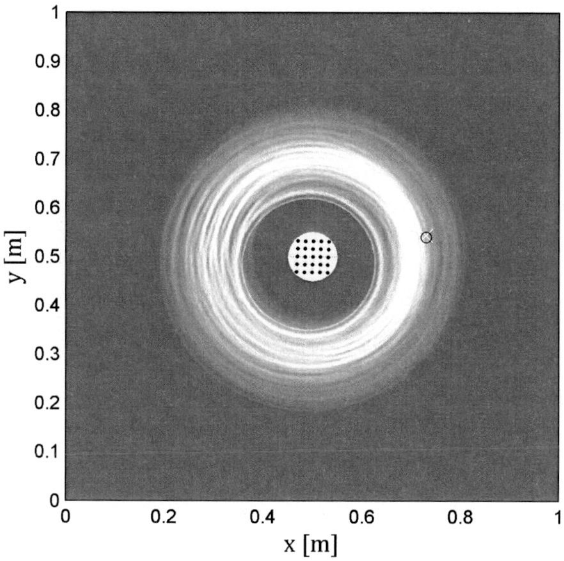

Figure 30. Localisation result for square 5 × 5 configuration with transducer spacing s =15 mm.

In this investigation a distributed array comprising of nine sensors was taken into conisderation. As previously the middle point was colocated with transducer being the source of waves. Large sensor spacing was chosen equal to 163 mm. This means that the edge of the configuration was 326 mm-long. Damage localisation result is presented in Figure 32. Characteristic ellipses can be seen. Their focii lie in the transdcuers position and at each sensing point. The point of intersection of the ellipses is more obvious that in the case of concentrated arrays. This interesection results in higher DI value that lies near the true poistion of the additional mass.

Test case - multi-site damage The usefulness of the energy summing algorithm was shown on the basis of vibrometer measurement. In this subsection the focus was on a more complicated case. More than one defect was considered. Research reported in this subsection was focused on square arrays used for four defects localization. The arrays consisted of nine sensors. Spacing between centres of these elements was equal to 5 mm. Middle sensor of the array was used for Lamb wave excitation. Remaining eight

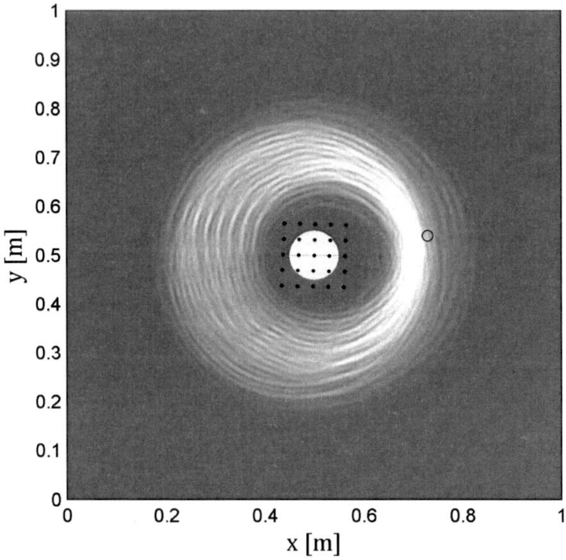

Figure 31. Localisation result for square 5 × 5 configuration with transducer spacing s =30 mm.

was registering propagating waves. The dimensions of the sensors were following: 3× 3 × 2 mm^3. Again investigation were performed on aluminium alloy AA5754 panel (1000 mm × 1000 mm × 1 mm). Damage was simulated in the panel by drilling two through thickness holes (Φ 6) and introducing two shallow notches, one 10 mm long and second 184 mm long. The notches were 0.2 mm wide. Considered arrays were attached to the panel in such way that the middle sensor of each array was at the centre of the panel. In order to ensure high sensitivity to small defects a relatively high frequency waves were used. The excitation frequency was chosen equal to 220 kHz. Results are showed in Figure 33. Information about damage presence was obtained. Moreover partial localisation was achieved because the radii on which the defects lie were found. In order to obtain complete localization an angular prediction is necessary. It was not achieved in ths case.

6.3 Summary

In this section an experimental study was described. Its main goal was to present energy summing algorithm. This special numerical algorithm

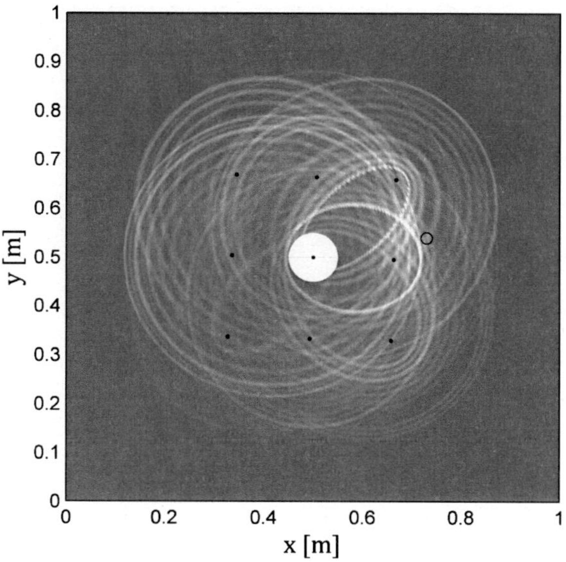

Figure 32. Localisation result for square distributed configuration with transducer spacing s =163 mm.

was used to process signals gathered be the sensors and to perform damage detection. The energy summing algorithm is a very usful approach in Structural Health Monitoring. It can be applied to any sensor arrays either concentrated or distributed.

In the investigation laser vibrometer mesurement points and piezoelectric transducers were used as sensors. The influence of sensor array parameters on damage localization result was investigated. The results were compared. In particular it was shown that increase in the sensor spacing s helps to remove the ambigiuity in damage localization.

In the test case subsection it was proven that the proposed damage localization algorithm allows to localize multiple damage in an isotropic structural element using proposed sensor array. It should be emphasized that the damage localization algorithm does not need baseline data in order to visualize DI. It means that an SHM system based on this idea can be used on a structure without the need for reference (undamaged or previous) measurements. The conducted research showed that directivity of the damage map is strictly correlated with transducer spacing. Better directivity will

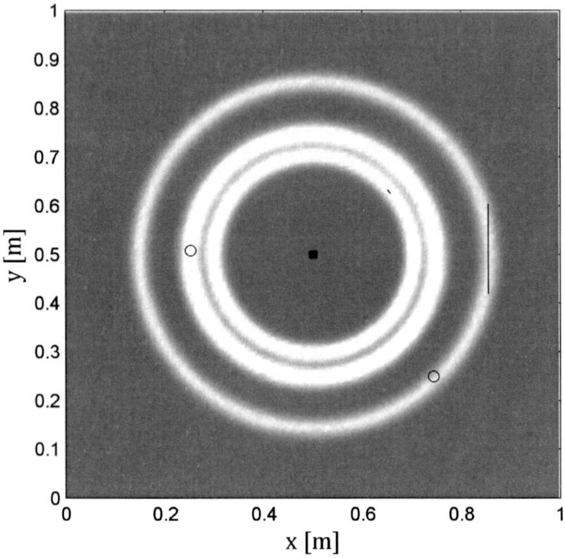

Figure 33. Four defects localization using a square sensor arrays. Circles indicate two holes. Lines indicate notches on specimen surfaces.

be obtained for larger transducer spacings. It was shown that by changing the distance between transducers it is possible to control the damage map directivity. It should also be mentioned that a larger distance between transducers is connected with the fact that a larger portion of the signal is interrupted by wave reflections that occur between transducers. As a result part of the signal connected with reflections between transducers must be rejected and, in effect, a certain area around the transducers cannot be monitored. The presented research showed that it is possible to differentiate the damage states of the structure. The results for one and four defects differ significantly. It means that the proposed DI is a very useful tool.

7 Beam forming algorithm

7.1 Basis for numerical algorithm

In this approach it is also assumed that any discontinuity on the elastic wave propagation path causes wave reflection. In the general approach the sensing configuration consists of N transducers that excite and sense elastic

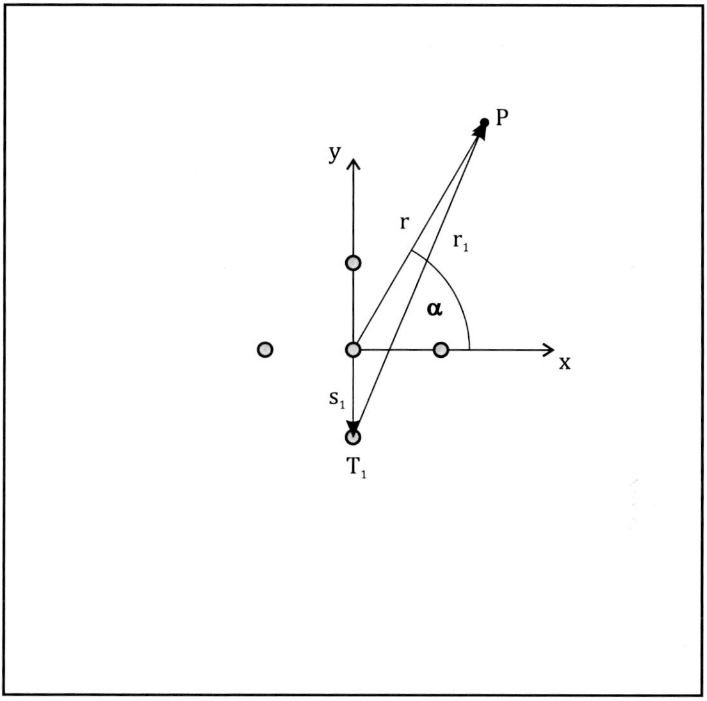

Figure 34. Sensor configuration with illustraation of the symbol for beam forming algorithm.

waves. In order to monitor a surface with these transducers they are placed in special manner in points $T_i (i = 1, .., N)$. The crucial point in applying this algorithm is that it makes use of the wave intefrence phonomenon. Assuming a harmonic wave at point $\vec{r}$ at instance t propagating with velocity c one obtain (Giurgiutiu, 2008):

$$f(\vec{r}, t) = \frac{A}{\sqrt{|\vec{r}|}} e^{j(2\pi f t - \vec{k} \cdot \vec{r})}, \tag{21}$$

where A - real amplitude, f - frequency, $\vec{k}$ - wave vector. Using a defined time delays:

$$\begin{aligned}\Delta(\alpha) &= \frac{1-\hat{r}}{c/|\vec{r}|} \quad (22)\\ \Delta(\alpha) &= |\vec{r}|/c - |\vec{r_i}|/c \quad (23)\\ \Delta(\alpha) &= t - t_i, \quad (24)\end{aligned}$$

and defined weights for signals registered by T_i sensors:

$$w_i = \sqrt{r_1} \quad (25)$$

a constructive interference for angle α is obtained (Figure 34). The symbol definition is follwing

$$\begin{aligned}\hat{r_1} &= \frac{|\vec{r_m}|}{|\vec{r}|}, \quad (26)\\ \vec{r_i} &= \vec{r} - \vec{s_i}, \quad (27)\end{aligned}$$

$\vec{s_m}$ is the position of the i-th sensor, t is the time of wave propagation from the origin to point P, while t_i is the time of wave propagation from sensor T_i to point P (Figure 34). In this way a N-times amplification of the sensor array response is obtained (Giurgiutiu, 2008):

$$S(\vec{r}, t) = N \times f. \quad (28)$$

The amplification can be obtained for wave generation and reception. Wave excitation in each transducer with the delay given by formula 24 results in a constructive interference for point P (Figure 35a). Form the other hand if point P is a source of wave, the delays defined by formula 24 can be applied to shift the signals registered by the sensors in an array (Figure 35b). This take into account the diffrence in time of arrival of the wave front at individual sensors. In the case of Structural Health Monitoring the sensors arrays are used both for wave generation and sensing, therefore the delay algorithm is used twice.

What is important it is not neccessary to excite wave with the mentioned delays to obtain interference in wave generation. Application of the delays can be realised in a post processing mode. This mean that waves can be excited sequentially in each transducers and registerd by the whole array of N transducers. Next the registered signals are processed by the delay algorithm to simulate the interfrence both in generation and recpetion.

In the conducted discussion point P was a source of wave, however in real

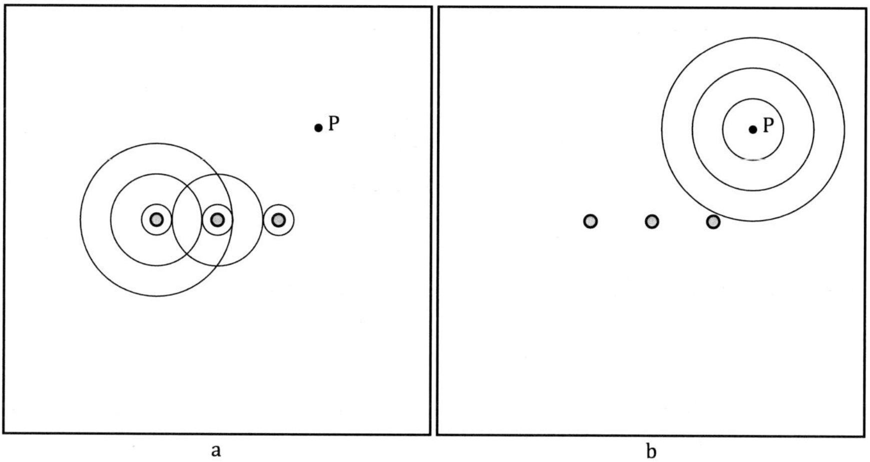

Figure 35. Sensor configuration working in a generation (a) and reception mode (b).

application the source of wave will be damage or other type of discontiunity (rivet, stiffener, etc.). Such discontiuiuty causes wave reflection, therefore it can be detected. If there is no wave reflection source at chosen point P the beam forming algorithm will not give any amplification. This will result in low damge index defined, as before, in the form of squared signal (20).

Characteristics of sensor arrays (phased arrays) Sensor arrays that utilize such beam forming algorithms are called phased arrays. It is important that in order to take advantage of the interference phenomenon the sensors should be placed with a carefully defined spacing. This spacing should be less or equal half of the wavelegth (Malinowski et al., 2008).

$$s \leq \frac{\lambda}{2} \tag{29}$$

This wavelength is of the conisdered wave, so in the case of Lamb waves the user should define which mode will be used A_0, A_1, S_0, S_1 or other. The Lamb wave dispersion should be also accountned for. Their wavelegth depends on the frequency, therefore if the excitation is broadband several wavelengths are excited and the phased array is tuned only to one of them. This may make damage localization more difficut and dispersion compensation algorithms could be helpful.
The spacing s (29) together with number of sensors N have also an influnce

on the precission of sending or reciving of the wave from a certain angle α. In the simplest case - a linear phased array, the dependence is given by the formula (Orfanidis, 2004):

$$W = 0.886\frac{\lambda}{Ns}. \tag{30}$$

W is the width of the main lobe steered at α angle. If it is narrow, arrays sends waves and is sensitive to the response only from this angle. As an example a directional characteristics of a 3×3 array was calculated. It was plotted in Figure 36. This array was steered at $\alpha = 60^o$ and the charateristic was normalised so its maximal value ie equal to one. If the condition given by (29) is fullfilled the charateristics of the array unambigious as in Figure 36a. Waves that are sent and recieved are related to angles indicated by the main lobe. In the provided example the main lobe is rather wide and in the case of discontiuities lying close to each other, wave reflection from them cannot be differentiated. Following the rule given by (30) an increase in the spacing results in a narrower main lobe. Provided example confirms this. Spacing equal to $s = 2\lambda$ (Figure 36b) ensures a very fine main lobe for $\alpha = 60^o$. However this was obtained by the cost of loosing the unambigious characristic of the array. By choosing spacing $s = 2\lambda$ the condition given by (29) was broken. As an effect one can notice six additional lobes. This phenomena did not facilitate the damage localization and should be avoided. If damage is present at an angle corresponding to any of the additional lobes the information about this will be obtained in the response for the angle $\alpha = 60^o$. This will result in false-positive indication of damage for this direction. Using an array with an ambigious characteristics (as for example this in Figure 36b) should be avoided.

7.2 Results

In this section concernig the beam forming algorithm the experimental example is the same as in the section 6, that concerned energy summing algorithm. Aluminium panel with dimensions of 1000 mm $\times$ 1000 mm $\times$ 1 mm was considered as it is presented in Figure 26. Piezoelectric excited waves were registered by the laser vibrometer. Square sensor array was conidered as presented in Figure 26a. Piezoelectric transducer with diameter of 10 mm and 0.5 mm thickness was used for excitation of the elastic waves. It was attached at the centre of the panel surface. Excitation signal was in the form of tone burst with frequency 16.5 kHz and 5 periods. As previously the same circular grid of points was chosen for result visualisation.

The first localisation example was obtained for an array of spacing $s = 3$

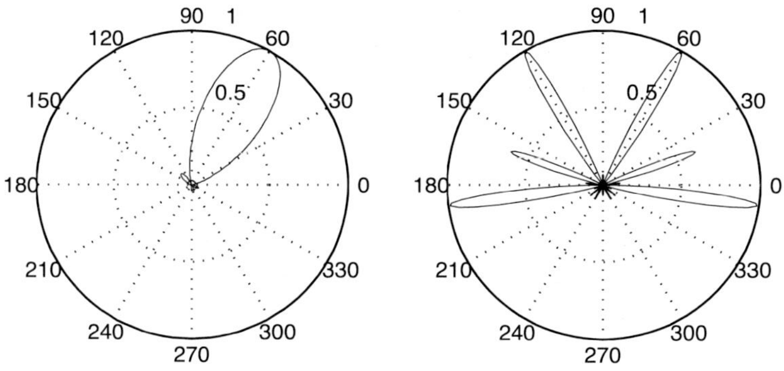

Figure 36. Phased array characteristic dependence on sensor spacing s; $s = 0.4\lambda$ (a),$s = 2\lambda$ (b)

mm (Figure 37). This result can be directly compared to the example illustrated in Figure 27 for the energy summing algorithm, because parameters (s and N) were identical to the case presented there. As one can notice, thanks to directional properties of the beam forming algorithm the damage localisation is more precise. Damage index value was amplified on the right from the array - in the region of simulated defect. Previous result (Figure 27) indicated only the distance from the array on which the discontinuity was placed. The result presented in Figure 37 can be rather comparable to the result of a 5 array with spacing $s = 30$ mm for which the directivity of the response is more apparent. Incresing the spacing by factor two $s = 6$ mm, results in narrower response for the angle correpsonding to discontinuity (Figure 38). However unwanted additional increase of DI appears symmetrical on the left hand side of the array. This is related with the increase in spacing. The case is similar to that presented in Figure 36. The increse of spacing reduced the width of main lobe but this also cased the amplification of the response from other angles. In order to avoid this

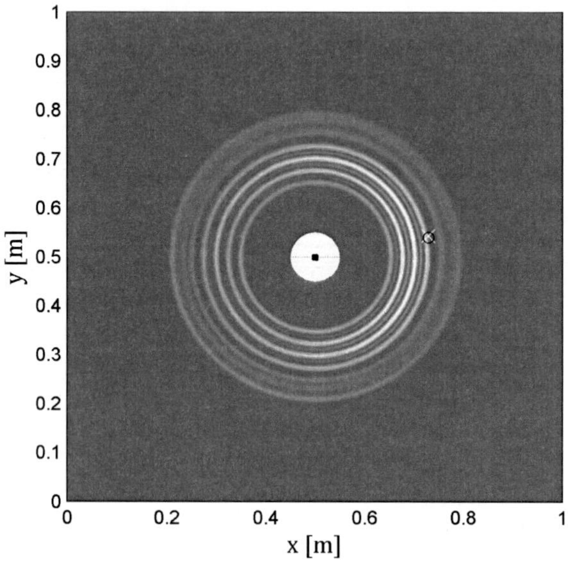

Figure 37. Localization result for a square 3×3 phased array wih spacing $s = 3$ mm.

phenomena the spacing should be kept constant and the increse of sensors will help to reduce the main lobe width. This rule is given by formula 30. The next example followed this rule. An array comprising of 25 sensors was selected with smaller spacing $s = 3$ mm. The result was presented in Figure 39. Unambigious and precise localization was achieved in this way. Although the parameters (s and N) were the same as for energy summing algorithm result presented in 29, this beam forming result is far more accurate.

Test case - CFRP panel The usefulness of the beam forming algorithm was shown on the basis of vibrometer measurement. In this subsection a CFRP panel with permanently attached piezoelectric transducers was investigated. This panel came from a AW139 helicopter and is stiffened by horizontal and diagonal stiffeners indicated in Figures 40 and 41. In order to perform health monitoring five transducers were placed in the middle of the panel forming a cross shaped array. The transducer separation was 2 mm. Rectangular multilayer piezoelectrics were used ($3 \times 3 \times 2$ mm^3). After

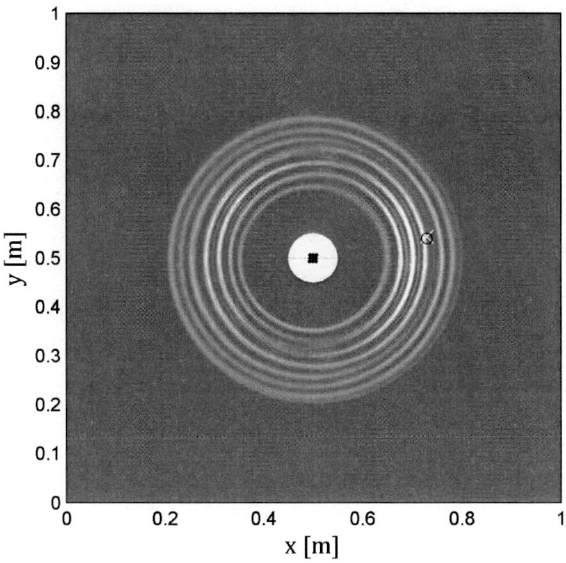

Figure 38. Localization result for a square 3×3 phased array wih spacing $s = 6$ mm.

the measurement for pristine plate damage was introduced by hollow metal block impact that was hit by a hammer. Obtained delaminated region was approx. 350 mm^2. Next the measurements were performed for the damaged panel and a beam forming algorithm was applied. The result for 140 kHz excitation is presented in Figure 40. The mesurments were also conducted for other frequencies (100,110,120,130 and 150 kHz) but the localization was less accurate. The maximum of damage index value (indicated by X in Figure 40) differed from the assumed centre of delamination by 43 mm.

The next step was the modification of the transducer array. Four more transducers were added to obtain a square array just like this in Figure 26a. In this case the damage index maximum was 45 mm away from the assumed centre of delamination Figure 41. It should be highlighted that in this case 120 kHz excitation gave better result than 140 kHz.

In comparison to cross-shaped array the indication is less accurate. However it is important to underline that in the case of the cross-shaped array baseline information was utilized and the signals were subtracted (damaged state - pristine state). In the subsequent case (square-shaped array) the

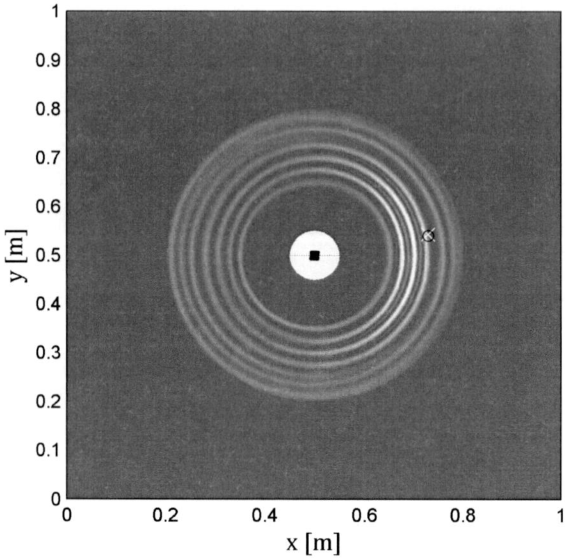

Figure 39. Localization result for a square 5×5 phased array wih spacing $s = 3$ mm.

panel was already damaged, so the baseline information was not available.

It should be underlined that measurement on the composite panel was made using only piezoelectric transducers. This approach resulted in more signals for signal processing algorithm because each transducer was sequentially used for wave excitation and the remining ones for sensing. Due to this mesurment scheme for cross-shaped array 20 (5 excitations $\times$ 4 sensors) signals were obtained. In the case of square-shaped array 72 (9 excitations $\times$ 8 sensors) signals were available. If the vibrometry mesurement had been concerned, it would have resulted in 5 (1 excitation $\times$ 5 sensing points) signals in the first and 9 in the second case.

The use of an array of transducers influences also the DI visualization (Figure 40 and 41). The DI values for an area around the array were not plotted due to numerous reflection of waves from the sensors. In the case of laser measurment the registered signals at the beginning contain direct waves propagating from the transducer to the sensing points. If the whole transducer array is considered, the signals will not only contain direct waves but also those reflected from the transducers. This initial part of the signals

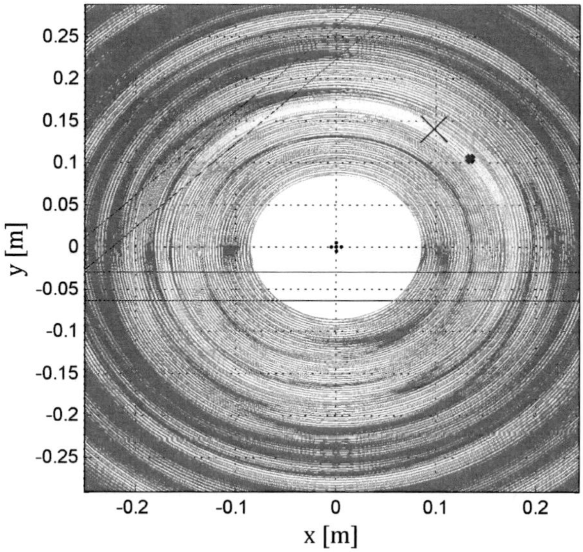

Figure 40. Damage localization result for a CFRP panel with an impact damage. Localization using square-shaped phased array.

was rejected in order not to disturb the vizualization of the DI.
One can notice that the results obtained for composite panel (Figure 40 and 41) are not so clear and pure as those obtained for vibrometer mesurements (Figures 37-39). One reason is that the laser vibroemetr mesurmements for mesuring the out-of-plane vibrations so the influence of S_0 mode was negleg-ible, while the piezoelectric transducers register both modes. The second reson is that the signal-to-noise ration for aluminium panel was higher due to lower attenuation of Lamb waves in aluminium in comparison to CFRP.

7.3 Summary

The beam forming algorithm is a very usfull to in the Structural Health monitoring techniques, that are based on propagation of guided elastic waves. It was shown, that successful application of the algorithm is only possible when a proper transducer spacing s will be chosen. The influence of the spacing and the number of senor is significant and should be carefullly controled. High precision of damage detection can be obtained, providing the main lobe width is sufficiently narrow. The spacing of a phased array is

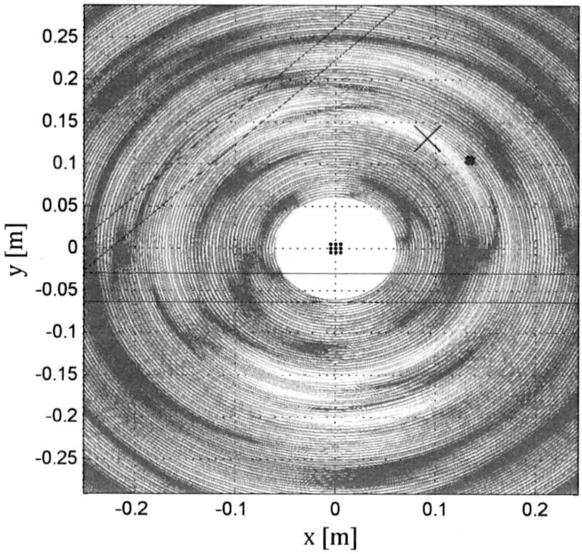

Figure 41. Damage localization result for a CFRP panel with an impact damage. Localization using star-shaped phased array.

chosen in relation to the wavelength (see formula 29). This means that the array should be tunned to previously chosen wavelength of either symmetric or anti-symmetric Lamb wave mode. According to the theory, for the same frequency, A_0 mode wavelength is shorter than S_0 wavelength. This should be taken int account. In the investigation conducted in this section the arrays were tunned for the A_0 mode, because in the laser vibrometry mesurments only the out-of-plane signals were taken for signal processing. Regarding the investigation related to aluminium panel it should be emphasized that the damage localization algorithm does not need baseline data in order to visualize DI. It means that an SHM system based on this idea can be used on a structure without the need for reference (undamaged or previous) measurements.

As far as composite panel is concerned, the investigation showed that using only 5 transducers 43 mm accuracy was achieved with baseline data. It was also shown that if there is no information about the healthy structure nine transducers are enough to indicate a defect with a 45 mm accuracy. Presented work proved that phased arrays that were used previously for

aluminium specimen may be successfully utilized for composite structures monitoring.
Obtained results showed high sensitivity of the method. It detected delaminated area of 0.12 % of the monitored area. For instance using vibration methods for delaminated beams only 5 % delaminations were detected (Zak, 2005). In work (Brunner et al., 2008) detection also of 5 % using two other methods: electrical impedance and ultrasonics.
In this chapter authors considered single phased arrays used for SHM. However it is also possible to use a combined approach, that employs few phased arrays working together and giving a cumulative damage index. In a research paper (Malinowski et al., 2011) authors investigated three phased arrays placed in vertices of an equilateral triangle. Each vortex contained a 2×2 array. It was shown that such configuration succesfully localizes discontiunities outside and inside of the triangle placed on a aluminium panel surface. Damage detection on the triangle edge was also possible.

Bibliography

J.-M. Berthelot, M.B. Souda, and Robert. J.L. Frequency response transducers used in acoustic emission testing of concrete. *NDT&E International*, 25(6):279–285, 1992.

A. J. Brunner, M. Barbezat, N. A. Chrysochoidis, A. K. Barouni, and D. A. Saravanos. Delamination detection in composites using active piezoceramic wafer and AFC sensor. In *Proceedings of 4th European Workshop Structural Health Monitoring (Krakow, Poland, 2-4 July 2008)*, pages 657–664, 2008.

F. Cau, A. Fanni, A. Montisci, P. Testoni, and M. Usai. A signal-processing tool for non-destructive testing of inaccessible pipes. *Engineering Applications of Artificial Intelligence*, 19:753–760, 2006.

Y. Ding, R.L. Reuben, and J.A. Steel. A new method for waveform analysis for estimating AE wave arrival times using wavelet decomposition. *NDT&E International*, 37:279–290, 2004.

S. Dixon and S.B. Palmer. Wideband low frequency generation and detection of Lamb and Rayleigh waves using electromagnetic acoustic transducers (EMATs). *Ultrasonics*, 42:1129–1136, 2004.

R. Gangadharan, G. Prasanna, M.R. Bhat, C.R.L. Murthy, and S. Gopalakrishnan. Acoustic emission source location and damage detection in a metallic structure using a graph-theory-based geodesic approach. *Smart Materials and Structures*, 18(11), 2009.

V. Giurgiutiu. *Structural Health Monitoring with piezoelectric wafer active sensors*. Academic Press, Elsevier Inc, 2008.

V. Giurgiutiu, L. Yu, J.R. Kendall, and C. Jenkins. In situ imaging of crack growth with piezoelectric-wafer active sensors. *AIAA Journal*, 45(11): 2758–2769, 2007.

K. Hongjoon, J. Kyungyoung, S. Minjea, and K. Jaeyeol. A noncontact NDE method using a laser-generated focused-Lamb wave with enhanced defect detection ability and spatial resolution. *NDT&E International*, 39:312–319, 2006a.

K. Hongjoon, J. Kyungyoung, S. Minjea, and K. Jaeyeol. Application of the laser-generated focused-Lamb wave for non-contact imaging of defects in plate. *Ultrasonics*, 44:1265–1268, 2006b.

J.-B. Ihn and F.-K. Chang. Detection and monitoring of hidden fatigue crack growth using a built-in piezoelectric sensor/actuator network: I. diagnostics. *Smart Materials and Structures*, 13:609–620, 2004.

J.-B. Ihn and F.-K. Chang. Pitch catch active sensing methods in structural health monitoring for aircraft structures. *Structural Health Monitoring, An International Journal*, 7 (1):5–19, 2008.

H. Jeong. Analysis of plate wave propagation in anisotropic laminates using a wavelet transform. *NDT&E International*, 34:185–190, 2001.

G.R. Kirikera, J.W. Lee, M.J. Schulz, A. Ghoshal, M.J. Sundaresan, R.J. Allemang, V.N. Shanov, and H. Westheider. Initial evaluation of an active/passive structural neural system for health monitoring of composite materials. *Smart Materials and Structures*, 15:1275–1286, 2006.

T. Kundu, S. Das, S.A. Martin, and K.V. Jata. Locating point of impact in anisotropic fiber reinforced composite plates. *Ultrasonics*, 48:193–201, 2008.

R. Lammering. Observation of piezoelectrically induced wave propagation in thin plates by use of speckle interferometry. *Experimental Mechanics*, 50:377–387, 2010.

C.M. Lee, J.L. Rose, and Y. Cho. A guided wave approach to defect detection under shelling in rail. *NDT&E International*, 42:174–180, 2009.

Y.S. Lee, D.J. Yoon, S.I. Lee, and J.H. Kwon. An active piezo array sensor for elastic wave detection. *Key Engineering Materials*, 297-300:2004–2009, 2005.

W. Lestari and P. Qiao. Application of wave propagation analysis for damage identification in composite laminated beams. *Journal of Composite Materials*, 39:1967–1984, 2005.

M. Lin and F.-K. Chang. The manufacture of composite structures with a built-in network of piezoceramics. *Composites Science and Technology*, 62:919–939, 2002.

P. Malinowski, T. Wandowski, and W. Ostachowicz. Multi-damage localization with piezoelectric transducers. In *Proceedings of 4th European Workshop on Structural Health Monitoring*, pages 716–723, 2008.

P. Malinowski, T. Wandowski, and W. Ostachowicz. Damage detection potential of a triangular piezoelectric configuration. *Mechanical Systems and Signal Processing*, 25:2722–2732, 2011.

M. Melnykowycz, X. Kornmann, C. Huber, M. Barbezat, and A. J. Brunner. Performance of integrated active fiber composites in fiber reinforced epoxy laminates. *Smart Materials and Structures*, 15(1):204–212, 2006.

M. Melnykowycz, M. Barbezat, R. Koller, and A. J. Brunner. Packaging of active fiber composites for improved sensor performance. *Smart materials and structures*, 19(1):015001, 2010.

J. E. Michaels, A. J. Croxford, and P. D. Wilcox. Imaging algorithms for locating damage via in situ ultrasonic sensors. In *Proceedings of the 2008 IEEE Sensors Applications Symposium*, pages 63–67, 2008.

R. Murayama and K. Mizutani. Conventional electromagnetic acoustic transducer development for optimum Lamb wave modes. *Ultrasonics*, 40:491–495, 2002.

F. Mustapha, G. Manson, K. Worden, and S.G. Pierce. Damage location in an isotropic plate using a vector of novelty indices. *Mechanical Systems and Signal Processing*, 21:1885–1906, 2007.

S. J. Orfanidis. *Electromagnetic waves and antennas.* `http://www.ece.rutger.edu/ orfanidi/ewa`, 2004.

W. Ostachowicz and P. Kudela. Experimental verification of the Lamb-wave based damage detection algorithm. In *Proceedings of 6th International Workshop on Structural Health Monitoring*, pages 2066–2073, 2007.

W. Ostachowicz, P. Kudela, P. Malinowski, and T. Wandowski. Damage localisation in plate-like structures based on PZT sensors. *Mechanical Systems and Signal Processing*, 23:1805–1829, 2009.

M. Palacz, M. Krawczuk, and W. Ostachowicz. The spectral finite element model for analysis of flexural-shear coupled wave propagation. part 1: Laminated multilayer composite beam. *Composite Structures*, 68:37–44, 2005.

I.-K. Park, T.-H. Kim, H.-M. Kim, Y.-K Kim, Y.-S. Cho, and W.-J. Song. Evaluation of hidden corrosion in a thin plate using a non-contact guided wave technique. *Key Engineering Materials*, 321-323:492–496, 2006.

P.X Qing, S. Beard, S.B. Shen, S. Banerjee, I. Bradley, M.M. Salama, and F.-K. Chang. Development of a real-time active pipeline integrity detection system. *Smart Material and Structures*, 18:1–10, 2009.

X.P. Qing, S.J. Beard, A. Kumar, H.-L. Chan, and R. Ikegami. Advances in the development of built-in diagnostic system for filament wound composite structures. *Composites Science and Technology*, 66:1694–1702, 2006.

J.L. Rose. *Ultrasonic waves in solid media.* Cambridge University Press, 2004.

D. Royer and E. Dieulesaint. *Elastic waves in solids II. Generation, Acousto-optics interaction, applications.* Springer, 1999.

A. Rytter. Vibration based inspection of civil engineering structures. Ph.D. Dissertation, Department of Building Technology and Structural Engineering, Aalborg University, Denmark, 1993.

K.I. Salas and C.E.S. Cesnik. Guided wave excitation by a clover transducer for structural health monitoring: theory and experiments. *Smart Materials and Structures*, 18:1–27, 2009.

M.H.S. Siqueira, C.E.N. Gatts, R.R. da Silva, and J.M.A. Rebello. The use of ultrasonic guided waves and wavelets analysis in pipe inspection. *Ultrasonics*, 41:785–797, 2004.

Z. Su, C. Yang, N. Pan, L. Ye, and L.M. Zhou. Assessment of delamination in composite beams using shear horizontal (SH) wave mode. *Composites Science and Technology*, 67:244–251, 2007.

N. Takeda, Y. Okabe, J. Kuwahara, S. Kojima, and Ogisu. T. Development of smart composite structures with small-diameter fiber Bragg grating sensors for damage detection: Quantitative evaluation of delamination length in CFRP laminates using Lamb wave sensing. *Composites Science and Technology*, 65:2575–2587, 2005.

A.B. Thien, editor. *Pipeline structural health monitoring using macro-fiber composite active sensors*, 2006. Ph.D. Dissertation, College of Engineering, University of Cincinnati.

G. Thursby, B. Sorazu, D. Betz, W. Staszewski, and B. Culshaw. Comparison of point and integrated fiber optic sensing techniques for ultrasound detection and location of damage. In *Proceedings of SPIE*, volume 5384, pages 287–295, 2004.

C. Valle and J.W. Jr. Littles. Flaw localization using the reassigned spectrogram on laser-generated and detected Lamb modes. *Ultrasonics*, 39: 535–542, 2002.

D. Wagg, I. Bond, P. Weaver, and M. Friswell. *Adaptive Structures. Engineering Applications.* John Wiley & Soons Ltd, 2007.

P.D. Wilcox, editor. *Lamb wave inspection of large structures using permanently attached transducers*, 1998. Ph.D. Dissertation, Imperial College of Sciences, London.

K. Worden and J.M. Dulieu-Barton. An overview of intelligent fault detection in systems and structures. *Structural Health Monitoring, An International Journal*, 3(1):85–98, 2004.

L. Yu, G. Santoni-Bottai, B. Xu, W. Liu, and V. Giurgiutiu. Piezoelectric wafer active sensors for in situ ultrasonic-guided wave SHM. *Fatigue & Fracture of Engineering Materials & Structures*, 31:611–628, 2008.

A. Zak. Non-linear vibrations of a delaminated composite beam. *Key Engineering Materials*, 293-294:607–616, 2005.

G. Zumpano and M. Meo. A new damage detection technique based on wave propagation for rails. *International Journal of Solids and Structures*, 43: 1023–1046, 2006.

Application for Wind Turbine Blades

Malcolm McGugan *

* Materials Research Division, Risoe DTU, Roskilde, Denmark

Abstract The following text and images are taken from the five lectures given by Malcolm McGugan in the CISM "New Trends in Structural Health Monitoring" Advanced Professional Training Course on the 20th-24th of June, 2011 in Udine, Italy.

1 A look at the Wind Energy Industry

Wind power is an ancient technology that has been used by human civilisation for many thousands of years. The Mycenaean sea-trade culture of 1700BC was driven by the wind energy harnessed by their ships. Wind mills dating to the period around 200AD have been confirmed from ancient Persia to China. These early examples of the technology were based around a vertical axis and driven by reed mat sails. Later developments included horizontal axis mills with cloth sails. The most common tasks for early wind powered mills were pumping water and grinding corn.

By the 14th Century the Dutch were using wind mills to drain otherwise unusable areas of the Rhine Valley. There were 500 years of development in this technology that improved the aerodynamic lift in the sails, introduced cambered leading edges, optimised spar placement, measured the centre of gravity to balance the mill, innovated brakes, spoilers, flaps, and so on. This machine was a major factor in pre-Industrial Europe; irrigating, grinding grain, milling timber, processing spices and commodities, and more. But by the 17th century the steam engine began to take over these and other tasks.

At this point the US takes over the story of wind power development with the construction of electricity producing wind turbines. The wind turbine generator constructed by Charles Brush was discussed in a special edition of the Scientific American Journal in 1860. It consisted of a disc (17m in diameter) made of slated wood that was turned out of the wind by a huge tail fin in order to protect it from wind speeds that were too high for the fragile structure. Brush utilised a stepped gearbox (50:1 ratio) for his turbine and in good conditions generated up to 12kW of Direct Current.

Figure 1. The Windmill De Kameel in Schiedam. Photograph by Martijn Janssen, GFDL.

The Brush wind turbine was limited, but in 1891 a Danish man, Poul La Cour, was showing that fast running turbines with fewer blades would generate electricity more efficiently, and by the 1920s there were electricity producing turbines of his design all over Denmark. But this first flowering of commercial wind powered turbines proved to be brief as the introduction of cheap, large, fossil fuel driven steam plants soon put them all out of business.

In the 1970s the events of the first oil crisis prompted governments around the world to support either the few remaining enthusiasts of wind power generation or the aerospace industry, according to availability in their respective countries. And while the aerospace designed turbines tended to be very fragile, the (mostly Danish) backyard industry succeeded by starting simply and slowly scaling up their designs and processes one step at a time in order to reach the sizes we see today. This means that the different designs which have existed earlier have now converged to what is often called the Danish design: three-bladed, upwind rotors, with a gearbox, on a tubular tower. And despite renewed enthusiasm for new and innovative wind powered devices, it is likely that the Danish design will be dominant for many years to come.

In order to stimulate the adoption of wind power, support schemes like the German Einspeisegesetz (Energy Feed Law) and similar laws in Denmark and Spain initially made these countries the top three for installed capacity. But by 2010 China, USA, and Germany are the countries with most installed capacity. 2010 was the first year where most of the new wind farm installations were in developing countries and emerging economies, led by China of course. This development shows that wind is not just a premium technology that can only be deployed at scale in rich countries.

Italy has almost 6GW of wind power installed at the end of 2010 which makes it the 3rd largest producer in Europe, and 6th in the world. Most of the new Italian capacity is to be found on Sicily, Calabria, and Molise. It could be noted that the only commercially exploited area that is not in the South is Liguria. Nationally there are 28,000 people employed in the Italian Wind Power Sector with 10,000 of these directly employed. 3.4 percent of Italys electricity requirement was met by the wind in 2010. In Denmark this figure is 24 percent. So far Denmark remains the country in the world with the highest penetration of Wind Power into the National Energy consumption.

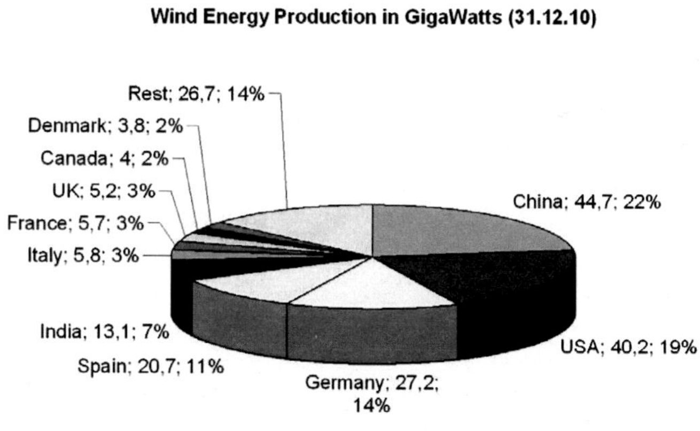

Figure 2. Pie Chart showing wind energy production by country - the global total is 200GW, figures from EWEA

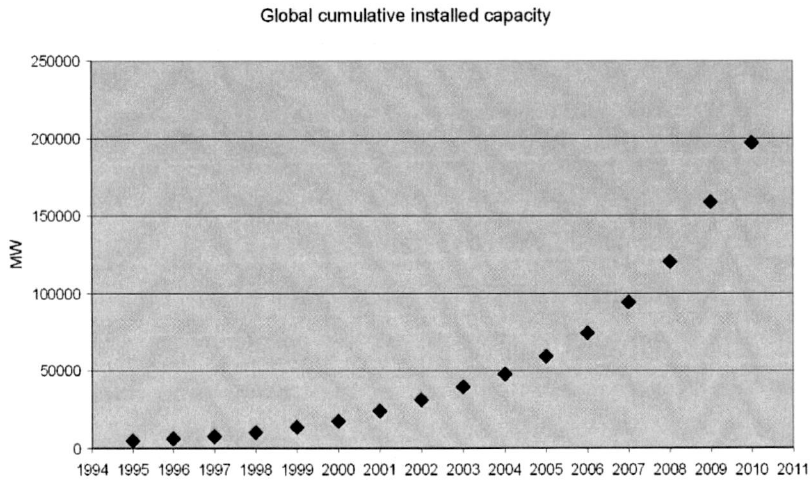

Figure 3. Graph showing installed global capacity from 1995 to 2010, figures from EWEA

Worldwide the level of installed wind power capacity approximately doubles every three years. While Europe is still the commercial area hosting the largest wind capacity, by the end of 2012 China will have overtaken that position. So the largest application focus for this industry is quickly moving East. So far, however, the new technology developments are still under development here in Europe. The challenge of offshore farms is being solved here (robust environment, scale, maintenance, deepwater, floating platforms, etc). The power network integration issues are being addressed here (transmission/distribution, smart networks, storage facilities, super grids). And the political targets for renewable energy are still most advanced here.

While the actions of the OPEC nations forty years ago were the impetus for the start of our modern drive to exploit wind power, the factors pushing us in this direction have not diminished. On the contrary, it can be easily proven that the worlds need for more wind energy is greater now than it has ever been. If we were to consider what these factors are we might start with the geopolitical perspective and list the following:-

- The world is facing an energy crisis
- Demand for power is increasing
- We now live in an era of energy uncertainty

- High oil prices are "here to stay"
- The effect of wind power as geopolitical leverage
- Climate change

The European nations need to ensure energy supplies has a major effect on the way we interact with our neighbours. Fossil fuel resources have lead to military confrontations and heavily skew the relationships the EU maintains with Saudi Arabia and Russia. On the 9th of March 2007 the EU Heads of State adopted binding targets for renewable energy (20% of the total energy mix by 2020, later increased to 31%). Since then several individual countries have chosen to publicly announce even more challenging targets as a way of maintaining their leading position in the new energy technologies; for example the Danish national target is now to achieve 50% of all power from renewable energy by 2020 and 100% by 2050. Within the next few decades it is only wind (among the many renewable sources available) that is mature enough and capable of contributing significantly to reaching these targets so quickly.

Of course political will on its own is not enough. Fortunately the technological developments in wind power generation encourage those in the Industry to agree that the targets set by politicians are challenging, but achievable. This technological perspective could be illustrated by pointing out that a modern turbine in 2010 produces 180 times more electricity than its counterpart could in 1990, and at half the cost. This improvement is expected to continue alongside corresponding improvements in wind farm control and power distribution networks.

Added to this are the many advantages that a "natural" energy source with no fuel (and therefore no waste products) has to offer society; Wind Energy is clean, free, indigenous and inexhaustible. Furthermore the globe is wealthy in wind resource. Wind power involves no issues regarding fuel, geopolitical risk, external energy dependence, energy imports, fuel costs, fuel price risk, exploration, extraction, refining, pipelines, resource constraints, air pollution, CO_2 emissions, or radioactive waste.

These geopolitical, technological, and societal perspectives support an industry that has quickly become a proven technology area and is now poised at a period of massive potential growth. Across Europe and the World, huge wind farm developments have been planned. The towers, turbines and blades themselves are all becoming larger (to catch more wind) as modern materials permit more ambitious designs to become reality. The locations of these new farms are increasingly remote (offshore), and this in turn promotes the strong interest in SHM technology within the industry.

The desire to place wind farms offshore is due to the better use of space (Europe in particular is a crowded continent) and the improved wind regime.

In general moving 10km offshore improves the wind resource by 70%, and a doubling of the average wind speed at a particular location gives a potential eight times higher energy output. However against these advantages is the fact that the offshore environment is incredibly tough and difficult (expensive) to visit.

Trends in the Wind Energy Industry

- an increase in installed capacity world-wide
- an increase in the physical size of the structures
- new materials and designs
- an increase in the size of proposed wind farms
- a tendency to place wind farms offshore
- higher requirements of reliability and easier maintenance

The operation and maintenance of wind farms today is presenting an increasing challenge due to the trends summarised in the list above. It is a set of practical problems that have increased the relative cost of maintenance within the industry, problems encountered when employing traditional scheduled-based inspection maintenance of turbines placed inconveniently on many tall towers located in remote/inaccessible areas. A growing problem particularly exists with the maintenance of offshore wind turbines and the operation of a centralised maintenance group covering wind farms located in several countries.

The solution to this problem involves the use of remote sensing to reliably report equipment faults and structural health information. This has the potential to hugely improve maintenance performance under these difficult conditions and so affect the profitability of the farm. Sensor and communication technology have advanced rapidly over the last decade allowing this ambition to become a realistic goal for the offshore wind farm operator.

Wind energy can be an important industry for establishing Structural Health Monitoring as an applied industry. The industry is growing rapidly and is under pressure to adopt remote sensing. The prospect for the extensive use of sensor technology is apparent. And a further important factor in identifying the wind energy sector as the best prospect for adoption of monitoring technology lies in the pioneering culture that exists within this young industry, in contrast to that which exists in more traditional areas of structural engineering. All stages in the rapid development of the wind turbine structure design have required new approaches to solve challenging engineering problems. This is a sector that encourages the rapid integration of new ideas and approaches. A far more conservative point of view exists among civil sector engineers due to the greater emphasis on minimising risks, generally tighter constraints on budgets, and a huge reserve of good engineering practice and already established material and design information.

2 Condition Monitoring of Wind Turbine Blades

The design and manufacture process produces two items, a structure and a set of general instructions for how that structure is to be maintained and repaired during the prescribed service lifetime. These generic maintenance guidelines are then applied to each structure. The frequency of inspections is determined by the guidelines provided by the manufacturer, the occurrence of special loading conditions (such as severe storms for instance), and the requirements of insurance or certification bodies. Not only the blades but also the maintenance of all the other structural components in an industrial wind energy installation is organised along similar principles.

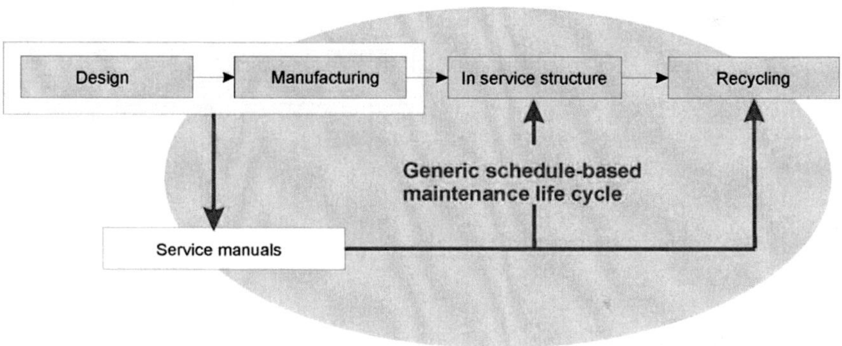

Figure 4. A traditional lifecycle paradigm

Also illustrated is a structure lifecycle controlled by a SHM system and we can compare this with the traditional paradigm. The most obvious difference is the presence of information feedback loops at every stage in the structure lifecycle. This structure response data, which is not available in traditional systems, is the means by which optimisation at each stage is made possible. A further key difference is that as all structures are sensorised, the data obtained is specific for each. This allows decision making based on the particular status of the structure, taking into account the manufacture, service life, repair history and so on.

Through the use of a successful SHM system it becomes possible to closely control the lifecycle of each individual structure. The user interface presents all the information necessary to make qualified decisions. The wind

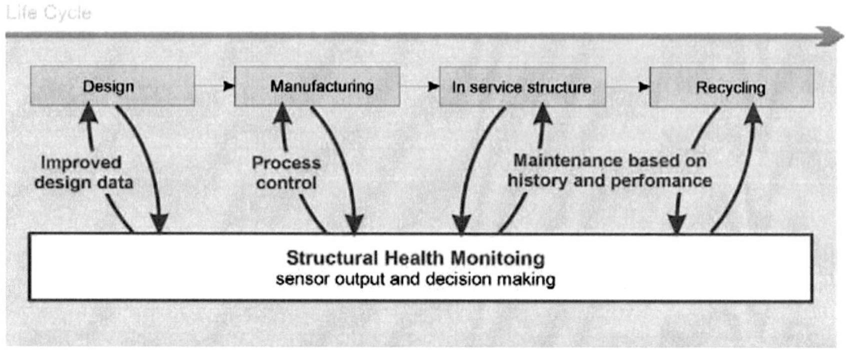

Figure 5. An SHM based lifecycle paradigm

energy industry has been identified as one of the best prospects for adoption of SHM principles that currently exists.

A Structural Health Monitoring System should:-

- remotely identify the presence of damage within a structure
- locate the damage within the structure
- assess the "severity" of the defect
- issue a prognosis of the damage development

In the list above Structural Health Monitoring (SHM) is defined as the ability to remotely (i.e. without direct inspection) identify the presence of damage within a structure. Furthermore, the application of SHM assumes a desire to further locate the damage within the structure, assess the "severity" of the defect, and issue a prognosis of the damage development in a way that allows a recommendation of action.

People have been building structures for thousands of years, and we can see that SHM, as the term is defined, has only become technologically feasible fairly recently due to remarkable advances in sensor development, data handling, networking, communication systems, computing power and by establishing a deeper understanding of the mechanisms involved in material degradation and failure. The adoption of a SHM approach to physical asset management however, is not dependent on an innovation within any single one of these technological disciplines, rather it is dependent on an exploitation of a set combination of them all with respect to a particular monitoring challenge. These technologies could therefore be said to represent the platforms upon which SHM systems can be created.

SHM is massively multi-disciplinary and working to develop this technology as an industrial application it can seem to be a very thin thread with

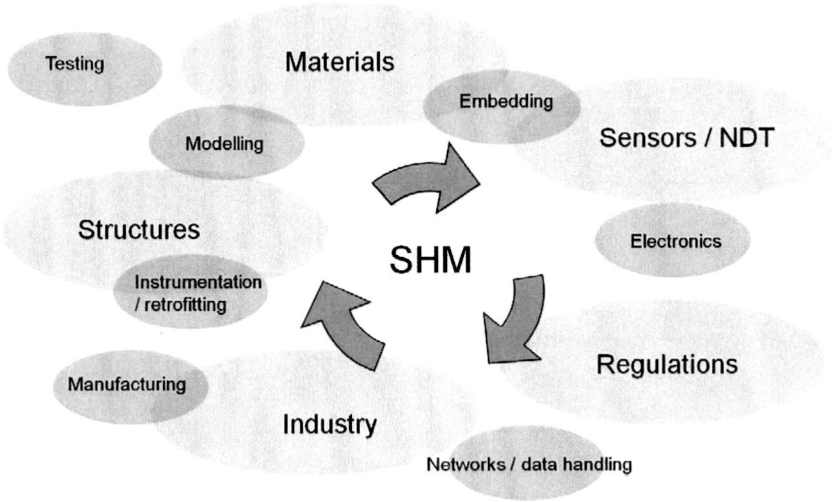

Figure 6. Technological platforms upon which an SHMS can be built

which to bind such large and independent research areas together. But in certain situations it becomes possible to position SHM as a central part of the strategies for these different groups. Where there is a need to demonstrate a high degree of integration and co-operation across disciplines then SHM can act as a kind of grand objective that leads to coherent research and allows a combination of resources to compete for new funding and new facilities.

When considering the specific issue of applying some form of SHM on wind turbine blades we need to consider the problem in relation to the current operation and maintenance procedures and ask what problems exist with traditional techniques?

Inspecting the condition of an in-situ wind turbine blade involves a considerable marshalling of resources. The timing of the work is critical and depends on favourable weather conditions. For this reason the inspection crew is often "on-standby" for when a suitable opportunity might present itself. When this occurs the turbine is place out of service and the blades are locked in the Y orientation, which presents the lowest arm for inspection. The most robust technique is visual inspection. Even with a highly skilled inspector, it is clear that this is a time consuming and inexact process, checking for damage/defects and cracks in the blade structure through the surface gel coat layer. Such inspections are considered worthwhile however as although the prospect of a catastrophic failure in a blade is remote, if it occurs it will certainly affect the remaining two blades, most likely the turbine machinery, and possibly even the entire tower structure. For this reason the default response to any detected defect is either to repair or replace the blade.

The inspection/maintenance procedure that is established for offshore wind farms will be crucial in determining the profitability of the industry. If the profitability is low then exploitation of the offshore resource will be compromised, and the national objectives with respect to adopting renewable energy cannot be made.

However there is encouragement to be found in the fact that the capability requirement for a practical SHMS for turbine blades in the wind energy industry is considerably lower than one that would be accepted for an aerospace application, for example. An aerospace application is always likely to demand something towards the very limit of what sensor and signal analysis is capable of providing in laboratory conditions with a millimetre precise mapping of laminate defects in the structural composite. Whereas a more technologically ready system, using commercial off the shelf components, that only provides a coarse set of information about structural condition is still likely to be commercially viable for wind turbine blades.

So let us consider the general idea behind how a modern offshore wind farm should be managed. The management system must integrate information from many different sources; an example of how this can be done is illustrated in Figure 7 where each component or system in the turbine can extract the relevant data from embedded sensor systems and/or other data sources and deliver this collated information to database and to a management system. The management program organises the many information streams from throughout the wind farm and presents daily reports, updates, alarms, etc. to the various users of the system. In such a modular system all that the Structural Health Monitoring System (for the turbine blade) needs to do is to provide output that fits into the pre-existing industrial communications framework. There is currently great activity inside the Wind Energy industry to ensure data tools conform to IEC standards in such a way those companies wanting to develop sensor inputs for wind farm operators are able to manufacture hardware that is instantly compatible in any wind farm.

The amount of information (data) generated by the monitoring systems in a wind turbine has always been increasing over time, and this is expected to continue. The graph in Figure 8 shows how as the amount of data increases, the options for control and optimisation of the turbine function also increases. This graph shows the measurement data generated for a

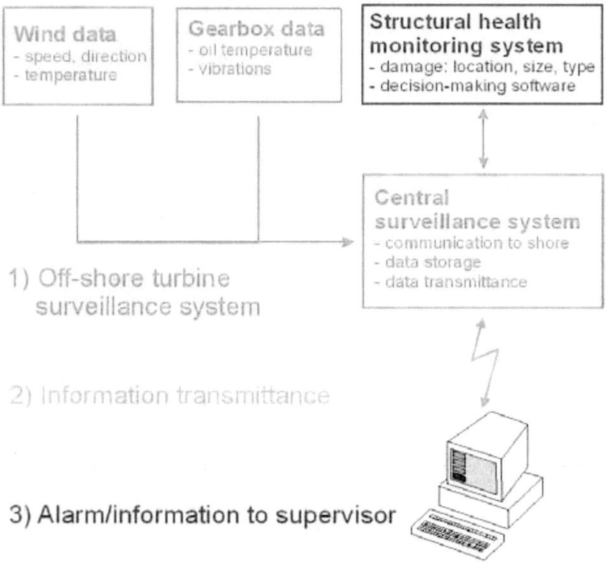

Figure 7. SHMS information embedded in a general wind farm surveillance system

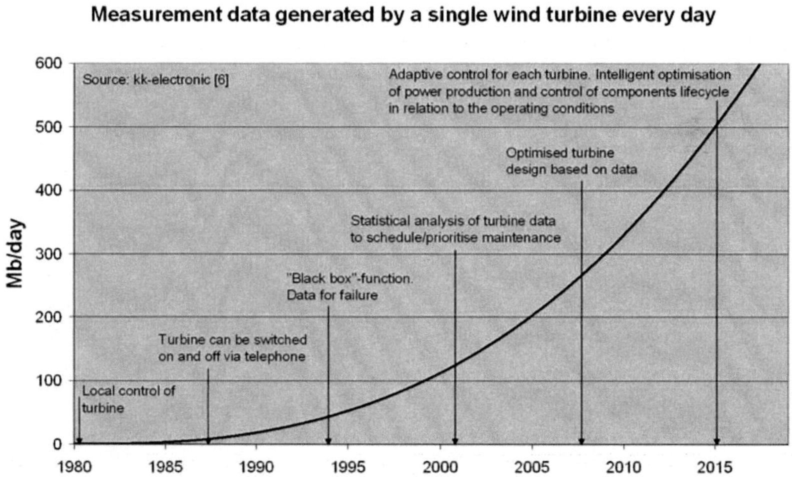

Figure 8. Graph showing increase in data generated by wind turbine - source KK-Electronics

single turbine, currently about 350MB per day. Projecting the trend only a few years into the future we see the electronic suppliers expect over 500MB of data per day (per turbine) and the possibility for "adaptive control for each turbine and intelligent optimisation of power production and control of components lifecycle in relation to the operating condition."

Most of the data represented in the curve shown in Figure 8 comes from wind and temperature measurements, pressure readings for the generator, gears and electronics, status checks on the brakes, relays and fire alarms and so on. Added to this will be some vibration and strain readings from the drive train components and tower structure. The physical network for gathering and controlling all these measurements and combining them into information that is relayed to a central control system is therefore already in place. And the development of a SHMS for the blade structure can then exploit this already present infrastructure to organise and transmit the data/information generated by the sensors in the blade. This is not something that can be expected for SHM implementations in all applications. It should also be noted that these "parametric" data streams are invaluable when interpreting the meaning of the sensors with respect to the operating conditions of the structure.

Later in this text the challenge of using sensors to understand the damage processes taking place in the polymer composite materials and turbine blade structures is discussed in more detail. As well as using specimen and component testing under laboratory conditions, the sensor data is also interpreted from measurements taken in full-scale structure tests, and finally from operating turbine blades. As the application of sensor systems moves from the small scale laboratory investigations (material properties), to full scale controlled loading of the entire structure (structure response), and then to in-situ measurement, the capability requirement of the system also changes. The practical application of sensors in a robust working environment that can be achieved at a financial cost low enough to be replicated across a fleet of similar structures is a significantly different challenge to the testing of material specimens and structural sub-components under laboratory conditions using the best available measurement and inspection technologies.

One of the conclusions of the research work addressing the challenge of moving research on the effect of defects/damage in materials and structures into an application technology has been that in the case of wind turbine blades the material damage processes and the structural geometry and loading conditions are so complex and challenging that it is not likely that a single sensor type will be sufficient to give a complete understanding of the "health" of the entire structure. In order to have confidence about the remote condition assessment from a wind turbine blade it will be necessary to create a synergy by combining different sensor/measurement technologies in one implementation.

Sensor fusion is the term used to imply a combination of disparate data sources (rather than direct fusion of information from several similar SHM sensors for example) to create a level of information that is somehow "better" (more accurate, more reliable, more complete, synergistically developing an emergent view) than any of the sources considered alone could generate. These other sources of data could include other types of sensors mounted on the blade (or elsewhere) as well as parametric data (including environmental and operational data), historical data, a priori information (e.g. about structural/material characteristics), and soft/virtual sensors (from software algorithms).

Considering only the data from sensors mounted on the blade we can characterise different types based on their structural range as follows:-

POINT sensors - This type includes strain gauges returning information about the material response at a precise point on the structure, or crack detectors in the leading or trailing edges of the blade that indicate the presence (or absence) of a crack.

LOCAL sensors - This type includes a passive Acoustic Emission sensor that returns information about the level of stress wave activity in a volume of the blade structure, a distributed strain fibre optic that returns strain measurements along the length of a fibre, or an active acoustoultrasonic instrumentation that interrogates the condition of a "hot-spot" and give warning of new structural discontinuities.

GLOBAL sensors - This type includes system identification via information from accelerometers and rotation sensors determining modal properties of the entire structure and variations due to the presence of damage.

The fact that there are many different damage types to be characterised and many sensor types under investigation suggests that no single sensor can be utilised to deliver all the information about potential damage of interest that can occur throughout the structure. A combination of sensor inputs is required, and this input can be optimised based on the best available understanding of the structure responses and damage types of interest. The sensor combination chosen is therefore based on the capabilities of the different techniques, "fusing" the best characteristics of each technique in such a way that they complement each other and their limitations are compensated for.

Note that a definitive matrix of sensor technique advantages/disadvantages with respect to polymer composite structures does not yet exist so predicting the ideal sensor fusion for this application is not straightforward.

So, an array of sensors is available within a linked communication network. Each sensor is unique and supplies local data. In a monitoring system for a large offshore wind farm there will be thousands of unique sensors available in the array. The raw data from these sensors can be processed by three modules to give information to Management, Maintenance and Design functions concerned with the structure operation. This information is interpreted within an expert system to aid decision making. The interpretation of this data requires correlation between the sensor data within the frame of an expert system.

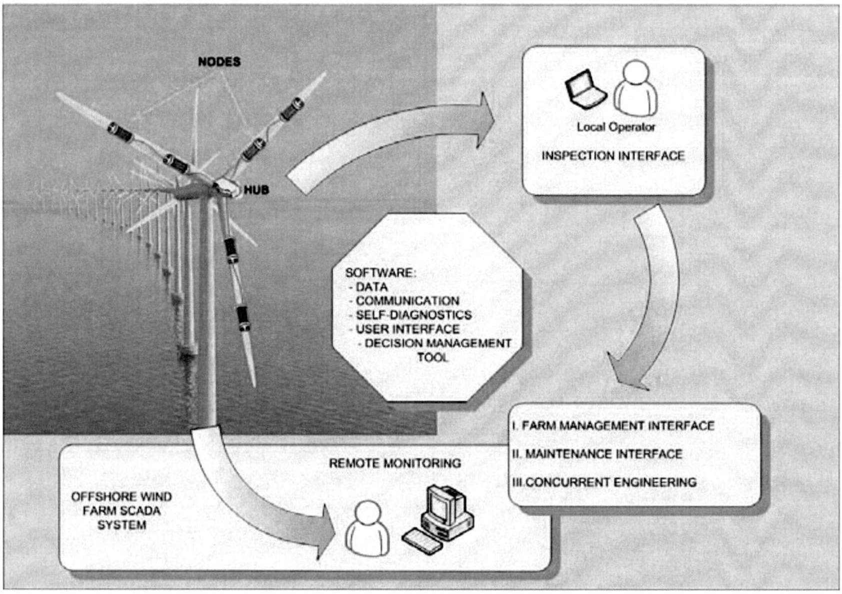

Figure 9. Sensor data extracted (locally and remotely) to three functional modules

The same sensor array is available to each module, but each module uses the data to generate different information. This information is utilised (possibly within expert systems / decision making systems) by the Management, Maintenance and Design functions.

The first module generates information on the operating parameters of the wind farm. This information will be utilised within management tools seeking to optimise output from the farm and from individual turbines. The second module generates structural health information for key plant in the wind farm. This information is primarily available in order to improve the maintenance/repair functions associated with the structure. The third module permits a concurrent engineering approach by generating information on structural response for comparison with the responses predicted by design. This information is primarily available to designers of components within the structure in order to continually improve their designs.

The benefits of such an organisation includes the fact that a huge amount of available "data" can be presented in a user relevant "information" interface and that several independently running management systems can be integrated into one. As the specific device manager software is open source, newly developed sensor types and/or analysis techniques can be quickly implemented. The data communication system is open and on-line at all times, therefore all structural/environmental changes and management/maintenance updates are visible in real-time. The remote interface with system output and controls (remote warnings and confirmations) is possible via all types of media. The in-service response data improves subsequent production run or design. Maintenance scheduling is improved by early fault recognition, localisation and diagnosis. And an improved management feedback and control allow profit optimisation.

The trends in the wind energy industry make it increasingly attractive to implement full condition monitoring of the blade in operation as the consequences of a failure is increasing for large turbines placed offshore and there is a need to integrate condition monitoring of the blades into the overall monitoring standards. Although many monitoring systems are now being evaluated, there are still challenges to overcome before a system is ready and these include the integration of different systems, the issue of system robustness/reliability, and avoidance of false alarm indications.

The concept of damage control in wind turbine blade structures that are monitored needs to be addressed now as the ability to remotely detect the occurrence of damage assumes a strategy for assessing the severity of detected changes and acting accordingly. Condition monitoring systems for operating wind turbine blades could be expected to provide the following capability:-

1. identify that damage evolution takes place
2. identify the location of damage
3. identify the damage type
4. identify the size of damage

In this way the need for manual inspection for operating turbines can be reduced, and the maintenance resources available can be more effectively deployed to respond to specific identified problems in the structure condition.

Various sensor types can be imagined that could contribute to such a remote damage detection/assessment system including point sensors (e.g. strain gauges), area or local sensors (e.g. acoustic emission sensors), and global sensors (e.g. accelerometers).

The data combination provided by the sensor network allows the remote surveillance system to provide an alert that this particular structure requires attention. This attention could include an on-site inspection that would give the full diagnostic result for that structure. And by combining the full structural information thus achieved with models of material/structural behaviour determined by previous experimentation, the final prognostic step can be taken that provides a suitable response by the operator. Illustrated in Figure 10, such a concept presupposes that the damage state can be assessed at a stage where the correct response will prolong safe structural life and/or prevent structural failure.

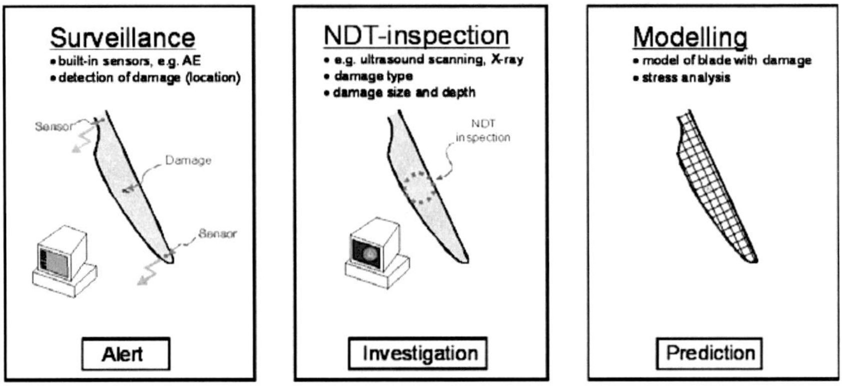

Figure 10. The Alert, Investigation, and Prediction stages in this damage control concept

Figure 11 shows the anticipated response of the structure under normal operation against the projected response (and premature failure) of a structure containing an unrecognised damage. Note that the y-axis places the sensor detection limit earlier than the visual, and of course earlier than the critical, damage state.

Figure 12 projects the effect of two possible responses to a detected damage, either a full repair to return the structure to the designed/expected damage evolution, or a change in the aeroload experienced by this particular structure in order to achieve the expected service life with no further intervention necessary.

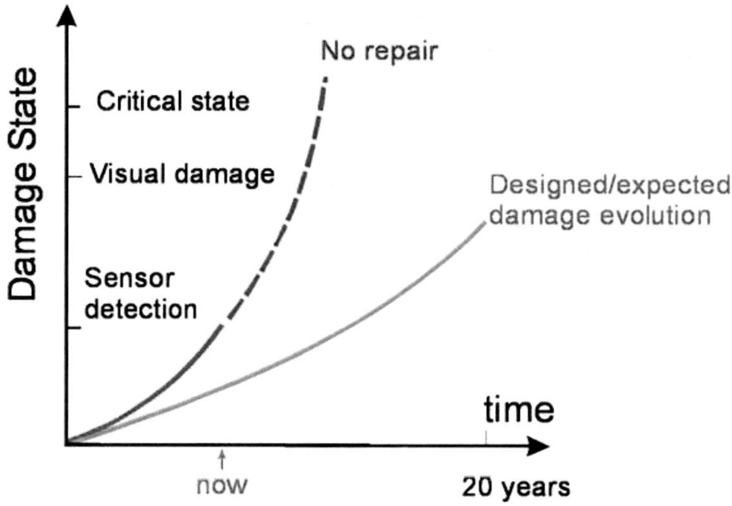

Figure 11. Damage state development for a healthy and a damaged structure

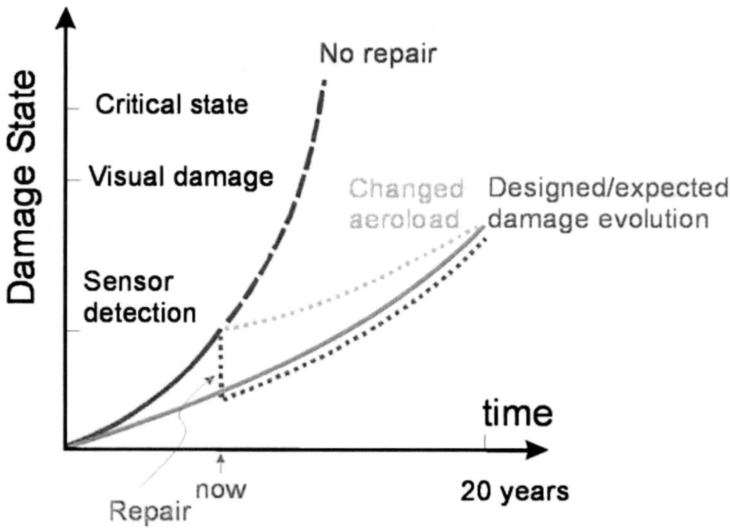

Figure 12. The effect of repair and of changed operation in achieving a full life cycle for the damaged structure

The possibility of making such decisions in the future is dependent on two things. Firstly the demonstration of sensors that can remotely indicate the presence and development of critical damage types, and secondly models that integrate the local material damage evolution within the global structural response.

Once present these two techniques allow knowledge-based decision making such as:-

1. The damage is stable therefore continue use of blade
2. The damage growth is slow therefore change the aerodynamical loads
3. The damage growth is fast therefore repair the blade
4. The damage is large (un-repairable) therefore replace the blade

This concept of turbine blade damage control for large remote wind farms requires a suitable interface relevant for the "non-technical" operators aiming to maximise the productivity from their assets. A relevant example could be the sensors and diagnostic processes in a modern car. Here also we find a network of sensors based on advanced technologies that are "hidden" behind a simple user interface. The underlying knowledge that controls the sensor network is not relevant to the user who requires only a clear indication of the overall damage state (or condition). Warning indications result in expert intervention where the system can provide a more detailed level of information relevant to the particular repair requirement.

3 Damage in Polymer Composite Materials

The wind turbine blades being produced have become larger over the years and this creates several technical challenges.

Figure 13 shows a scatter plot of the lengths of various wind turbine blades against their weight. As the length of the blade design is increased it can be seen that the weight of the blade increases more sharply. The scaling factor for aerodynamical forces is L^2, while the scaling factor for gravitational forces is theoretically L^3 (in practise $L^{2.25}$). The fact is proven that with larger lengths of wind turbine blades, (saving weight) becomes increasingly important in order to break the trend seen in the figure and achieve blade lengths of 80m, 100m, or more.

Another consequence of the increasing size of wind turbine blades is found in the manufacturing reproducibility and cost of fabrication. Blades are assembled from relatively few, but increasingly large, parts glued together. A large blade part has a high cost value it is therefore unattractive to be forced to discard such a part if it is sub-standard in any way. For this reason quality control becomes more important as blade size increases.

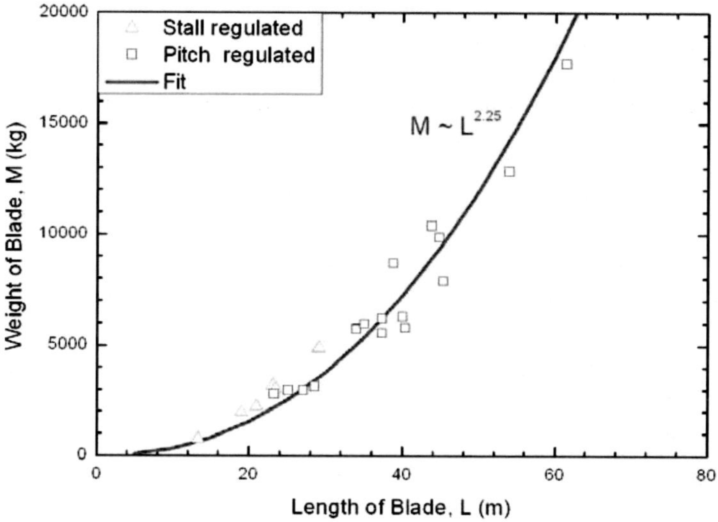

Figure 13. Blade length vs weight (LM Glasfiber A/S)

As the largest wind turbines are most likely to be placed offshore, the access to these structures is difficult and costly. Two further technical challenges are therefore to reduce the requirement for on-site maintenance, and develop reliable modelling tools for damage assessment and repair.

Looking at the response to these technical challenges we can see that in order to save weight we must develop better (stronger, lighter) materials and develop better structural design solutions. In order to improve quality control we must increase our understanding (and control) of the manufacturing process and the ways that different production defects can develop into damage (propagating cracks/fracture). To reduce the need for visual inspection of the composite structure we must investigate the response of various embedded SHM sensors to the initiation and growth of damage, as well as continuing to develop the correct Non-destructive evaluation hardware. And to provide correct modelling tools we need to further investigate the properties that govern the behaviour of cracks in these complex materials.

These are the considerations that underpin the research focus for the polymer composite material mechanics (AFM-KOM) research group at Risoe DTU. They are also shown in Figure 6 as the technological platforms to the top and the left hand sides of the illustration.

Within the KOM group the behaviour of fibre reinforced polymer matrix materials can be studied (and modelled) at all length scales using test facilities capable of testing single fibre specimens under microscopic analysis, mesoscale testing of all standard mechanical properties as well as specially designed test fixtures for more precise measurements of response or for studying fracture mechanics under various conditions, Sub-component testing, and full-scale blade testing.

Of particular interest is the response of the material (or structure) in the presence of various types of defect or damage. The way that damage initiates and grows can be revealed during the testing using sensors and scanners in order to correlate with the fracture mechanical models.

The laws governing the behaviour of the material is the same at all scales, stress-strain, and traction-separation.

Looking more closely at modern wind turbine blades, we see they are structurally advanced constructions utilizing sandwich design composite materials, gelcoat films and adhesive joints. Although there is a variety of wind turbine designs (reflecting different manufacturing processes and material selection), the functionality of wind turbine blades can be understood by considering the blade as a load-carrying beam enclosed by a shell. The primary purpose of the shell is to give the blade an aerodynamic shape, creating aerodynamic forces that cause the wind turbine blade to rotate and thus extracts energy from the wind to make electrical power. The aerodynamic forces are transmitted to the wind turbine hub through the load-carrying beam of the blade. Thus, the primary load-carrying part of a wind turbine blade is the load-carrying beam, sometimes called the main spar.

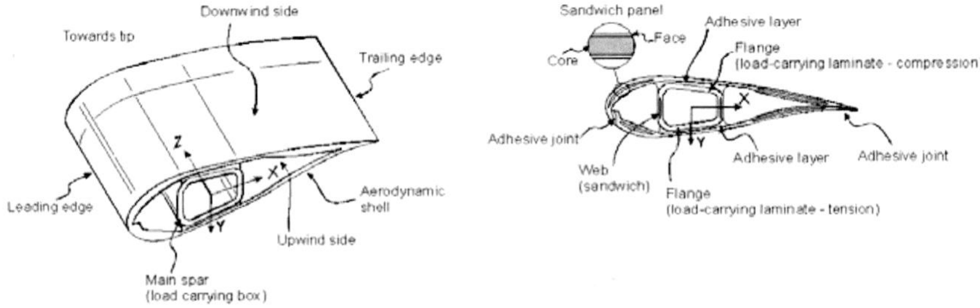

Figure 14. Cross section of a typical wind turbine blade

Looking at a typical cross section, we can conclude that wind turbine blades are made of several structural parts and can potentially fail in a number of ways. The details of damage evolution can easily differ from one blade design to another. However, the assumption that many failure modes may develop before a blade actually fractures remains plausible irrespective of different design details of different blade manufactures.

An overview of the types of damage observed in one wind turbine blade tested to failure is given by Debel (2004) and categorised as follows.

Type 1: Damage formation and growth in the adhesive layer joining skin and main spar flanges (skin/adhesive debonding and/or main spar/adhesive layer debonding)

Type 2: Damage formation and growth in the adhesive layer joining the up- and downwind skins along leading and/or trailing edges (adhesive joint failure between skins)

Type 3: Damage formation and growth at the interface between face and core in sandwich panels in skins and main spar web (sandwich panel face/core debonding)

Type 4: Internal damage formation and growth in laminates in skin and/or main spar flanges, under a tensile or compression load (delamination driven by a tensional or a buckling load)

Type 5: Splitting and fracture of separate fibres in laminates of the skin and main spar (fibre failure in tension; laminate failure in compression)

Type 6: Buckling of the skin due to damage formation and growth in the bond between skin and main spar under compressive load (skin/adhesive debonding induced by buckling, a specific type 1 case)

Type 7: Formation and growth of cracks in the gel-coat; debonding of the gel-coat from the skin (gel-coat cracking and gel-coat/skin debonding).

And illustrated by Sorensen et al. (2004) in the following sketches:-

It is clear that any attempt to detect the initiation and development of such a wide range of potential damage types along the entire length of a large wind turbine blade would require sensor instrumentation so extensive as to be scarcely feasible. In order to guide and focus the sensor installation therefore, it is necessary to work with the current research on initiation and growth of different material damage types and the structural failure modes this damage can cause. Only in this way is it possible to establish the critical damage types and the critical structural locations that will focus the sensor instrumentation effort required to monitor and assess structural health in a particular blade.

It requires a lot of understanding to assess the severity of damage in a wind turbine blade. Such an understanding does not exist at present. Never the less, it is of great importance to develop approaches that will allow an

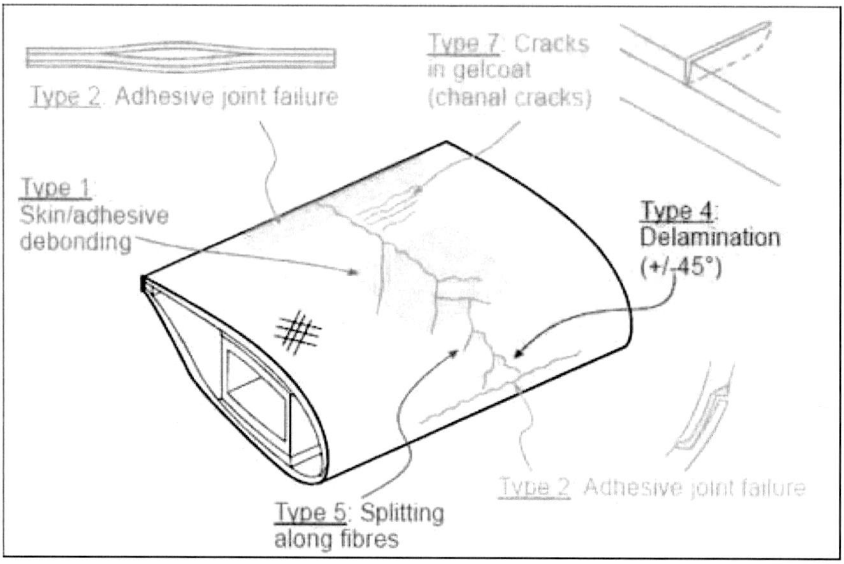

Figure 15. Damage types identified on the suction/compression face of a wind turbine blade tested to failure

inspector of a wind turbine blade to evaluate how serious any observed damage is. It is thus a long-term goal of the AFM-KOM group to establish a framework that enables reliable analyses of any set of damage given the position, size and type of that damage.

This framework can begin by taking the observed damage categories listed previously and assigning them to a general structural failure mode and a location within the structure.

Taking this approach further it is possible to consider each structural failure mode in terms of the basic material damage modes that control it. At first it may seem that this has uncovered an increase in the complexity of the problem of characterisation as the term adhesive joint failure or laminate failure both cover a range of different basic damage modes. However the table in Figure 18 shows that several structural failure modes actually belong within the same family of basic damage modes. This implies that the same concepts and criteria (once understood in the laboratory) can be used for characterising a range of different structural failure modes.

This then is the research approach taken by materials research groups like AFM-KOM at Risoe DTU. The observed damage is categorised and assigned a specific structural failure mode which is then rendered into the

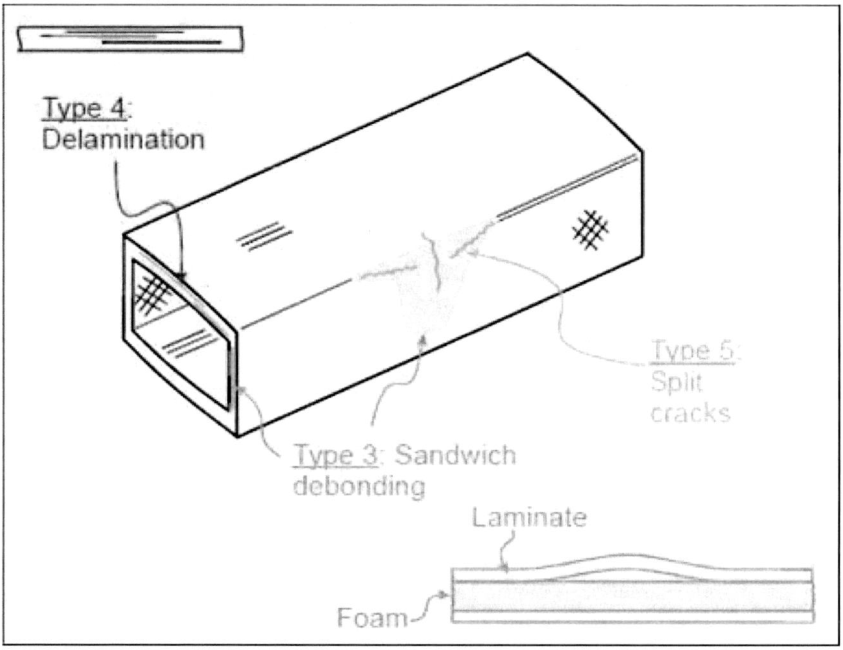

Figure 16. Damage types found in the main spar section of a wind turbine blade tested to failure

Structural failure mode	**Location in wind turbine blade**
Adhesive joint failure (type 1 and 2)	upwind/downwind skins at leading and trailing edges skin/main spar joint
Sandwich face/core delamination (type 3)	skins webs in main spar (e.g. buckling driven delamination)
Laminate failure (type 4 and 5)	skins main spar (e.g. buckling driven delamination)
Gelcoat/skin delamination (type 6)	upwind/downwind skins
Gelcoat cracking (type 7)	upwind/downwind skins

Figure 17. Table showing structural failure modes and their position in the wind turbine blade

Structural failure mode	Basic damage modes
Adhesive joint failure	crack in adhesive layer laminate/adhesive interface cracking interface cracking with fibre bridging
Sandwich face/core delamination	interface cracking interface cracking with fibre bridging
Laminate failure	tensile failure (fibre fracture mode) - damage zone compressive failure (fibre fracture mode)- damage zone tensile failure (matrix fracture mode) - cracking shear failure splitting crack (crack parallel with fibre direction) delamination crack between plies
Gelcoat/skin delamination	interface cracking interface cracking with fibre bridging
Gelcoat cracking	thin film cracking

Figure 18. Table showing the basic damage modes assigned to the observed structural failures

basic damage modes involved. It is these basic damage modes which are investigated in the test laboratories to establish the material properties that control the initiation and growth of these modes and provide the micro and meso scale models that determine their behaviour within a structural context.

So let us now consider the critical damage modes. Fibre reinforced polymer composites are damage tolerant. This implies that a structure manufactured with these materials will not fail as the result of the growth of a single crack (such as is the case with brittle materials like glass and ceramics). Rather, structural failure in a polymer composite wind turbine blade involves the progression of several damage mechanisms with the global collapse occurring when a damage mode at a particular structural location has reached a critical size (at which point the fracture growth becomes unstable). Considering this damage tolerance property and the size of wind turbine blades it is clear that research into the damage modes of interest for these structures focuses on understanding the mechanisms by which different damage modes can progress (crack propagation - is it stable, unstable, or arrested/stopped?) rather than the micro-structural details that describe how/why the damage formed in the first place (crack initiation - from existing flaws/defects).

A wind turbine blade under operation is subjected to varying aerodynamic loads due to the fluctuation in the local wind speed, plus gravitational and aerodynamic effects resulting from the turbine rotation. When characterising the critical damage modes for structural material used in the blades it is necessary to consider two types of loadings: Very high static loading ("an extreme wind load") and various types of cyclic loads. The wind turbine blade must be able to survive a few extreme loads (however the number of extreme wind load events is small), but also a huge number of cyclic loads (where the load level is relative low however).

Therefore, researchers consider two cases for damage evolution: static growth and fatigue (dynamic) damage. The material properties which describe damage evolution under static loading and under fatigue loading are not the same.

Damage mode	Loading regime	Material property of interest
Damage zone	Static	Tensile strength σ^{+}_{11u}
		Compressive strength in the fibre direction σ^{-}_{11u}
		Composite shear strength τ_{12u}
	Fatigue	Maximum applied stress σmax
		Number of cycles to failure *Nf*
Crack propagation	Static	Critical mode I stress intensity factor *KIc*
		Critical mode I energy release rate G*Ic*
	Fatigue	$\frac{da}{dN} = A\left(\frac{G_{max}}{G_{Ic}}\right)^n$
Crack experiencing fibre bridging	Static	Mode mixity Gc= Gc(ψ)
	Fatigue	? Not yet entirely clear... See Meeking et al. (1990), Cox & Marshall (1991), Cox (1993), Sørensen & Jacobsen (2002), and Sørensen 2011.

Figure 19. Table showing the modes of damage in polymer fibre composite laminates and examples of the corresponding material properties governing growth criteria under static and fatigue loading regimes

<u>Static</u>

For a damage mode that develops a damage zone (where the failure process zone constitutes a considerable volume) the damage is usually described in terms of a critical stress value, i.e., by a maximum stress criterion (tensile or compressive strength).

The onset of growth of a crack (crack propagation) can be described in term of a maximum stress intensity factor (fracture toughness) or a maximum energy release rate (fracture energy).

A crack experiencing fibre bridging requires modelling of the bridging fibres. This can be done by a cohesive law (a stress-opening law). A cohesive law describes the relationship between the local crack opening and the local stress across the failure process zone. In the general case, the cohesive stresses can be normal and shear stresses; we then talk about mixed mode cohesive laws.

Fatigue

For materials that fail by a damage zone, such as composites loaded in the fibre direction, the strength properties under cyclic loading can be described by a so-called S-N curve, which is the relationship between the maximum applied stress, and the number of cycles to failure.

For materials that fail by cyclic crack growth, the crack growth rate, da/dN, should be measured as a function of the applied energy release rate.

For materials experiencing large-scale-bridging under cyclic crack growth, the situation is more complicated. The cohesive laws that operate under cyclic loading are likely to be different from those present under monotonic crack opening. Thus, the cohesive laws should be characterised as a function of the number of cycles. Precisely how this should be done is not quite clear, although a few ideas have been developed (see McMeeking and Evens (1990), Cox and Marshall (1991), Cox (1993), Sorensen and Jacobsen (2002) and Sorensen (2011)).

For a review of the materials testing guidelines, standards and techniques that can generate these material properties relevant to certifying and repairing new wind turbine blade designs see Holmes et al. (2007).

Many parameters that affect the growth of damage in a wind turbine blade are known with a high degree of accuracy (structural dimensions and the materials elastic properties for example) while others are subject to uncertainties. Among these uncertain parameters are naturally varying aerodynamic forces. But fracture properties of the materials are another uncertain element when considering damage growth, partially due to scatter in material properties, but also because the many possible modes of fracture failure are not fully characterised and modelled.

The table in Figure 19 shows that damage in composite structural materials is considered as two separate classes. The first is a distributed damage within a volume of material that can be described in terms of a critical stress value (tensile or compressive strength) and number of cycles. The second is a distinct crack present in the structure (either within the laminate structure or at an adhesive interface) that can be described in terms of a maximum stress intensity factor (fracture toughness) or a maximum energy release rate (fracture energy) governing its' propagation.

In general the first class is better understood by materials scientists and structural engineers and is a well defined consideration in the design and maintenance procedures for wind turbine blades. The second however is a topic of current research aiming to describe and model the growth patterns of various types of cracking possible in complex laminate composite components, and the effect of the presence of such defects in a dynamically loaded structure. For this reason such failure modes represent a much greater risk for the owner of a wind turbine blade and should be the focus of useable SHM hardware development. This in turn helps us define the laboratory testing required; to optimise our monitoring effort to detect interface failures at an early stage in key structural locations.

One very significant damage event in sandwich composite materials is skin/core debonding, where the exterior layer of laminated composite material becomes detached from the lightweight core material. Once initiated, such damage has the potential to grow quickly and compromise structural performance. The speed at which this damage grows depends on many material and structural factors such as where the damage is located in the structure, how large it is, the external loading, the material quality and so on. Clearly one of the most significant factors determining the growth rate is the energy uptake required to advance the crack front.

The interface between skin and core in a sandwich composite structure is often characterised as a "resin rich layer", where there is little or no reinforcement of the polymer resin matrix by the glass or carbon fibres. Consequently, if damage is initiated between the skin and core, the energy uptake required to propagate the damage through this layer is low. This is known as debonding of the skin from the core.

There is a laminated structure within the skin material formed by the layers of fabric reinforcement held in place by the polymer resin matrix. Damage can also form between two layers of the composite material and this is known as delamination. By comparison with skin/core debonding however, there is likely to be far more fibre reinforcement between two laminate layers in the skin material itself than there is between the skin and core. This fibre bridging between skin laminate layers is the reason that the energy required to propagate a delamination is generally far greater than the energy required to propagate a debond.

There is a practical challenge regarding the inspection and maintenance

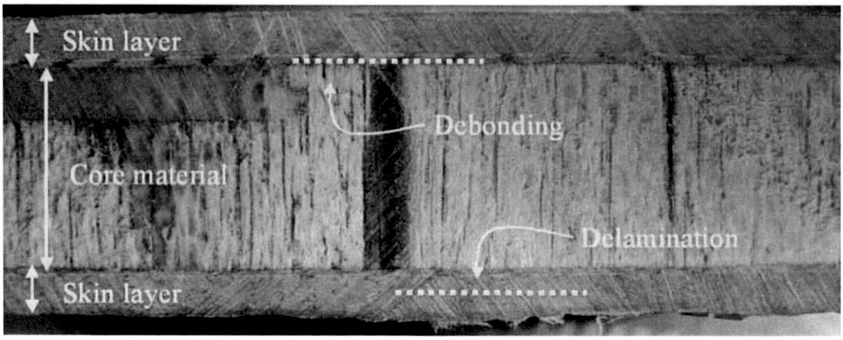

Figure 20. Through thickness profile of a sandwich core panel (GRP skin/Balsa wood core)

of these structures. This might be difficult (or impossible) to detect during a visual inspection, but could have serious implications if the damage grows over time until the structure/panel is at risk of failing under operational loads.

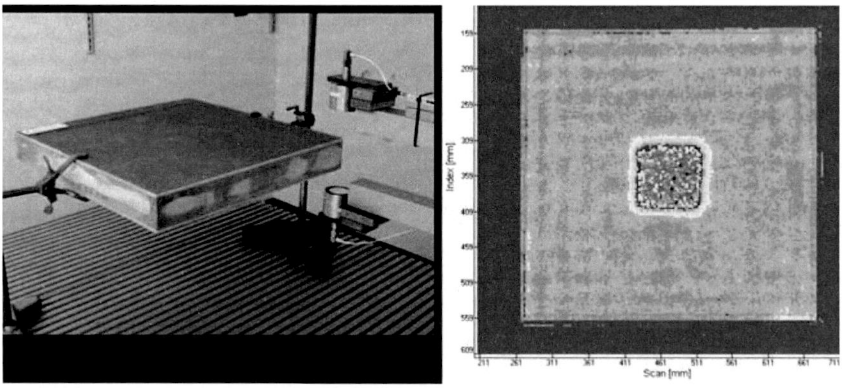

Figure 21. Inspecting a damaged sandwich core GRP panel with through transmission, air-coupled Ultrasonics

Damage of this kind (both delamination and debonding) has traditionally been detected during routine inspection using non-destructive techniques such as pulse echo ultrasonic inspection. However, such an inspection is not always able to distinguish between debonds and delaminations as both create a similar "loss of transmission" in the returned ultrasound signal and the difference in depth between a possible debond or delamination can be a matter of millimetres. But making such a distinction is not always a priority however, as a common philosophy for maintenance of sandwich FRP structures is that all detected damage should be repaired.

Over time, there has been an adoption of damage tolerance approaches to the maintenance of large GRP structures, in order to extend service life and to avoid invasive repair procedures. A successful damage tolerance assessment requires detailed information, including the energy uptake of the damage. These two paths of crack advance (debond and delamination) have significantly different fracture resistances. Obviously this difference has a critical influence on any damage tolerance assessment based on fracture mechanics, therefore, distinguishing between a debond and a delamination becomes necessary.

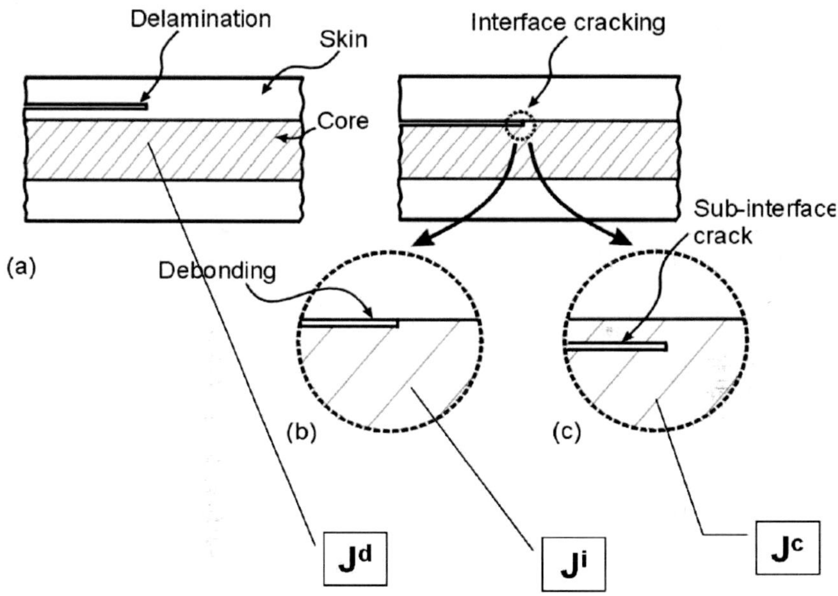

Figure 22. Crack propagation in sandwich GRP. J^d (delamination) J^i (interface/debond) and J^c (core)

The high stiffness to weight ratio of fibre reinforced polymer sandwich panels, and the ease with which this stiffness can be predicted, has contributed to their popularity. A more complex area that designers need to deal with however, is the relationship between the growth of damage/defects in the skin (or skin/core interface) and the presence of shear stresses in the panel.

Cracking in FRP sandwich structures occurs along weak planes. The criterion for crack growth is that the energy release rate reaches a critical value, the fracture energy of the crack front. The cracking occurring in these materials is mixed mode, meaning there is both normal and shear stresses present at the crack tip. The fracture resistance is affected by the degree of mode mixity.

As illustrated in Figure 22, crack propagation can occur in three different places; As delamination, as an interface crack, and in the core material. It is important to distinguish between them because the resistance to fracture can be very different for the three cracking planes.

In some systems, crack growth in the laminate and at the interface is associated with crack bridging. This is fibres or fibre bundles bridging the two crack faces. Crack bridging results in substantial increases in fracture Energy and can range from J_0^d (no fibre bridging) to J_{ss}^d (fully developed bridging zone). These are considered material properties in fracture mechanics calculations.

So it follows that the "severity" of a crack in a sandwich structure does not only depend on the size of the crack, but also on factors like position (both in the structure and through the thickness of the sandwich), the level of mode mixity at the crack tip, and the degree of fibre bridging associated with the crack front.

In order to determine the cohesive laws that govern the propagation of cracks in materials it is necessary to obtain data from stable crack growth experiments. Most test methodologies involving shear loading at the crack tip give unstable crack growth, however the DCB-UBM (Dual Cantilever Beam - Uneven Bending Moment) specimen configuration used at Risoe DTU allows stable crack growth in any mixity from Mode I (simple opening) to almost pure Mode II (in-plane shear).

The use of a wire and roller arrangement ensures that identical forces exist in the loading wire, but that the moment can be controlled by changing the distances in the loading arms. A support allows the specimen to move up and down in the test machine, but not to rotate. This ensures no axial forces as the crack propagates.

By controlling the mode mixity at a crack front in a sandwich core test specimen it is possible to control the J integral for the damage growth and even the amount of fibre bridging behind the crack tip. This allows the researchers to generate experimental data that can be used in validating models describing crack growth under different conditions, and direct measurement of specific crack characteristics with monitoring sensors.

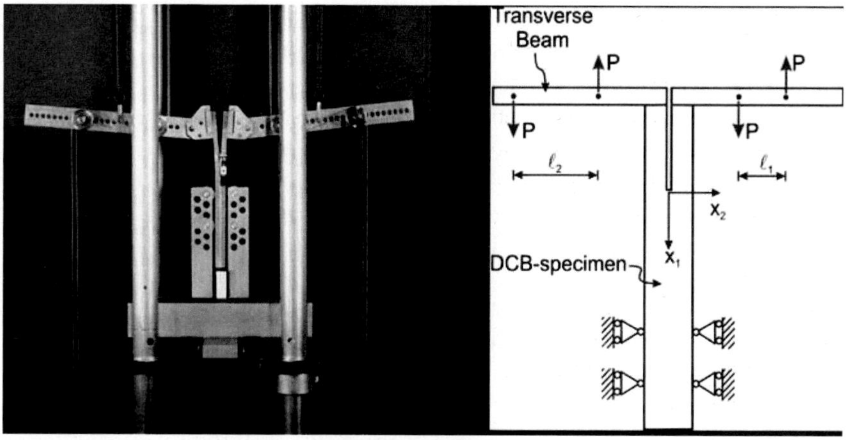

Figure 23. The DCB-UBM specimen configuration – Even bending moment

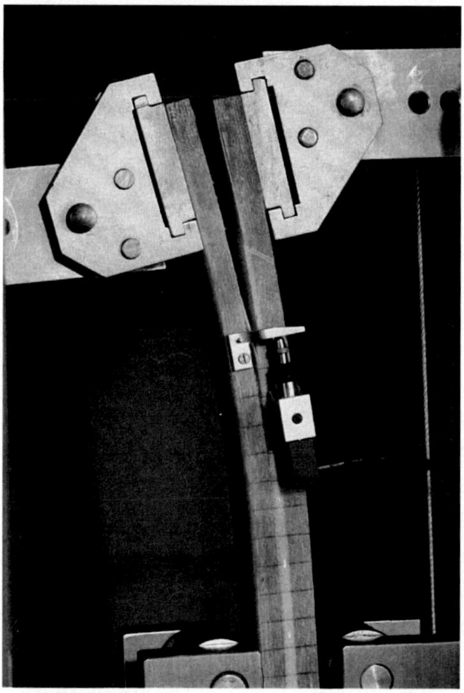

Figure 24. The DCB-UBM applying a mixed mode (uneven) bending moment

One of the techniques used to investigate damage in fibre reinforced polymer composites is AE (Acoustic Emission). In Laboratory testing AE can be used to mark the initiation of damage, localise and characterise the crack growth activity, and identify the failure mechanisms. In large scale structure testing AE is used to provide feedback during loading, focus inspection at "active" areas of the structure, and ultimately provide a more efficient structure test. While for in-situ measurements AE can provide an "exception analysis" that identifies problem structures and locations at an early stage.

In the crack growth tests conducted with the DCB-UBM machine a two sensor AE linear array localises the crack growth and the extent of the crack bridging zone behind the crack tip.

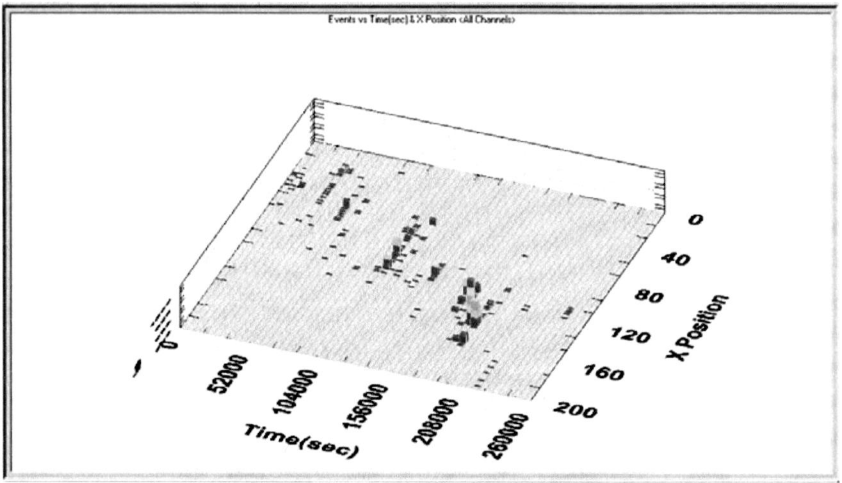

Figure 25. AE localisation showing the crack advance over time and extent of cohesive zone behind the tip

A test series was conducted to assess the extent of fibre bridging present behind the crack tip during loading using only AE output. Essentially this would be a way of estimating the J value in real time. Identical DCB specimens (GRP skins, polymeric foam core) were tested, and by using the DCM-UBM test machine it was possible to grow a crack along the skin/core interface; J^i. But by careful alteration of the mode mixity it was possible to vary between the J_0^i (no fibre bridging) and the J_{ss}^i (full scale fibre bridging) values of the crack front. In this way it was possible to vary the amount of fibre bridging present and assess its significance for the damage propagation.

H1309					
AE Hits	36102				
Cum counts	859580	Average NRG/hit	17.08		
Cum NRG	616714	Average counts/hit	23.8	**Average NRG/count**	**0.718**
Time (s)	415				
H1310					
AE Hits	23853				
Cum counts	530724	Average NRG/hit	22.20		
Cum NRG	529508	Average counts/hit	22.25	**Average NRG/count**	**0.998**
Time (s)	310				
H1311					
AE Hits	1324				
Cum counts	40589	Average NRG/hit	40.09		
Cum NRG	53083	Average counts/hit	30.66	**Average NRG/count**	**1.308**
Time (s)	180				

Figure 26. Table showing the AE data for different degrees of fibre bridging behind the crack tip

The AE data recorded in the three test specimens varied due to the different energy uptake values for the crack growth, both in the general activity during the test and also in the characteristics of the recorded waveforms. Differences in the general AE activity recorded during the tests reflected the type of crack growth taking place in each specimen. A huge number of low intensity hits were recorded for H1309 (Fully developed bridging zone, micro cracking) and the crack growth was controlled and grew only very slowly. By comparison, H1311 (crack in the resin rich layer, no fibre bridging, macro-cracking) had far fewer AE hits but each individual hit contained far more energy. The crack growth here was less stable (due to resin plasticity) and progressed entirely during two sudden "jumps" forward, the AE trace reflected this. H1310 was an intermediate case with an intermediate number of hits, energy per hit and length of test/crack growth. And although the AE trace (and crack growth) during the test was generally stable, it could be seen from the AE that the crack growth was in fact via a series of "mini-jumps".

A key observation is that simply observing total AE activity or cumulated AE energy release is not sufficient to determine the "severity" of a defect in a sandwich GRP material. The crack growth involving fibre bridging (H1309) generates a huge number of AE events, and far more energy is released overall than in the other specimens. However this does not correlate with the length of crack growth in this specimen. More significant is the amount of energy released per event detected, and the energy released per count (a function of the event duration).

It was proposed that by calculating a real time figure for average energy per count during the course of the test, it would be possible to estimate the relative amount of fibre bridging in any similar specimen at any time during the crack growth; lower ratios for average energy per count indicating a greater degree of fibre bridging. In a DCB test where the crack type changes (from skin/core debonding to skin delamination or vice versa) due to a change in applied moment, we suggested we could immediately detect this event using only AE sensor data.

This kind of monitoring could be used (initially) to automatically control the loading arms on the DCB-UBM test machine in order to maintain a certain crack pattern in a specimen, or perhaps in the future to adjust a structure loading pattern in order to prevent damage.

4 Structural Testing of Wind Turbine Blades

Wind turbines in Denmark must be certified in accordance with the Danish approval system at an accredited test centre. The rotor blades are the most critical parts of a wind turbine. Failure of a rotor blade in service often involves damage of the entire turbine. The economical consequences of a rotor failure can be devastating to a wind turbine project, wherever situated on the globe. Although the blade design and loading calculations are simulated thoroughly on computers, there is still a strong need to verify these calculations by full-scale blade tests. In addition, vital information concerning manufacturing issues can be brought to light by these tests.

The static tests specified for accreditation involve a determination of various physical properties of the blade (weight, centre of gravity, natural frequencies), a flapwise proof test, an edgewise proof test, a torsional proof test and a visual inspection. The fatigue tests involve instrumentation by many surface strain gauges followed by a static load calibration check. Then a fatigue loading (usually produced by an eccentric mass exciter) for a pre-determined number of cycles. Every 1,000,000 cycles the static load calibration check must be performed again. An optional post-fatigue static test to determine residual strength can also be included as a final step. These tests enable certifying bodies to compare measured data to the calculated data of the design criteria. The entire test process is reported using a standard format and the tested design is accorded a pass or fail based on these specified measurements.

Static blade tests are performed in order to determine the structural properties of a blade including stiffness data and strain distribution. Static tests may be carried out as a single point load application, but more usually via multipoint load application from carefully fabricated loading clamps. The load may be applied in horizontal or vertical direction depending on the test facility.

Dynamic blade tests are performed in order to determine the fatigue properties of the blade; the duration of a test may be several months. Surveillance, inspection and calibration of the test setup will take place at regular intervals.

Resistance strain gauges applied to the blade skin and internal blade structure provide response data in static load condition and over the duration of the dynamic testing. Non-linearity of the response in strain gauges positioned along the length of the blade can indicate that buckling of the blade skin is taking place. The slope of the strain graph as load increases is also essential information for the verification of the structural blade design. And the length-wise strain distribution under static load can be used to reveal any high gradients that will reduce the fatigue life of the blade.

The condition monitoring techniques specified by current certification authorities are surface strain measurement, displacement transducers, the output of the load cells, and visual inspections. However, in order to extract more data from these (rather expensive) full-scale tests, blade manufacturers have begun to use more and more sophisticated condition monitoring techniques and to perform more demanding and complex loaded tests on their new blades. This includes non-standard testing designed to get extra information about the design and manufacturing issues for the structure and to check actual response against models that can predict the onset of non-linear behaviour.

The increase in the production rate (and scale) of wind turbine blades has generated a need for improved and faster quality inspection at manufacture and during/after structural testing. A wide range of techniques are available and the most common are summarised by Drewry and Georgiou (2007) and Amenabara et al. (2011). The simplest and most robust technique available is a visual inspection by an experienced employee. A manual inspection of this kind can include the use of miniature cameras or endoscopes, and the use of tap testing (manual and/or computer aided) to investigate hidden structural irregularities such as skin/beam disbonding and so on.

The results of tap testing can be used to justify the application of an ultrasonic inspection, or to confirm the findings of such an inspection. Ultrasonic inspection can reveal hidden defects effectively; it is a reliable and repeatable technique and the most widespread non-destructive composite inspection method in the industry. The main advantage is its ability to locate and characterise hidden defects (or damage) like delaminations and laminate dry spots. The main disadvantage is usually the issue of applying the technique quickly over the large surface areas of a modern wind turbine blade. Commercial services offering blade inspection tend therefore to focus on the use of arrays or automated robotic systems to optimise the technique applicability, for example the NDT systems used by FORCE A/S technology (www.forcetechnology.com).

Infrared thermography is also a technique in widespread use for detecting flaws in adhesive joints and research/innovation groups are currently optimising hardware for commercial applications in wind turbine blade manufacture, for example the Fraunhofer-Allianz Vision groups Thermographic Testing of Rotor Blades in Wind Turbine Generators (www.fraunhofer.de).

Commercial Shearography systems are available that are designed for use of wind turbine blades, for example Dantech Dynamics Wind turbine Blade Inspection using Shearography (www.dantecdynamics.com).

Back scatter X-ray technology is also a system being developed for this application, for example InnospeXion ApS back-scatter x-ray technology for the detection of sub-surface defects accessible from one side (www.innospexion.dk).

Many of these NDT techniques (with some practical modifications) can also be applied to blades that are undergoing proof testing in order to document the initiation and development of damage, or even applied to search for problems on blades mounted on the towers. However the disadvantage of transporting personnel and hardware to carry out full structure inspection in such harsh conditions is the driving force for the development of embedded systems that can give early warning of problems and focus inspection effort where it is needed.

There are some key differences between NDT (non-destructive test) hardware and SHM (structural health monitoring) hardware.

- NDT hardware is used consecutively on a series of different structures. The cost of the system relative to the cost of the structure can be high.
- SHM hardware is used continuously on a specific structure for the remainder of that structures operating life. The cost of the system relative to the cost of the structure must be low.

Most SHM hardware is trialled for effectiveness in detecting damage initiation and growth during blade tests. The more successful techniques can in some cases be modified and made suitable for long term demonstration on operating blades (that is made cheap and robust). The next step in the development of such SHM hardware that is successful in both controlled tests and normal operation would be a prototype product development and integration with the relevant communication/data tools and SCADA systems.

For example, Acoustic Emission techniques have been used to good effect to document structural integrity and any damage initiation/development in connection with full scale blade tests during extreme static loading by Jorgensen et al. (2004), in fatigue series by Wei and McCarty (1993) and Beattie (1997), and for standard certification tests (AEGIS project, 1998-2002). The perspectives in mounting AE sensors on an operating turbine blade have also been a research topic investigated for the industry by Blanch and Dutton (2003).

Risoe DTU has a long history of investigation in this area (SHM for wind turbine blades) and the summary reports by Sorensen et al. (2002) and McGugan et al. (2008) detail the sensor types investigated (Acoustic emission, Fibre optic, and accelerometers), the monitoring approach (Ultrasonic and X-ray), and the laboratory and full-scale testing undertaken.

More recently, static loading tests at the Risoe DTU Blade test Facility (established 2008) have showcased the effectiveness of various measurement techniques in characterising the deformation behaviour and damage evolution of wind turbine blade structures. These techniques include Digital Image Correlation, Vacuum hood shearography, acoustic emission, ultrasonics, resistance strain gauges, and mechanical displacement measurements. See Jensen et al. (2008) and Nielsen et al. (2010).

Improvements in Optical Backscatter Reflectometry techniques now permit resolutions at which a single fibre optic can provide a distributed strain measurement along its' entire length with obvious advantages over multiple point measurement (whether from traditional resistance strain or fibre Bragg gratings). See Guemes et al. (2010).

Large EU sponsored projects such as UPWIND (www.upwind.eu) have included instrumentation for all wind turbine structures, including fibre optic strain systems for the operating blades, in a dedicated work package looking at condition monitoring (WP7). And indeed this is the technology type that it is most common to find offered as an extra feature in commercial blades, the inclusion of embedded (in fact bonded) fibre Bragg-grating strain systems and crack detectors.

It is clear that many different techniques are used in the industry but that simple and robust inspection methods are preferred for quality control purposes with the most successful NDT equipment (ultrasonic inspection) hampered by problems covering the large surface areas and complex geometries of the structure.

In-situ inspection is even more limited in application, and embedded sensor technology hardly used at all. In general the application of these technologies is most widespread in structural testing to understand structural response and damage progression; however modified forms of portable inspection and robust sensor systems have been used in blades, generally as a response to a known but uncharacterised problem. The extent of research in the area suggests that the industry focus on more reliable (and larger) structures will prompt new technological solutions involving combinations of sensor systems that can begin to provide remote structural data and hence maintenance optimisation.

So the idea that by concentrating the use of sensors and monitoring systems during structure testing, where the demand for them is apparent, we can also develop the techniques that will be most efficient when applied as SHM on operating blades. For example, stress wave analysis or acoustic emission gives the possibility of locating small cracks and imperfections in the structure. By use of the acoustic emission detection system during fatigue test and static test it becomes possible monitor the initiation and development of defects, and stop the test before damage extends to a size that will cause the blade to fail.

The acoustic emission detection system is a set of piezo-electric transducers positioned on the blade in a pre-defined pattern. The sensors are connected to a data acquisition system that samples the transducer signal. Knowing the position of the sensors, it becomes possible to locate the area of the signal source (the damage). Hence the system is set up to register "first hit" among all sensors; any signal will be determined to originate from the zone of a single sensor. With a closer pattern of sensors it becomes possible to make accurate determinations of the origin of the damage using time of flight information between two (linear location) or more (planer location) sensors.

An important objective for a development engineer working to develop SHM is to first demonstrate the practical benefits accruing to the test centre from using sensor technology as simply an aid for the test personnel when carrying out their regular work. This includes the following points.

- Increased safety
- Instinctive sensor feedback
- Minimise "unwanted" damage events
- More focused "faster" inspections
- Identify onset of different damage modes during testing
- Inspections closer to ultimate failure load
- Improved ratio of "working" to "resting" time in the rig

When carrying out tests on such large structures it is paramount that all personnel must kept safe. For this reason, the test operatives have no way of visually checking the blade during loading for evidence of damage development. Video links and stress-strain curves have previously been the only information available. The audio feedback from the AE sensors and the clear visual displays showing where "damage" is developing in the blade give the operator a much more meaningful indication of or "feel for" how the test is proceeding. Errors such as unwanted damage are much less likely to go undetected. This allows alterations to be made, if necessary, to ensure a good test. The process of visually or ultrasonically inspecting a 50m long blade for damage following a proof load becomes much less time consuming when a good indication already exists of where to begin looking. Different damage types occurring during the course of the test can often be identified during the actual load process. The predictive use of AE activity during load holds means that a full NDT inspection of the structure can be carried out at a damage level almost immediately before failure. Often this will give much more information about the root causes of the failure than is possible from only inspecting the destroyed structure. A net result of this will be that the ratio between "working time" and "resting time" while the blade is in the rig will be improved, thus saving money.

An example of a static blade test where AE measurements were used to follow the damage condition and modify the load application is given in Jorgensen et al. (2004) and summarised here.

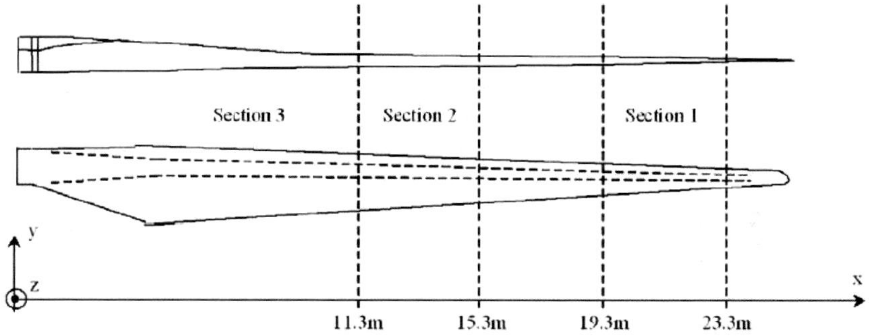

Figure 27. A Blade to be tested statically in three separate sections

The blade was tested in three sections, giving the opportunity to investigate three independent failures. The sections to test were decided based on knowledge of compression strength and geometry and these sections were loaded and supported in such a way that ultimate failure would occur in a particular area and in a particular way (one that involved the central spar "bulging" as predicted by new non-linear modeling analysis).

A step loading approach was taken whereby the load was increased and held for a short time, then either unloaded and the blade inspected or the load increased another step depending on the AE response from the attached sensors.

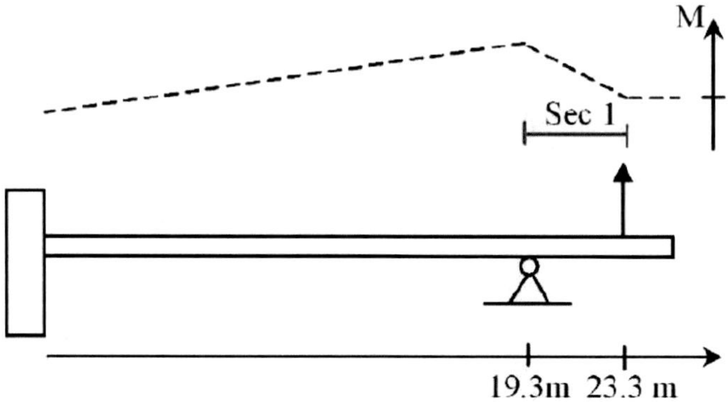

Figure 28. The Load schematic for blade section 1

Strength calculations for the section 1 loading scheme predicted blade failure at the 20.1m chord of the blade (x = 20.1m). AE sensors were used to monitor the level of activity at the 19.3m holding yoke and the activity at the 23.3 loading yoke in order to give a visual indicator to the test operatives that the loading was proceeding as expected and that no unwanted damage types were occurring. An array of sensors was also positioned on the compression face of the blade around the 20.1m chord in order to generate an activity localisation around the predicted failure.

The blade was loaded as shown in Figure 29 several times with AE activity recorded for each load. During these multiple loadings a form of the Kaiser effect was observed, in which acoustic emissions are not observed during the reloading of a material until the stress exceeds its previous high value. When the blade was loaded to failure it was observed that the fracture line at the 20.1m chord corresponded to the AE localisation data obtained as the structure approached failure.

Figure 31 shows that a slightly more complex loading scheme was used on blade section 2 with two "loading" yokes and one "clamping" yoke. This time the strength calculations predicted failure at the 12m chord of the blade. So guards were placed at the "clamping" yoke marked 1 (at 11.3m) and at the "loading" yoke marked 2 (at 15.3m). And the localisation sensors positioned at 3 (12m chord).

However, at the expected failure point (12m chord), almost no acoustic emission activity was detected, whereas guard sensor 2 which had been

Figure 29. Static blade test deflection

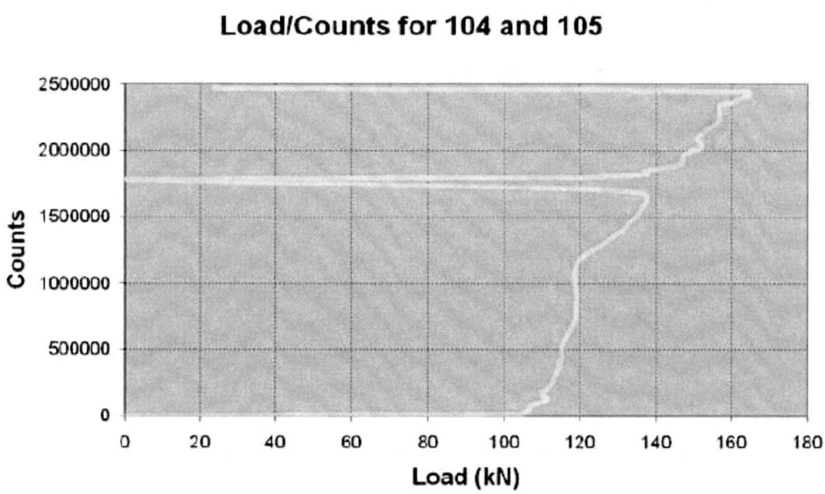

Figure 30. Cumulated AE activity against load showing the Kaiser effect

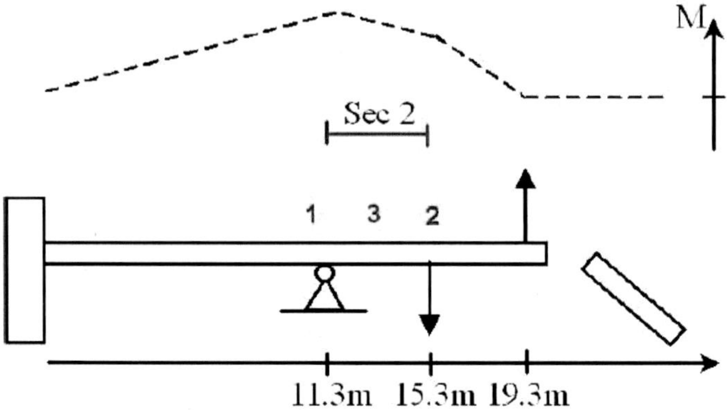

Figure 31. The Load schematic for blade section 2

positioned at the 15.3m loading yoke successfully warned, very early in the test, that failure would in fact occur here. This was an example where the model prediction was incorrect.

The static load was applied in a stepwise manner in order to assess the amount of acoustic emission and decide if the load should be increased further or the blade unloaded. Figure 32 shows that after each increase in the load ("step up") there is a burst of acoustic emission activity. As the load is held the activity level dies down and the load can be increased again to the next "step". At each step up and load hold the amount of AE increases and the time taken for the activity to die down becomes longer. This is a common technique when using acoustic emission to monitor a structural test.

On the final step load in test 201 (Figure 32) the AE activity is so great and takes so long to die down again that is was decided to stop the test and inspect the area. In this way it was possible to characterise the damage condition in the blade very close to failure load by using the sensor output to predict how close the loaded structure was to failure. In this case the test operators were able to predict failure load for this section with a less than 2% error as on the subsequent reload of the blade failure occurred as soon as the load was increased above the previous "step" level; see Figure 33.

The fact that the blade failed at section 2 by the 15.3m loading yoke was unwanted but, due to the warning from the AE sensors, not a surprise. At the moment of failure the broken blade "sprang back" against the clamping yoke at 11.3m. At this point it was observed that the guard sensor placed at the clamping yoke signalled AE activity, where it had previously been

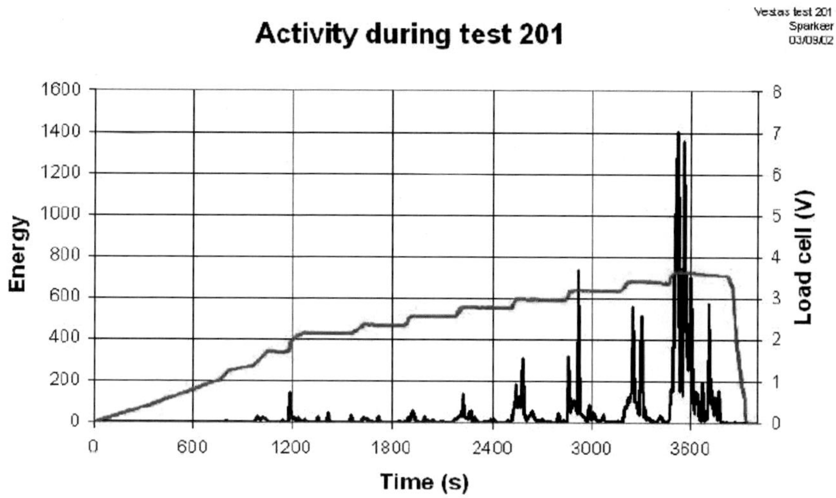

Figure 32. Showing bursts of AE activity at each "step" of test 201

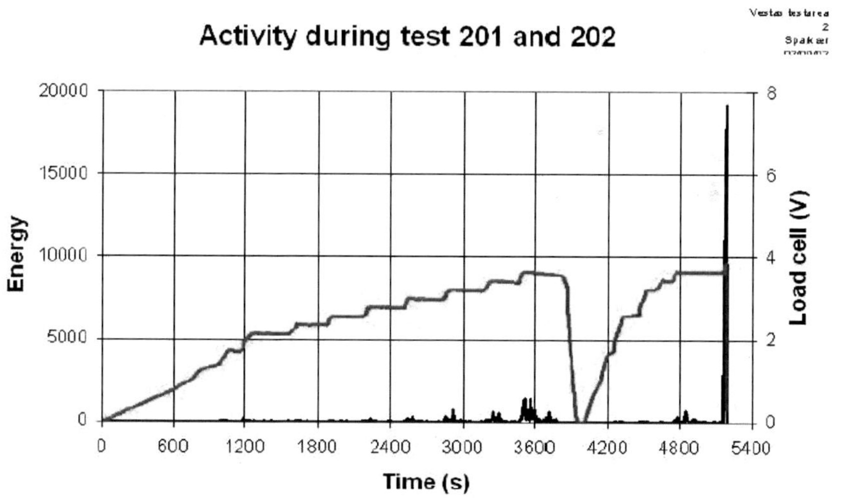

Figure 33. Showing bursts of AE activity at each "step" of tests 201 and 202

silent. This prompted a close inspection of the blade at this area. From outside the blade nothing could be seen.

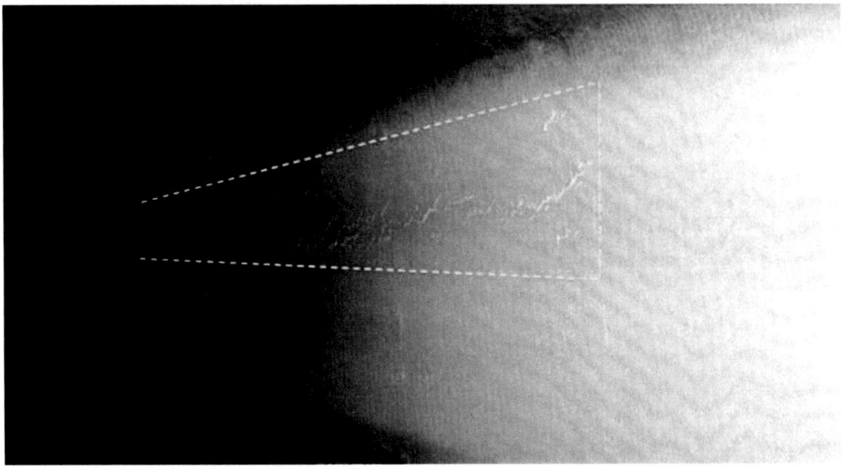

Figure 34. Image from inside the blade showing damage

But by climbing inside the blade central spar and crawling out to the 11m chord it was possible to see lateral cracking on both sides of the spar that were certainly caused when the blade "sprang back" against the clamping yoke at the moment of failure out at 15.3m.

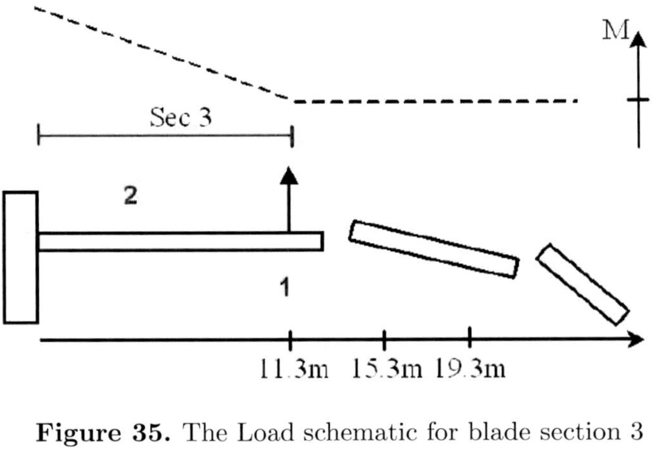

Figure 35. The Load schematic for blade section 3

This damage became very significant in the third and final blade section test. As shown in Figure 35 this was a simple static loading from a single loading yoke. The greatest bending moment is found at the root of the blade, but due to the thickness of the structural material at the root it was predicted by the model that the failure cord would be at 5.5m. AE monitoring was positioned at this predicted failure point and at the loading yoke (11.3m).

Figure 36. Alternate test configuration with reinforced loading section

The test began as expected but suddenly a great deal of AE activity registered at the loading yoke and the blade was quickly unloaded. An attempt was made to load up again but this time there was immediate activity (even at very low loads) from the AE sensor at the 11.3m loading yoke. By marking the extent of the lateral cracking inside the blade it was shown that this damage was growing and it was clear that the blade will not break as planned, the test as it was currently configured must be abandoned.

As the structure had been saved from premature failure by the AE monitoring it was possible to enact a change in the test configuration and try the loading again.

It was decided to change the loading angle (see Figure 36) and reinforced the blade around the loading yoke by using a solid plug. When loading again with this new configuration the AE sensor at the 11.3m loading yoke was silent (the lateral cracking now longer growing) and activity began at the 5.5m instrumentation indicating that the test is proceeding correctly. Note that there was no other way to predict where the blade was going to break during loading (only info is from the bending calculations made beforehand) and without the AE detecting the unwanted damage, and confirming it was active when reloaded, it is certain the test would have broken the blade at the loading point instead of at S2, and the results from this structure test would have been worthless.

5 Future work; Smart blades?

Wind turbine blades manufactured today deflect in such a way that the worst of extreme wind blast loading is avoided. This is simply due to the passive properties of the construction material, anisotropic, non-symmetrical laminates, that can be utilized to create a controlled bending and twisting under normal wind loading.

In addition to this passive response, there are also benefits to be gained from the performance of a wind turbine blade that can actively adapt its' aerodynamic profile to the conditions. Much as a bird of prey can hang motionless in the air due to slight wing adjustments, an adaptive blade can draw more energy out of the available wind and avoid peak loading from potentially damaging gusts. This saves materials and at the same time increases the potential operating lifetime as the mechanical loads on the wind become smaller. An adaptable blade will be less buffeted in a gusting environment.

In order to do this the trailing edge profile of the blade (possibly fabricated from tough rubber material) should be altered using piezoelements or pneumatic actuators linked to sensor feedbacks measuring the characteristics of the wind and the turbulent flow immediately approaching the blades. Such measurements could be realized by further development of the LIDAR (LIght Detection And Ranging) systems used in the windscanner project (www.windscanner.dk).

The structural design of wind turbine blades is also an area where significant changes will be made as the trend towards still larger structures continues. Figure 37 illustrates the multi-scale approach to testing of wind turbine blades. The smaller images inside the triangle are of tests that have been developed and carried out by Risoe DTU over the last few years. At present the certification agencies regulating structural materials used in the wind energy sector require documentation only for the top and bottom levels of this pyramid; that is small test specimens measuring material properties and full-scale blade tests measuring structural response. It is the objective of the Risoe DTU groups collaborating on blade materials and structures to systematise the sub-component testing (the levels in-between specimen and full scale) in such a way that it can also be used in the design and certification process.

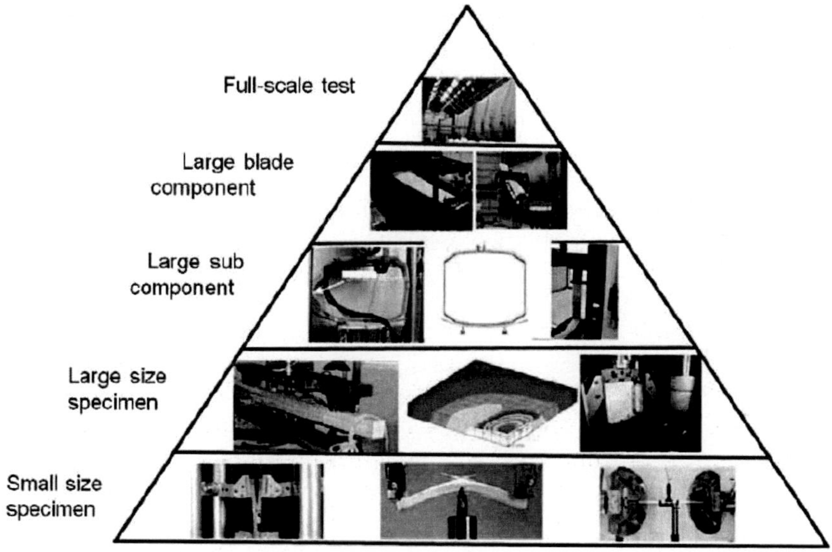

Figure 37. Test triangle for materials and structures at different scales

In order to be more complete this diagram should also include a sub-level corresponding to mechanical testing of specimens at the micro scale (perhaps approaching the nanoscale). Testing of this kind, that measures fibre pull-out and other interface effects, are carried out at Risoe DTU and the models for behaviour at this scale are also of importance when taking such a multi-scale view.

The breadth of scale under consideration here is not simply a practical issue of best mechanical testing methods. To characterise the responses at different scales we see different modelling processes implemented, and different sensor types and configuration adopted. The challenge posed by considering such a multi-scale triangle/pyramid is to develop a solution where the effect of small-scale local deviations in material properties (whether caused by the presence of damage/defects or due to design) distributed around the structure can be integrated with the full-scale models in a consistent way thus providing a tool for identifying and characterising material failure modes and locations that are critical for any given blade design. Such a tool can be used in the design and qualification stages of new blade types, and could later be used to assist maintenance decision making for the blades once in operation by predicting damage growth patterns and calculating the remaining life of any particular turbine blade based on its current health status and its operating environment.

A major ambition for the future is the establishing of test facilities at DTU that are capable of carrying out testing of sub-component structures under realistic boundary conditions. This requires the use of sensor data and NDT input to characterise the response of the sub-component in terms that can then be fed into models describing the effect of the sub-component response within the framework of the structure it will be a part of. This model is then used to recalculate the boundary conditions around the sub-component and hence correct the loading profile.

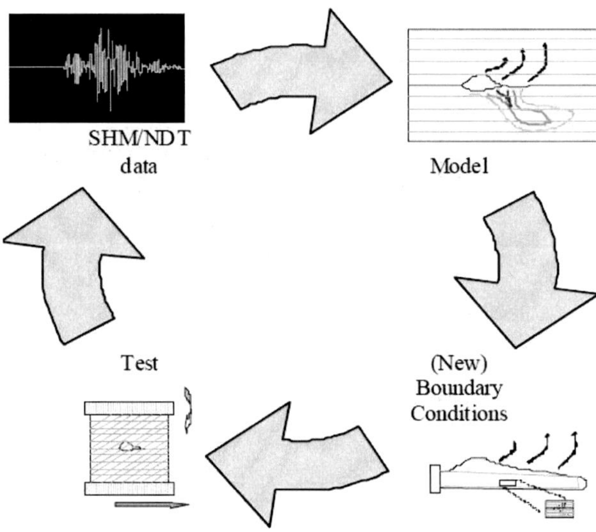

Figure 38. Hybrid testing of a sub-component with sensor feedback and boundary control

Such a facility would allow examples of the type of "smart structure" ambitions of designers who wish to create operating structures that can use embedded sensor data to inform the structure on how to automatically respond to operating loads; whether it be an adaption in order to optimise operating output under different conditions, a safeguarding of the structural integrity by avoiding potentially damaging load conditions and thus maximise total operating life, or as input to a condition based maintenance regime.

Figure 38 illustrates how an eventual hybrid test might proceed with sensor data from the test feeding into damage growth models that then update a structural response model, which supplies new boundary conditions for application on the sub-component test actuators, and so on. This gives a more realistic investigation of the sub-components mechanical properties in its' structural context.

Very often the emphasis on future developments is dominated by current research, technical progress and the possibilities that these offer to current procedures. However these advances are not enough in and of themselves to enable a change in an established industry. The argument in this section has been that the changes identified as specific to the wind energy industry allow better opportunities for new ideas like Structural Health Monitoring to be adopted than has been the case in other industries (Aerospace, civil, and military) that have followed the technology with interest for many years.

One positive way to consider this situation is with reference to the ideas first put forward by Christensen (2003) in the Innovators Dilemma. Here one can read about the possible need for an "enabling application" which successfully uses a lower capability type technology, and thereby promotes that technology's State of the Art (SOTA) in such a way that it later becomes competitive at the higher technology capability level. There are difficulties in successfully identifying Christensens enabling and disruptive type applications before the fact, but a good case can be made for offshore wind turbine blade monitoring currently being one of the best candidates for an SHM enabling application. Key terms include:-

- Sustaining technologies - improve product performance
- Disruptive technologies - "different value proposition"
- Enabling technology - "pilot" the technology, demonstrate reliability, validate cost benefits, reduce "program" risks prior to wide-scale application on an emerging program.

Sustaining technologies are those that improve the performance of an established product or procedure. The improvement can be large, but it is always evolutionary in nature (doing something "better"). The vast majority of all technological improvements (also in the aerospace structural maintenance industry) are sustaining in nature.

Disruptive technologies, on the other hand, "bring to a market a very different value proposition than had been available previously. Generally, disruptive technologies underperform established products in mainstream markets. But they have other features that a few fringe (and generally new) customers value."

Christiansen goes on to suggest that once established in a few fringe or niche markets, disruptive technologies can then quickly go on to become fully performance competitive in the wider market, where initially they underperformed with respect to established procedures and/or products.

The question could then be asked if SHM is disruptive with respect to current NDT and Inspection? If so, then in assessing technological possibilities for SHMS, we can use Christensens findings to judge the way that these new technological possibilities are most likely to be enabled.

The repair and maintenance procedures for (for example) military aircraft are well established and highly optimised. Advances in the procedures are without doubt of a "sustaining" nature as described by Christensen. However, there are a few areas where difficulties in the maintenance procedures are recognised.

For example:-

- Physical access to areas of inspection
- Adoption of new materials
- New operational requirements

The sustaining technology approach to the difficulties listed above might include the decision to alter designs to allow easier access to structural components in question, performing qualification trials for the materials and specifying in-service structural assessments, and so on.

The general approach identified by a SHMS approach would require "hot spot" monitoring of only the structural components in question, rather than attempting an ambitious "general" structural health assessment. The precise nature of the reason for adopting this hypothetical "hot spot" monitoring in an operational platform is not made clear, but it would most likely be due to one of the reasons listed above.

It should be recognised that the output of the "hot spot" monitoring from an SHMS does not need to equal the quality of output expected from using traditional inspection techniques on the same structural area. The reason for this is that the SHMS is offering a different "value proposition". A successful SHMS would allow operators the confidence of focussing traditional inspection effort (which is relatively expensive and time consuming) only on the platforms that contain structural areas that require it.

These characteristics would seem to cast SHM as a "disruptive" technology in the area of physical asset management with respect to traditional inspection techniques.

A true enabling application is one that involves mass production and associated economies of scale. One possible argument against the idea that successful adoption of SHM on aerospace structural "hot spots" will inevitably lead to full structure monitoring on aircraft, is that the mass production (of sensors) would not be achieved here. Similarly in other aerospace, the emphasis is on sustaining technology advances and optimisation rather than whole structure sensorisation, with only small potential applications for SHM in suspect structural areas, e.g. around the tail rivets on commercial airliners or on the insulation tiles on the Space shuttle. These small applications however may not require the mass production of sensors necessary to qualify for the epithet, "enabling" as it is defined by Christensen.

A brief consideration can made of other potential "enabling" applications for SHMS in other industrial sectors.

The application of sensor technology in civil engineering is progressing only very slowly, due to the well-established and optimised principles of building being employed. Where new materials and/or designs are adopted, there is a slightly greater interest, also where older buildings with historical value are monitored or buildings are placed in earthquake zones. Many new buildings are constructed with an in-built wireless communication system, and sensors/actuators designed to deliver automated energy savings. So although the damage detection technology is not employed, relevant SHM "sensor communication infrastructure systems" are common in modern civil engineering.

In the early 90s, helicopter-monitoring systems developed very rapidly when safety problems in the industry (serious problems which almost crippled the industry) required an immediate advance in the technology employed. In particular, the advances in vibration monitoring sensors were strong and allowed these sensors to later become widely applied in all (industrial) rotating machinery, including wind turbine gearboxes and drive trains. However, the development ceased when the required level of safety through monitoring and data-fusion had been achieved. No further SHM technology developments are expected here in the short to medium term.

SCADA systems for offshore wind farms are under pressure to deliver a far more comprehensive and integrated information flow to operators, including live structural health updates on turbine machinery, foundation piles, blades, etc. The repair and maintenance strategy adopted for the forthcoming round of large offshore installations has been identified as being a decisive factor in their financial success (or otherwise) and hence of the renewable energy strategy for Europe. The scale of this construction operation and the demands (industrial and political) for extra reliability in such a treacherous ocean environment make this a good prospect for a SHM enabling application in the next decade. This would also be a bonus for European competitiveness as this continent is ahead in wind power development and would therefore benefit most from such a SHM breakthrough.

Other possible enabling application could be computer (games) technology and/or human/machine interface technologies like augmented reality.

In the United States, sensor research and development is generally more advanced than in either Europe or Asia, but what this high technology American industry has so far lacked is an application that demands significant numbers of their products. The most common application for structure monitoring sensors is in Aerospace, where the volume of demand is not high, and within civil infrastructure, where a high level of confidence in design and materials have restricted the use of the relatively expensive sensors.

Adoption of SHM technology within the European wind energy industry would create a great demand for the hardware manufacturers and create a network of supporting industry and consultants on this continent. This concentration of expertise would form the base for the expansion of the technology into other sectors in Europe and around the world, re-exporting the "know-how" into diverse applications. The economies of scale would reduce hardware costs and the confidence from a successful application would win over more cautious engineers (in civil sectors for example).

Industries poised to follow the example of a pioneering European wind energy sector and embrace SHM technology include, Aerospace, Transport, Civil infrastructure, Earthquake construction, Historical building preservation and more.

It is the general feeling among experts, such as Patrick Goggin of Boeing Phantom Works (Goggin et al. (2003)), that a breakthrough application ("enabling application") is necessary in order to stimulate a more general adoption and development of SHM ideas and techniques throughout industry. The rapid manufacturing growth within the wind energy sector, and the trend towards large offshore farms, make this the most promising global candidate for fulfilling the requirements of an SHM enabling application.

With reference to Figure 39; a technology can be monitored thru time, and the CAPABILITY of the technology can be compared to the CAPABILITY REQUIRED for different applications. The technical requirements for SHM on aerospace is currently "not available". Hopefully the rate of improvement in the technology is rising faster than the requirement for the application but this cannot be known. Perhaps it never reaches the required level and arrives at the future scenario of SOTA2. But if Christensen is correct then by "discovering" an application that can match with the current level of SHM capability, the SOTA development curve is improved to such

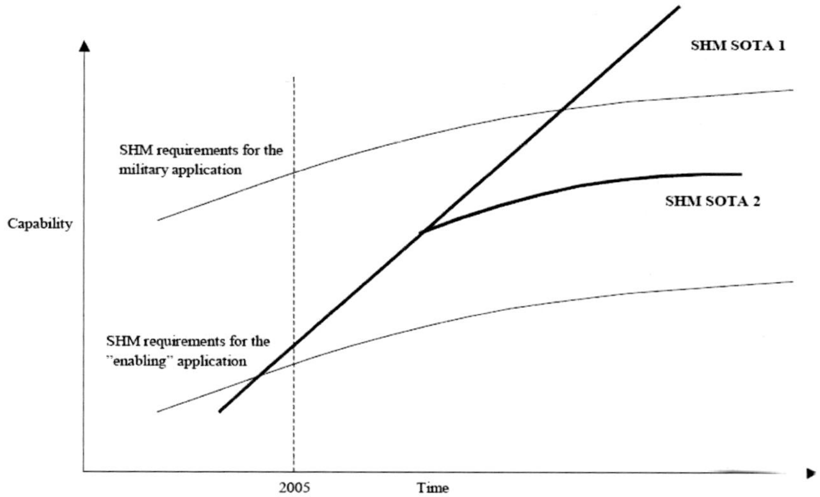

Figure 39. Clayton Christensens "Innovators Dilemma", Christensen (2003)

an extent that we do in fact manage to overtake the high capability requirement in the future; SOTA1.

A final consideration that may be fruitful when considering the best way to promote the use of Structural Health Monitoring could be biological systems; in particular the evolutionary approach to sensor development. In many presentations of the technology of Structural Health Monitoring it is common to compare the function of a sensor network in a biological phenotype (a human being) to that which is being proposed in the SHMS. However the consideration of how such a system came to be expressed (that is, the history recorded in the genotype) is not made.

It is possible that useful insights can be found by considering how small (but measurable) improvements in the performance of biological sensors through an evolutionary process are made in a competitive environment; particularly where the ultimate objective is a "high capability" system that cannot be achieved without demonstrating small but incontrovertible improvements at no (or limited) cost.

To summarise we can list the following concluding statements:-

- Developments in the wind energy industry demand new approaches
- Challenges exist for the effective structural performance of larger blades
- Opportunity exists for SHM and other "smart" solutions to speak to

these challenges

- Modular monitoring systems are envisaged to harness the synergy required to cover all significant damage types
- Implementation strategy will be critical in any successful adoption of SHMS
- Structural Health Monitoring is a form of Information Technology
- Management and Organisation of the developments are as much an issue as any technical challenges

Bibliography

I. Amenabara, A. Mendikutea, A. Lpez-Arraizaa, M. Lizaranzub, and J. Aurrekoetxea. Comparison and analysis of non-destructive testing techniques suitable for delamination inspection in wind turbine blades. *Composites Part B: Engineering*, 42:1298–1305, 2011.

A.G. Beattie. Acoustic emission monitoring of a wind turbine blade during a fatigue test. *Sandia National Laboratory USA*, 1997.

M.J. Blanch and A.G. Dutton. Acoustic emission monitoring of field tests of an operating wind turbine. In *Proceedings of the 5th International Conference on Damage Assessment of Structures (DAMAS)*, 2003.

C. Christensen. The innovators dilemma. *Harper Business Essentials*, 2003.

B.N. Cox. Scaling for bridged cracks. *Mechanics of Materials*, 15:87–98, 1993.

B.N. Cox and D.B. Marshall. The determination of crack bridging forces. *International Journal of fracture*, 49:159–76, 1991.

C.P. Debel. Identification of damage types in wind turbine blades tested to failure. *Risoe R-1391(EN) ISBN 87-550-3178-1 ISBN 87-550-3180-3 ISSN 0106-2840*, 2004.

M.A. Drewry and G.A. Georgiou. A review of ndt techniques for wind turbines. *Insight*, 3, 2007.

P. Goggin, J.Huang, E.White, and E.Haugse. Challenges for shm transition to future aerospace systems. In *Proceedings of the International Workshop on SHM 2003*, pages 30–41, 2003.

A. Guemes, A. Fernandez-Lopez, and B. Soller. Optical fiber distributed sensing - physical principles and applications. *Structural Health Monitoring*, 9 (3):233–245, 2010.

J.W. Holmes, B.F. Sorensen, and P. Brondsted. Reliability of wind turbine blades: an overview of materials testing. In *Proceedings of Wind Power Shanghai*, 2007.

F.M. Jensen, K. Branner, P.H. Nielsen, P. Berring, T.S. Antvorskov, M. Nielsen, M.M. Lindahl, B. Lund, C. Jensen, J.H. Reffs, R.F. Nielsen, P.H. Jensen, M. McGugan, K.K. Borum, R.S. Hansen, C. Skamris, F. Sorensen, R. Schytt-Nielsen, J.H. Laursen, M. Klein, A. Morris, A. Gwayne, H. Stang, J. Wedel-Heinen, J.P. Dear, A. Puri, and A. Fergusson. Full scale test of a ssp 34m box girder 1. data report. *Risoe-R-1622(EN) ISBN:978-87-550-3574-4*, 2008.

E.R. Jorgensen, K.K. Borum, M. McGugan, C.L. Thomsen, F.M. Jensen, C.P. Debel, and B.F. Sorensen. Full scale testing of wind turbine blade to failure - flapwise loading. *Risoe-R-1392(EN)*, 2004.

M. McGugan, G. Larsen, B.F. Sorensen, K.K. Borum, and J. Engelhardt. Fundamentals for remote condition monitoring of offshore wind turbines. *Ris-R-1639(EN), ISBN:87-55-03662-7*, 2008.

R.M. McMeeking and A.G. Evens. Matrix fatigue cracking in fibre composites. *Mechanics of Materials*, 9:217–27, 1990.

M. Nielsen, F.M. Jensen, P.H. Nielsen, P. Berring, K. Martyniuk, A. Roczek, T. Sieradzan, V. Roudnitski, P. Kucio, R. Bitsche, P. Andresen, T.V. Lukassen, Z. Andrlova, K. Branner, C. Bak, B.S. Kallesoe, M. McGugan, J. Wedel-Heinen, T. Lindby, F. Sorensen, Jensen C, H.W. Knudsen, U. Uldahl, A.B. Rasmussen, and J.J.A. Rasmussen. Full scale test of ssp 34m blade, edgewise loading ltt : Data report 1. *Risoe-R-1718(EN), ISSN:0106-2840, ISBN:978-87-550-3794-6*, 2010.

B.F. Sorensen. Cohesive laws for assessment of materials failure: Theory, experimental method and application, doctor of technices thesis. *Risoe-R-1736(EN) ISBN 978-87-550-3827-1 ISSN 0106-2840*, 2011.

B.F. Sorensen and T.K. Jacobsen. Crack bridging in composites: Connecting mechanisms, micromechanics and macroscopic models. In *Vol.2 of the Proceedings of the International conference on new challenges in mesomechanics*, pages 599–604, 2002.

B.F. Sorensen, L. Lading, P. Sendrup, M. McGugan, C.P. Debel, O.J.D. Kristensen, G.C. Larsen, A.M. Hansen, J. Rheinlnder, J. Rusborg, and J.D. Vestergaard. Fundamentals for remote structural health monitoring of wind turbine blades - a preproject. *Risoe-R-1336(EN)*, 2002.

B.F. Sorensen, E. Jorgensen, C.P. Debel, F.M. Jensen, H.M. Jensen, T.K. Jacobsen, and K.M. Halling. Improved design of large wind turbine blade of fibre composites based on studies of scale effects (phase i) - summary report. *Risoe-R-1391(EN) ISBN 87-550-3176-5; ISBN 87-550-3177-3 (internet) ISSN 0106-2840*, 2004.

J. Wei and J. McCarty. Acoustic emission evaluation of composite wind turbine blades during fatigue testing. *Wind Engineering*, 17, No 6:266–274, 1993.